Tropentag 2021

International Conference on Research on Food Security, Natural Resource Management and Rural Development

Tropentag 2021

International Research on Food Security, Natural Resource Management and Rural Development

Towards shifting paradigms in agriculture for a healthy and sustainable future

Book of abstracts

Editor: Eric Tielkes

Reviewers/scientific committee: Folkard Asch, Jan Banout, Mathias Becker, Michelle Bonatti, Marc Cotter, Christoph Gornott, Emiel De Meyer, Mizeck Chagunda, Thomas Daum, Anja Fasse, Falko Feldmann, Bernhard Freyer, Irmgard Jordan, Brigitte Kaufmann, Gudrun B. Keding, Holger Kirscht, Bohdan Lojka, Radim Kotrba, Lisa Murken, Joachim Müller, Sharvari Raut, Eva Schlecht, Maria Wurzinger

Impressum

Bibliografische Information der Deutschen Nationalbibliothek
Die Deutsche Nationalbibliothek verzeichnet diese Publikation in der Deutschen Nationalbibliografie; detailierte bibliografische Daten sind im Internet über http://dnb.ddb.de abrufbar.

Tropentag 2021: Towards shifting paradigms in agriculture for a healthy and sustainable future, Tielkes, E. (ed.) - Witzenhausen, DITSL

Nonnenstieg 8, 37075 Göttingen
Telefon: 0551-54724-0
Telefax: 0551-54724-21
http://www.cuvillier.de

ISBN: 978-3-7369-7573-6
eISBN: 978-3-7369-6573-7

Online-Version: https://www.tropentag.de/

Preface

Tropentag is the largest interdisciplinary conference in Europe on development-oriented research in the fields of tropical and subtropical agriculture, food security, natural resource management and rural development. Normally, the *Tropentag* takes place annually. However, for reasons that by now have become obvious, the past two years have been particularly challenging. We are therefore, delighted that the University of Hohenheim managed to host a hybrid version of the conference from 15th to 17th September 2021. Being a hybrid conference, it was pleasing to note that people did not only gather in one of the lecture theatres at the University of Hohenheim but also in one of the state-of-the art seminar rooms at the Czech University of Life Sciences in Prague. The rest, of course, attended via Zoom meetings being streamed on YouTube channels using the Whova online platform. Corona has taught us many things. One of the lessons that we learnt from the fully virtual *Tropentag* 2020, is that the poster sessions can be organised differently. We took this over into *Tropentag* 2021 and we are happy to say the response has been very positive.

Just like in the preceding years, the *Tropentag* being a development-oriented and interdisciplinary conference, in 2021 it addresses issues of resource management, environment, agriculture, forestry, fisheries, food, nutrition and related sciences in the context of rural development, sustainable resource use and poverty alleviation worldwide. Specifically, the theme was, "Towards shifting paradigms in agriculture for a healthy and sustainable future". Agriculture needs to be perceived in the minds of the people as in line with nature. As such we need to shift paradigms to a positive relationship between humans, environment, ecology and nature. Only products produced in line with the need and requirements of the planet should be positive in the minds of the people. If we achieve that, we may be able to still save the planet for a healthy and sustainable future. *Tropentag* 2021 reviewed recent research results addressing these challenges from various points of view with different approaches around this theme through discussions in plenary and thematic sessions, poster sessions, and workshops. We hope that this conversation triggered some behavioural and mindset changes that will continue beyond *Tropentag* 2021.

For the conference we received 564 contributions related to the theme of which 429 were presented either as theatre or poster presentations and are now available in these proceedings. Apart from the oral presentations, which were in 24 sessions, there were 29 poster sessions of 3–5 minutes video clips with moderated discussion. Further, there were 10 pre- and post-conference workshops. The International Livestock Research Institute (ILRI) was the CGIAR Centre of the Year.

The four plenary keynote contributions to this years' *Tropentag* gave us a very good overview of the understanding and application of paradigm shifts in agriculture towards a healthy and sustainable future.

Dr. Jimmy Smith, the Director General of ILRI started by highlighting that the demand for food is driven by three main drivers – human population size, income, and demographics. He further indicated that in the 2050s between 60–70% more food will be needed. Dr. Smith highlighted that food supply is less predictable, as it is influenced by many more variables such as crop, livestock and fish productivity, increasing resource constraints (land and water in particular), climate change and variability, and global as well as local economic and political conditions. To respond to this challenge, Dr. Smith proposed three paradigm shifts: (a) Agriculture must become a growth pole with equity, including for women and youth; (b) Research and innovation must respond to the needs of the agricultural population who are mostly small and medium scale farmers and entrepreneurs; (c) shorten and professionalise agriculture supply chains to reflect a focus on local and regional markets instead of export orientation.

Continuing with this theme, Prof. Dr. Ingo Grass of the University of Hohenheim, taking examples from Colombia, South Africa and Indonesia, illustrated potentials of agricultural diversification and ecological intensification in modern-industrial farms to promote environmental and economic sustainability. He emphasised that while individual methods hold great potential to reduce negative impacts, combining multiple methods of ecological intensification and diversification can create truly regenerative agricultural systems that have great potential to sustainably enhance food production and biodiversity.

In her presentation, Prof. Dr. Hale Ann Tufan of Cornell University, USA demonstrated the importance of gender responsive innovations in agricultural research and development. Prof. Tufan emphasised the need in this paradigm shift not only in primary production but also in capacity and strategy development, curriculum development and training, and in priority setting and market development.

In the last, but by far not the least in importance, Dr. Chikelu Mba of Food and Agriculture Organization of the United Nations (FAO), made a case for technologies that promote farmers' access to quality plant and animal seeds from varieties and breeds that are well-adapted, productive and nutritious. Dr. Mba demonstrated that this paradigm shift requires the safeguarding of the widest spectrum of genetic resources in the seed system for enhanced food security and agriculture.

In addition to the plenary keynote contributions, we had thematic session invited speakers who provided an excellent overview to the specific subject matter. In the session "Trees for people", Dr. Patrick Worms of CIFOR-ICRAF, illustrated the vital benefits of multi-crop agriculture involving trees in an Agroforestry setup. Further, Dr. Worms highlighted the importance of science-policy interface in dealing with issues of land degradation, environmental legacy and conflicts. In the session on "The role of wildlife", Prof. Dr. Louw Hoffman of the University of Queensland, Australia and Professor Emeritus, Stellenbosch University, South Africa demonstrated the importance of having a wider view of biodiversity in agriculture in order to ensure human food and nutritional security.

Tropentag 2021 has clearly shown that scientific exchange and discussions needed to keep science alive are possible and fruitful even in a hybrid conference. As part of the organising team, we have learnt some lessons in this hybrid conferencing that may particularly be suitable to increase *Tropentag*'s outreach and we will be willing to pass them on to the next team of organisers.

Special thanks go to Eric Tielkes and his team for his very valuable support in organising this hybrid *Tropentag*, getting the IT setup sorted and distributing the scientific debates to the likewise decentral *Tropentag* family around the globe. We thank the student reporters for keeping the blog and reports 'alive'. Thanks to ATSAF for all the guidance and for maintaining the institutional memory, thanks to the University of Hohenheim and all the hygiene brigade. Particular thanks to our longstanding donors (listed on the back cover) for their unwaning financial and in-kind support. Thank you all for participating - you made it happen and you showed that the is indeed urgent need to move "towards shifting paradigms in agriculture for a healthy and sustainable future".

Indeed, *Tropentag* is normally a whirling pool of people interacting and listening to each other, learning new things, building and refreshing networks, and enjoying science and coffee. We hope we can return to the in-presence format soon. As of now, let me wish you all, all the best and hope to see you at *Tropentag* 2022 in Prague.

On behalf of *Tropentag* 2021 organising team.

Mizeck Gift G. Chagunda

Hohenheim, September 2021

Contents

6. Institutions

Transforming Food Agriculture to Feed the People and Save the Planet

Jimmy Smith

International Livestock Research Institute, Kenya

The theme of Tropentag 2021 is 'Towards shifting paradigms in agriculture for a healthy and sustainable future'. I have taken the liberty of interpreting agriculture as 'food agriculture' and I will confine my remarks on 'the future' to developing countries where both opportunities and challenges for agricultural transformation are significant.

Demand for food is driven by three main drivers – human population size, income and demographics. Robust modelling of population growth, incomes and demographics show that demand for food will likely plateau in the 2050s, but by then between 60–70 % more food will be needed than is currently produced. However, food supply is less predictable, as it is influenced by many more variables such as crop, livestock and fish productivity, increasing resource constraints (land and water in particular), climate change and variability, and global as well as local economic and political conditions. Our quest must be about both food and nutritional security – along with addressing the current and future food demand sustainably – environmentally, socially and economically.

To respond to this challenge in developing countries three paradigm shifts are proposed: (a) Agriculture must become a growth pole with equity, including for women and youth; (b) Research and innovation must respond to the needs of the agricultural population who are mostly small and medium scale farmers and entrepreneurs; (c) shorten and professionalise agriculture supply chains to reflect a focus on local and regional markets instead of export orientation.

Should these paradigm shifts take place, the food agriculture sector would respond to meeting future food and nutritional needs sustainably, contribute to gainful and equitable employment, as well as to growth and development in developing countries.

Contact Address: Jimmy Smith, International Livestock Research Institute, P. O. Box 30709 00100 Nairobi - Kenya, 00100 Nairobi, Kenya, e-mail: J.Smith@cgiar.org

Enabling Farmers to Exploit Genetic Gains for Sustainable Crop Production Systems

Chikelu Mba

Food and Agriculture Organization (FAO), Italy

Global efforts to accomplish the onerous task of a 50 percent increase, over the 2012 figures, in the production of food and other agricultural products sustainably by 2050 are confounded by the impacts of climate change and other drivers. To underscore the enormity of the constraints to attaining universal food security and nutrition by 2030, a commitment of the Sustainable Development Goals, one in every 10 persons globally or one out of every five persons in sub-Saharan Africa did not have enough nutritious food to eat in 2019. Still more worrisome, the number of the food insecure and malnourished has been increasing steadily over the last six years. With 80 % of our food being plant-based, a significant component of the solutions to these untenable conditions must be sourced from crop production systems – which produce more with fewer inputs. Towards this end, the case is made that farmers' access to the quality seeds and planting materials of the well-adapted, productive and nutritious crop varieties which are resistant to myriad biotic and abiotic stressors must be enhanced. This requires the safeguarding of the widest spectrum of plant genetic resources for food and agriculture, the use of their inherent variations in breeding progressively superior crop varieties and the agency of responsive seed systems that cater especially to resource-poor farmers of food security crops in vulnerable parts of the world. The normative and operational work of the Food and Agriculture Organisation of the United Nations and its partners in this regard is reviewed and future perspectives shared.

Contact Address: Chikelu Mba, Food and Agriculture Organization (FAO), Rome, Italy, e-mail:

Reconciling Agricultural Production with Biodiversity Conservation through Ecological Intensification and Diversification

INGO GRASS

University of Hohenheim, Inst. of Agric. Sci. in the Tropics (Hans-Ruthenberg-Institute), Germany

More than seventy percent of the land surface is transformed by humans, and 1 million species of animals and plants are threatened with extinction. Intensification of agriculture is a major driver of this dramatic biodiversity decline. Modern-industrial agriculture has spurred considerable yield increases in the second half of the 20th century, however, the current agrichemical model of agriculture undermines itself and much else besides. Hence, a paradigm change in agriculture is urgently needed. Ways forward include the (re)diversification of agriculture from field, farm to landscape scales. In addition, ecological intensification, that is, the replacement of conventional intensification practices through ecological processes to increase crop production, has been proposed. Taking examples from Colombia, South Africa and Indonesia, I illustrate potentials of agricultural diversification and ecological intensification in modern-industrial farms to promote environmental and economic sustainability. While individual methods hold great potential to reduce negative impacts, combining multiple methods of ecological intensification and diversification can create truly regenerative agricultural systems that have great potential to sustainably enhance food production and biodiversity

Contact Address: Ingo Grass, University of Hohenheim, Inst. of Agric. Sci. in the Tropics (Hans-Ruthenberg-Institute), Stuttgart, Germany, e-mail: ingo.grass@uni-hohenheim.de

Crops

Crops and cropping systems

A Synthesis of Twelve Years of the "Long-Term Farming Systems Comparisons in the Tropics (SysCom)"

David Bautze, Gurbir Bhullar, Noah Adamtey, Laura Armengot, Eva Goldmann, Amritbir Riar, Johanna Rüegg, Monika Schneider, Beate Huber

Research Institute of Organic Agriculture (FiBL), Department of International Cooperation, Switzerland

There is substantial evidence that carrying on with the prevalent agricultural management practices is not a sustainable option. As an alternative system to conventional agriculture, organic farming precludes the use of synthetic pesticides and fertilisers and seeks to produce food while improving environmental and human health. It is, however, legitimate to question whether organic agriculture can contribute to sustainable development as it is often perceived to be based on ideology. Comparative long-term studies were often conducted under temperate environments, and thus several questions remain open regarding the performance of organic systems under tropical conditions. The SysCom programme (https://systems-comparison.fibl.org) has addressed this gap by over a dozen years of long-term farming systems comparisons research on different organic and conventional agricultural production systems in Kenya, India, and Bolivia.

The results from the programme show that crop productivity of organic systems can match those of conventional systems yet varies depending on the type of crop and management practices. The variation in yields of annual organic crops can primarily be explained by the slow release of nutrients from applied inputs or their lower nutrient content and yield losses from pest and disease pressure. This is often the case if organic systems try to mimic conventional methods and only substitute conventional inputs. In perennial production systems, the results show that the system's complexity (monoculture vs. agroforestry) has a more significant influence on the total productivity than the type of farming system (organic vs. conventional). The profitability of all the organic systems was found to be mainly influenced by crop productivity. However, by employing a systemic approach and implementing good agricultural practices, organic systems could be managed successfully and profitably. Organic systems also offer additional benefits to society as well as to the environment when compared to conventional systems, e.g., i) buildup of soil fertility over the long-term, ii) reduction of pesticide residues in soils, crop products, and run-off water, iii) higher concentrations of vital elements in food crops, iv) enhanced flora and fauna diversity and abundance, v) a reduction of non-renewable energy resources used, and vi) increased resilience.

Keywords: Long-term comparative research, organic farming, tropics

---

Contact Address: David Bautze, Research Institute of Organic Agriculture (FiBL), International Cooperation, Ackerstr. 113, 5070 Frick, Switzerland, e-mail: david.bautze@fibl.org

Crop Diversification as a Coping Strategy to Changing Environment

Aksana Zakirova

Eberswalde University for Sustainable Development (HNEE), Center for Econics and Ecosystem Management, Germany

Humanities' ever increasing desire for consumer goods is resulting in mounting pressure on natural resources. The increasing consumer demand is not only related to food products, but also to clothing, the production of which is becoming unsustainable and encouraging farmers to engage more in cotton cultivation. Such an increase in demand negatively affects the small-scale cotton producing countries such as Tajikistan in Central Asia, where production resources are limited and the continued intensive cotton production will further exacerbate environmental problems of the country.

In Tajikistan, cotton monoculture practices were established during the Soviet era, and have left an indelible mark on its environment through the intensive use of land and water resources. Since the disintegration of the Soviet Union in 1991, the situation for cotton monoculture in Tajikistan has changed, and there has been a slow shift towards crop diversification in support of food security. While crop production patterns have not shifted completely, it is interesting to observe how has the shift from cotton monoculture to other crops started and is currently taking place in Tajikistan. It is important to understand whether farmers can move completely away from cotton production, and if so, what social, economic and environmental factors influence this transition?

A field research conducted during the cotton harvest season of 2019 in Khatlon region, one of the major cotton producing regions of Tajikistan, has shown that the majority of small-scale farmers prefer to gradually move away from cotton mainly because of declining yields, which inherently expose them to the high risks associated with low harvest and poor sales price. One of the reasons for declining yields is linked to a deterioration in land quality, manifested as increased level of soil salinity resulting from the long history of the excessive irrigation used during the Soviet era. Moving away from cotton production towards crop diversification is perceived by farmers as a coping strategy that addresses environmental changes such as soil degradation, increasing temperatures and decreasing water availability.

Keywords: Cotton monoculture, crop diversification, environmental change

---

Contact Address: Aksana Zakirova, Eberswalde University for Sustainable Development (HNEE), Center for Econics and Ecosystem Management, Schicklestr. 5 (Haus 22), 16225 Eberswalde, Germany, e-mail: aksana.zakirova@hnee.de

Smallholder Farmers' Perspective on Sustainable Agriculture Intensification in the Guinea Savannah Agro-Ecological Zone of Ghana

Ebenezer Boateng[1], Benjamin Kofi Nyarko[1], Christine Fürst[2], Martin Schultze[2], Fifii Amoako Johnson[3], Simon Mariwah[1], Ishmael Mensah[4]

[1]*University of Cape Coast, Geography and Regional Planning, Ghana*

[2]*Martin Luther University Halle-Wittenberg, Inst. for Geosc. and Geography, Germany*

[3]*University of Cape Coast, Population and Health, Ghana*

[4]*University of Cape Coast, Hospitality and Tourism Management, Ghana*

Sustainable agriculture intensification (SAI) has emerged as an important concept worth considering whiles attempting to achieve food security with minimal environmental impact. Several studies have focused on the conceptualisation, and developing indicators and drivers of SAI. However, the perspective of impoverished Sub-Saharan Africa smallholder farmers on SAI has not been considered. Therefore, this study was conducted within the Guinea Savannah Agro-Ecological Zone of Ghana and aimed at assessing smallholder farmers' perspectives on sustainable agriculture intensification. Within the study area, 698 smallholder farmers were interviewed from 25 communities. The data gathered was analysed using exploratory factor analysis, Kendall coefficient and logistic regression. The results showed that there is a high internal consistency of the Likert scale used in assessing SAI. However, it was identified that the respondents are practising a weak SAI because the environmental dimension was less prioritised. In addition, variables such as community, marital status, religion, level of education, income, household size and farm size were identified as significant (between 90 % to 99 % confidence level) predictors of SAI. In order of importance, the study found that drivers such as inadequate capital, low soil nutrients, farm size, limited access to modern farming technologies, and population growth hinder SAI. To enhance SAI, smallholder farmers need to be encouraged and supported to protect the environment. Together with non-governmental organisations dedicated to SAI, Ministry for Food and Agriculture and other stakeholders such as the Environmental Protection Agency (EPA) could strengthen efforts to educate smallholder farmers on the importance of SAI and its contribution to the broad SDGs.

Keywords: Ghana, Guinea savannah, smallholder farmers, sustainable agriculture intensification

---

Contact Address: Ebenezer Boateng, University of Cape Coast, Geography and Regional Planning, University Avenue University of Cape Coast Ghana, 00233 Cape Coast, Ghana, e-mail: ebenezer.boateng@stu.ucc.edu.gh

Quality Seeds in Traditional Systems: Evidence in Household Consumption of Indigenous Crops in Peru

Nancy Pierina Benites Alfaro
National University of Engineering, Economic Engineering, Peru

The traditional Peruvian production systems, which concentrate the greatest agrobiodiversity areas, and constitute one of the centers of origin of crops, have gained international relevance in the last two decades as providers of food security and strategies for adaptation to climate change: The International Year of Potato (2008) and International Year of Quinoa (2013), two native Peruvian crops, were declared by FAO as examples of that; also, the 75 % of Peruvian family farms have subsistence farming, concentrate in the Andes, and uses their own seeds as the most relevant input in their fields. However, the inclusion of this system in agrarian policies, particularly in agrarian laws of seeds is still incomplete: the agriculture sector has a technological transformation posture, based on high yields per crop, and high demand in urban and international markets. The legal framework promotes the use of certified seeds. For these reasons, traditional or ancestral production systems have been replaced progressively, based on the premise that it only shows low yields per crop. The regulatory Peruvian framework on seeds does not fully integrate traditional production systems as suppliers of quality seed and doesn't include all native crops, especially some crops with high domestic consumption in the Andes. For these reasons, the study analyzes the historical changes (2000–2020) in the yields of high native crops that exclusively use seeds from traditional production systems, which are not being included in Peruvian seed regulation, and that not counted with certified seed (national or imported), but they are native seeds widely consumed in local and national markets: Peruvian pumpkin (*Cucurbita ficifolia*), Peruvian squash (*Cucurbita moschata*) and Peruvian hot pepper (*Capsicum pubescens*). The analysis has shown that the seeds worked under traditional systems, providing sufficient yields that allow guaranteeing food security and resilience supported by their sustained sowing in subsistence family agriculture. In other words, the inclusion of the traditional system and domestic native crops needs their inclusion in agricultural policies.

Keywords: Formal seed system, native crops, Peruvian Andes, quality seeds, traditional production systems

Contact Address: Nancy Pierina Benites Alfaro, National University of Engineering, Economic Engineering, 391 Mezarina Street Limatambo. Lima - Peru., lima51 Lima, Peru, e-mail: pierinabenitesalfaro@gmail.com

Seedball Technology Development and its Application for Pearl Millet Production in West African Sahel

Charles Ikenna Nwankwo[1], Hannatou Moussa Oumarou[2], Ali Maman Aminou[3], Ludger Herrmann[1]

[1]*University of Hohenheim, Soil Chemistry and Pedology, Germany*

[2]*National Agronomic Research Institute of Niger, Crop Production, Niger*

[3]*FUMA Gaskiya, Crop Production, Niger*

In quest for an effective local material-based innovation for West African Sahelian smallholder farmers, the seedball technology was developed. Seedball is a cheap seed pelleting technique that combines sand, loam, seeds and optionally wood ash or mineral fertiliser (NPK) as an additive to enhance pearl millet seedlings growth in chemically infertile soils. To form pearl seedballs, a mixture of 80 g sand+50 g loam+25 ml water+2.5 g seeds serves as the standard recipe; either 3.0 g wood ash or 1.0 g NPK can be added as effective nutrient compounds. About 10 seedballs of 2.0 cm diameter size could be formed from the standard recipe. This study summarises about 7 years of seedball technology and its application findings in greenhouse and field trial conditions. Our greenhouse results showed that seedball increases shoot (height, leaf count, dry matter and stem diameter) and root (length, density, diameter, fine root and dry matter) variables of pearl millet seedlings in different folds, depending on the soil nutrient condition (low and normal) and seed size (high and low TGM); the lower and smaller the soil nutrient status and seed sizes, respectively the higher the enhancement effect of seedball most likely associated with nutrient supply to seedling by seedball. The effects of seedball were often relatively higher in the local seed variety. The field trial results showed that seedball does not suppress seedlings emergence. An average of about 30 % panicle yield increase was observed, with a slight decline from 2014 to 2020 trial years, attributed to differences in production practices and annual rainfalls over the years. Panicle yield comparison was made with respect to (i) seedball type – wood ash- vs NPK-additive, (ii) sowing time – wet vs dry, (iii) weed management – complete vs partial, (iv) local-soil type (v) cropping system - sole vs mixed and (vi) gender. Seedball-derived plants produced fewer but denser panicles. The average panicle yield of NPK-amended seedball was relatively higher. Wet sowing, partial weeding, sole cropping, and male farmers produced relatively higher panicle yield. Seedball technology seems to have favourable adoption conditions in the Sahel. A recommendation is the socio-economic evaluation of the seedball technology as an innovation.

Keywords: African Sahel, dry farming, local farmers, low-income farmers, millet, peasants, WAS farmers

Contact Address: Charles Ikenna Nwankwo, University of Hohenheim, Soil Chemistry and Pedology, Emil-Wolff-Str. 12a, 70599 Stuttgart, Germany, e-mail: charsile2000@yahoo.com

High Chloroplast Diversity of Cacao (*Theobroma cacao* L.) in Western Amazonia

Helmuth E. Nieves-Orduña[1], Markus Müller[1], Konstantin Krutovsky[1,2], Oliver Gailing[2,1]

[1]*Georg-August University of Göttingen, Dept. of Forest Genetics and Forest Tree Breeding, Germany*

[2]*Georg-August University of Göttingen, Center for Integrated Breeding Res., Germany*

The cacao tree, *Theobroma cacao* L., is cultivated in the tropics mainly in agroforestry systems for the production of seeds, the raw material for the chocolate industry. Thus, conservation and use of cacao genetic resources in breeding programmes are vital for the cacao economy. Wild cacao populations have a wide geographic distribution in their native range along the Amazon basin. However, the phylogeographic structure of natural cacao populations is not well known. In this context, chloroplast DNA (cpDNA) polymorphisms can be used as informative markers to study geographic variation of cacao populations and geographic origin of planting material used in cacao farms and breeding programs. Here we show how chloroplast genetic diversity is geographically distributed in Amazonia, and how it is represented in common cacao genotypes by using cacao chloroplast microsatellite (simple sequence repeat - SSR) markers to genotype 235 cacao samples. Western Amazonia has high cpDNA variation. In total, 23 chloroplast haplotypes were identified. Two observed haplotypes were widespread across the Amazon basin suggesting seed dispersal patterns from west to east in Amazonia, likely influenced by humans. We also observed a common haplotype in selections representing cacao farms in west Ecuador. When we compared the observed haplotypes with previously identified cacao genetic groups based on nuclear DNA we observed specific chloroplast haplotypes within eight genetic groups. Our results can help to determine the chloroplast diversity of cacao accessions, and using them together with agronomic characteristics geographically distinctive varieties can be selected in breeding programs. The analysis of different cacao populations with wide geographic distribution in the Amazon basin may reveal new haplotypes. Furthermore, the detection of new haplotypes will expand our understanding of patterns of geographic genetic variation and dispersal in cacao. This information will help to guide the conservation of cacao genetic resources.

Keywords: Chloroplast haplotypes, chocolate, crop dispersal, geographic origin

Contact Address: Oliver Gailing, Georg-August University of Göttingen, Center for Integrated Breeding Res., Buesgenweg 2, 37077 Göttingen, Germany, e-mail: ogailin@gwdg.de

Twenty Years of Agroecological Practices on a Family Farm in Pinar del Río, Cuba

Raymundo Vento Tielves[1], Victor Jorge Díaz García[2], Eduardo Cabrera Carcedo[2], Evelyn Pérez Rodríguez[1], Bettina Eichler-Löbermann[3]

[1]*University of Pinar del Río, Center for Environment and Natural Resources Studies, Cuba*

[2]*Soil Institute of Pinar del Río, Siol, Cuba*

[3]*Rostock University, Agricultural and Environmental Faculty, Germany*

The research is integrated with the DiveCropS project supported by DAAD, Germany, and shows that experience of more than twenty years of agroecological practices in a family farming, located in an area of slate heights in the Guaniguanico mountain, in the municipality of Los Palacios, province of Pinar del Río, Cuba. The studied is located in areas of fragile agricultural ecosystems with lithosol-type soil, which for many years have been subjected to poor management and intensive use, as a consequence they have been degraded by erosion and compaction, limiting their productive capacity. In the experience carried out, the effect of the system of agroecological practices implemented for the management of soils that allows them to improve, conserve and increase their productivity over time is evaluated. The study has a surface area of 5 hectares where they applied technologies for the improvement and conservation of soil, crop rotation, minimum tillage and agroforestry management. The results achieved show a reduction in erosion, by improving the chemical, physical and biological properties of the soil, increasing its fertility and the biodiversity of the agro-ecosystem, on the basis of reaching an increase in organic matter in the soil with values of 0.97 to 1.96 %, achieves a greater cation exchange capacity, with an increase in structural porosity and soil permeability, contributing to a 30 % increase in yields. These results allow a greater stability of the family, improve their quality of life and guarantee their permanence in the environment without the need to emigrate to other areas

Keywords: Agroecology, agroforestry, biodiversity, family farming, soil

Contact Address: Raymundo Vento Tielves, University of Pinar del Río, Center for Environment and Natural Resources Studies, Borrego Avenue Building 108 Apt C-9 Brothers Cruz, 20100 Pinar del Río, Cuba, e-mail: tielve@upr.edu.cu

From Top to Bottom, How Meristem Position Affects the Regeneration Capacity of Sweet Potato after Cryopreservation

Hannes Wilms[1], Natalia Fanega Sleziak[2], Maarten Van der Auweraer[1], Edwige Andre[1], Bart Panis[1,2]

[1] *KU Leuven, Dept. of Biosystems, Belgium*

[2] *Bioversity International, Belgium Office, Belgium*

Sweet potato (*Ipomoea batatas*) provides, together with other tubers such as cassava and potato, more than 1000 million people a daily meal with a production of more than 112 million tons of sweet potato each year. Continuously new challenges such as diseases, pests and drought due to climate change arise, upon which farmers need to be able to react. Introducing improved varieties is a part of the solution, but to be able to do this, farmers and breeders need access to safely stored germplasm.

Cryopreservation could play an important role, since it allows the storage of genetic resources for an indefinite amount of time. The development of a cryopreservation protocol of sweet potato meristem already started in 1992 by Towill and Jarret and more protocols with ranging success rates have been developed over time. However most of these reports used apical meristems without considering the axillar alternative. These two types of meristems differ greatly in their composition, both hormonal and physiological, often resulting in different outcomes after cryopreservation as has also been shown in other plants.

This project aimed to optimise different parameters of a droplet vitrification protocol with a special focus on the meristem type. Three different cultivars CIN, IBA and CMR were cryopreserved with either apical or axillar meristems, which resulted in a plantlet regeneration rate of respectively 27.7 % and 56.6 %. The meristem type has thus a significant impact on the regeneration rate. This effect was found to be slightly cultivar dependent, since the CIN cultivar showed no significant improvement.

This improved protocol was subsequently tested on seven other cultivars. Since different cultivars of the same species vary greatly in their reaction towards cryopreservation. Cryopreservation of this material resulted in an overall regeneration rate of 49.7 %, with cultivar regeneration rates varying between 9.5 and 83.9 %. These results shows that the protocol is applicable to large parts of existing collections and that other protocols could also benefit from using axillar meristems.

Keywords: Conservation, cryopreservation, *in vitro*, sweet potato

Contact Address: Hannes Wilms, KU Leuven, Dept. of Biosystems, Willem de Croylaan 42, 3001 Leuven, Belgium, e-mail: Hannes.wilms@kuleuven.be

Forgotten Crops and the Vulnerability of Rural Livelihoods: The Case of Enset in Ethiopia

Ashenafi Duguma Feyisa, Yann De Mey, Miet Maertens
KU Leuven, Dept. of Earth and Environmental Sciences, Belgium

This study examines the role of enset or false banana, one of the many identified forgotten crops, in the livelihood of rural households in Southern Ethiopia. Like many other forgotten crops, enset has a high resilience to extreme environmental conditions and a high nutritional value. Enset matures in 2 to 3 years but can stay in the field for up to 12 years and can be harvested any time of the year. Some enset-derived fermented food products can be stored for a long period of time. With these characteristics, enset could play a vital role in supporting rural livelihoods and reducing their vulnerability to shocks but this is poorly understood. In this study, we analyze the contribution of enset to household income, food security and vulnerability to shocks. We use data from a large comprehensive household survey, covering 684 rural households in 46 kebeles in Southern Ethiopia. We rely on the sustainable livelihoods framework to guide the empirical analysis and use a stepwise regression approach in line with this framework. We find that household income increases with the number of enset plants a household has. We find no significant effect of enset cultivation on household food security but owning enset plants significantly reduces the negative impact of shocks on food security. Moreover, we find that households with more enset plants are less vulnerable to shocks and perceive less risk. We conclude that enset cultivation plays a crucial role in reducing household vulnerability, rather than being a food security strategy.

Acknowledgments: This research is funded by VLIR-UOS, under the Interuniversity Cooperation Programs (IUC 2017 Phase 1 AMU). We thank Zonal, District, and Kebele level administrative bodies in Ethiopia for unreserved support during data collection.

Keywords: False banana, orphan crops, risk, rural livelihoods, vulnerability

Contact Address: Ashenafi Duguma Feyisa, KU Leuven, Dept. of Earth and Environmental Sciences, Celestijnenlaan 200e - box 2411, 3001 Leuven, Belgium, e-mail: ashenafiduguma.feyisa@kuleuven.be

Investigating the Change in the Pattern of Crop Dominance in Iranian Agricultural Ecosystems

Jafari Leila[1], Sara Asadi[2], Asgari Ashkan[3]
[1] *University of Hormozgan, Horticultural Science Department, Iran*
[2] *Ferdowsi University of Mashhad, Fac. of Agriculture, Iran*
[3] *University of Hormozgan, Minab Higher Education Center, Iran*

Human have made fundamental changes in his environment in many aspects, and agriculture is no exception. We are witnessing the simplification of agricultural ecosystems, which has affected the biodiversity of agricultural products in different parts of the world, and ultimately human behaviour has changed the natural patterns of several thousand years and its effects are the instability of agricultural ecosystems. In order to investigate the predominance of some species in Iranian agricultural ecosystems in terms of biodiversity, a study was conducted in the period 1991–2018. We used time series of crop area data from 1991 to 2018 for Iran. Also, some biodiversity indicators, including Shannon, Simpson, dominance and uniformity, were used to determine the status of biodiversity. The results showed that wheat and barley species are cultivated in almost all parts of Iran (97 % of agricultural areas). Wheat and barley in Iran have about 5.4 and 1.45 million hectares of cultivated area (61.83 % of the total area under cultivation of crops and orchards), respectively. The trend of Simpson dominance index fluctuated in the study period but was generally affected by the area of wheat and barley. Also, the Shannon diversity index was strongly influenced by the area of wheat ($b = -0.41$, $r = -0.51$ **) and barley ($x = -0.09$, $r = -0.34$), so that by increasing the area of wheat and barley, Shannon index decreased. The results show the dominance of these plants and the prevalence of monoculture in Iran, which results in instability and vulnerability of agricultural ecosystems. Due to Iran's arid and semi-arid climate, the dependence of these crops on seasonal rainfall has increased the vulnerability of the agricultural system and food security. During the occurrence of drought in some years, the amount of wheat and barley production in Iran decreased significantly. In general, the biodiversity trend of crops in Iranian agricultural ecosystems was a function of the amount of dominance or area of wheat and barley cultivation. In other words, when the area of these two crops in this period decreased due to climatic or political reasons, the amount of diversity indices increased.

Keywords: Barley, diversity, food security, sustainability, wheat

---

Contact Address: Sara Asadi, Ferdowsi University of Mashhad, Fac. of Agriculture, Mashhaf, Iran, e-mail: saraasadi.sai@gmail.com

Genetic Diversity in Napier Grass (*Cenchrus purpureus*) Collections and Progeny Plants: Potential-Duplicates and Unique Genotypes

Meki Shehabu Muktar[1], Tadelech Bizuneh[2], Besufekad Wolde[3], Yilikal Asefa[1], Chris S. Jones[4]

[1]*International Livestock Research Inst. (ILRI), Feed and Forage Development, Ethiopia*
[2]*Ethiopian Inst. of Agricultural Research, Holeta Agricultural Research Centre, Ethiopia*
[3]*Ashoka University, Department of Biology, India*
[4]*International Livestock Research Institute (ILRI), Kenya*

Napier grass is an important perennial tropical grass grown as a forage and energy crop, however, it has a limited global diversity mainly due to its vegetative propagation. In this study, 345 Napier grass genotypes composed of 199 progeny plants raised from 13 seed-setting accessions, 19 accessions from USDA-ARS, 25 accessions from ICRISAT, 42 accessions from EMBRAPA, and 60 accessions from ILRI were investigated with the aim of enhancing the genetic diversity and increasing the population size of the Napier grass collection in the ILRI forage genebank. The plants were genotyped using the DArTseq genotyping by sequencing (GBS) method. A total of 96,321 SNP and 96,454 SilicoDArT genome-wide markers were generated. Of these, 85,619 (89 %) SNP and 73,573 (76 %) SilicoDArT markers were mapped onto the fourteen chromosomes of the Napier grass genome. Over 46 % of the SilicoDArT and 35 % of the SNP markers had a polymorphic information content (PIC) value above 0.25. Genetic diversity analysis using a subset of independent and highly polymorphic SNP markers detected moderate genetic variation among the collections, particularly between the progeny and the four collections, and indicated the presence of unique diversity in the progeny plants. The progeny plants clustered separately from the collections in multivariate analysis (UPGMA and PCA) and scattered across the PCA plot showing the high genetic variability within progeny plants. Variance components calculated by analysis of molecular variance (AMOVA) were also significant both among and within populations, however, more variation (90 %) was within than among populations (10 %). Multilocus genotype (MLG) analysis using Nei's genetic distance identified 33 potential duplicates (0.02 threshold) suggesting the remaining 312 genotypes as distinct. Of these 312, 81 genotypes were selected as unique genotypes (0.26 threshold) by MLG analysis. The unique genotypes were mainly from progeny plants while the ILRI collection had most of the potential duplicates. The results of this study provide a useful information for the Napier grass breeding strategy and enhancement of genetic diversity in the ILRI collection. The results also provide a useful guide for the management and conservation of the collection in the ILRI forage genebank.

Keywords: Duplicate genotypes, elephant grass, genetic diversity, progeny plants, unique genotypes

Contact Address: Meki Shehabu Muktar, International Livestock Research Inst. (ILRI), Feed and Forage Development, Addis Ababa, Ethiopia, e-mail: mshehabu@gmail.com

Rationale and Motivation of Rural Farmers in Adopting Floating Agriculture in the Haor Region, Bangladesh

KHONDOKAR HUMAYUN KABIR, MOHAMMED NASIR UDDIN, SOURAV SARKER

Bangladesh Agricultural University (BAU), Agric. Extension Education, Bangladesh

Floating farming, a climate-smart practice, is a response to the challenges facing agriculture in wetland areas due to climate change. However, the motivation in adopting floating agriculture in wetland areas (also known as Haor) in Bangladesh is slow. This study aims to identify the factors that motivate and barriers that inhibit the adoption of floating agriculture in the Haor region of the Kishoreganj district, Bangladesh. To achieve the objective, Roger's five-stage innovation-decision model is used. Data from a sample of 120 Haor rural farmers is collected using a pre-tested structured questionnaire via a personal interview. Focus group discussions, key informant interviews, and secondary sources were used to gather additional data. A binary logistic regression is used to identify the factors that predict farmers' motivational actions in adopting floating agriculture. In addition, the order of rank is used to identify the obstacles that prohibit farmers from implementing floating agriculture. The findings demonstrate that education, agricultural training related to floating agriculture, credit received, prior conditions, communication behaviour, trialability and observability, complexity in practicing floating agriculture of the Haor farmers are the motivating factors in adopting floating agriculture. The result also demonstrates that the physical and natural barriers that inhibit the adoption of floating farming. Farmers' dependence on full-time farming and engagement with new methods may be a motivating force in the adoption of floating farming, but their continued practice of growing grain crops can demotivate them to do so. Finally, this study provides suggestions for increasing the motivation to adopt floating farming among the farmers in wetland areas.

Keywords: Adoption, Bangladesh, climate-smart agriculture, floating agriculture, Haor, Rogers model

---

Contact Address: Khondokar Humayun Kabir, Bangladesh Agricultural University (BAU), Agric. Extension Education, 2202 Mymensingh, Bangladesh, e-mail: kabirag09@bau.edu.bd

Smallholder Typologies in Pigeon Pea-Based Farming Systems: An Application to Rural Uganda

Dorothy Birungi Namuyiga, Till Stellmacher
University of Bonn, Center for Development Research (ZEF), Germany

Worldwide, over 500 million smallholders exist who are diverse and mostly operate in resource-constrained environments that are heterogeneous. Smallholders produce over 30 percent of the food supply and operate 24 percent of the global gross agricultural area, yet they are seldom recognised as critical entities contributing to food security globally. Pigeon pea is an orphan legume crop grown in the semi-arid tropics in Africa and Asia. Despite smallholders' importance, there is little scholarship regarding their diversity and heterogeneity. Against this backdrop, this study sought to cluster smallholders within the pigeon pea-based farming systems and identify the distinguishing characteristics. We gathered data through group discussions to develop a hypothesis about the farming system, in addition to a cross-sectional survey with 257 smallholder households in northern Uganda. Using multivariate analysis, we employed both Principal Component Analysis (PCA) and Hierarchical Cluster Analysis (HCA)), and construct six typologies. The six distinct smallholder clusters consisted of 55, 35, 43, 55, 43, and 28 members respectively. Cluster one, two, and six are more livestock reliant because much of their land is dedicated to livestock grazing in addition to owning high TLUs. Cluster three, four, and five are crop-reliant since much of their land is dedicated to crop production and own high units of farm assets. There are further distinctive differences between clusters in terms of farming experience, pigeon pea sales demonstrating a level of commercialisation and access to and use of farm remedies, providing evidence of the large differences between individual households. These results are evidence for targeted and differentiated programs and policies by the government and probable funders.

Keywords: Livelihoods, rural-Uganda, smallholders, targeting, typologies

Contact Address: Dorothy Birungi Namuyiga, University of Bonn, Center for Development Research (ZEF), 53113, 3, 53113 Bonn, Germany, e-mail: dbirungi8@gmail.com

The Seedball Technology also Enhances the Panicle Yield of Sorghum in the Sahel

Charles Ikenna Nwankwo[1], Hannatou Moussa Oumarou[2], Ali Maman Aminou[3], Ludger Herrmann[1]

[1]*University of Hohenheim, Soil Chemistry and Pedology, Germany*

[2]*National Agronomic Research Institute of Niger, Crop Production, Niger*

[3]*Fuma Gaskiya, Crop Production, Niger*

Sorghum and pearl millet are the major staples often produced by the Sahelian smallholder farmers. Following the favourable experience of seedball technology in pearl millet production, transferring the seedball technology to sorghum was frequently requested by farmers. Therefore, the objective of this study was to test the optimised sorghum seedball on-site in the Sahel region. In 2020 planting season, 57 on-farm trials were conducted in over 20 villages of Maradi region in Niger. Conventionally sown and seedball-derived pearl millet crops were grown using "farmer-optimised" simple split plot designs with three treatments, mainly: (i) farmers' practice as control i.e., conventional sowing and 3.0 cm diameter-sized seedballs (made from 80 g sand + 50 g loam + 25 ml water +0.9 g seeds as standard recipe), which contained either (ii) 4.5 g wood ash or (iii) 1.5 g mineral fertiliser (NPK) as effective nutrient compounds. Results showed that seedballs do not suppress seedling emergence in sorghum. Compared to control, NPK- (N = 21) and wood ash-amended (N = 36) seedballs increased panicle yield by 40 and 15 %, respectively. The yield effect was positive in several villages. However, there was a high variation in panicle yield ranging between 200 and 2730 kg ha^{-1}. Seedball technology appears suitable for sorghum production, and seems too have a high chance of adoption by the local farmers. However, in order to enable more precise recommendations for its application further on-farm trials are necessary that differentiate the effects of soil type, gender, post-emergence fertilisation, sowing time and other management practices.

Keywords: Maradi farmers, Sahel local farmers, sorghum seedball, sorghum yield, technology transfer

Contact Address: Charles Ikenna Nwankwo, University of Hohenheim, Soil Chemistry and Pedology, Emil-Wolff-Str. 12a, 70599 Stuttgart, Germany, e-mail: charsile2000@yahoo.com

Genetic and Morphological Stability of Autopolyploid *Thymus vulgaris* L. and Changes in its Anatomy and Physiology

Yamen Homaidan Shmeit[1], Pavel Novy[1], Rohit Bharati[1], Katerina Hamouzová[1], Jana Žiarovská[2], Ingrid Melnikovová[1], Eloy Fernández Cusimamani[1]

[1]*Czech University of Life Sciences Prague, Fac. of Tropical AgriSciences, Czech Republic*
[2]*University of Agriculture in Nitra, Dept. of Genetics and Plant Breeding, Slovakia*

Thymus vulgaris L. is a species of the Lamiaceae family endemic to the West Mediterranean Basin. It is well known as an aromatic and culinary herb, but thyme also has important medicinal properties. The plant produces essential oils rich in monoterpene phenols which are expressed by a wide range of chemo-diversity. The high chemical polymorphism of its essential oils provides various biological activities. Which allows their implementation as a biosphere-friendly and biodegradable substance for plant protection in agriculture. Variety selection and plant breeding techniques such as *in vitro* induced somatic polyploidisation can improve the quality and yield of essential oils in plants. This research aims at assessing the genetic and morphological stability as well as physiological and anatomical changes in a new tetraploid genotype (2n = 2x = 60) of *T. vulgaris* obtained from a diploid control (2n = 2x = 30) using *in vitro* induced polyploidisation. The genetic stability was tested using chromosome counts and repeating the flow cytometric analysis. Chlorophyll fluorescence analysis was performed to show the photosynthesis variability between the tetraploid and diploid plants. The antioxidative activity was determined by analysing the total polyphenol, flavonoids and phenolic acid content. Morphological characteristics such as plant height, number of shoots and plant/branch thickness, length, internodal distances, leaf length and leaf thickness were measured. The anatomical assessment was performed by wood anatomical analysis for the variability of xylem anatomy features in cross-sectional view. Stomata and guard cell size and stomatal density microscopic measurements were also performed. Preliminary results showed that the newly obtained tetraploid genotype of T. vulgaris proved to be genetically and morphologically stable. The physiological and anatomical changes varied between the tetraploid and its diploid counterpart. After a two-year assessment, we also aim to patent the new genotype and introduce it as a capable variety to satisfy the growing demand of *T. vulgaris* essential oil.

Keywords: *Ex vitro*, flow cytometry, *in vitro*, polyploidy, stomata, *Thymus vulgaris*, xylem

Contact Address: Yamen Homaidan Shmeit, Czech University of Life Sciences Prague, Fac. of Tropical AgriSciences, Dept. of Crop Sciences and Agroforestry, Kamycka 933, Prague, Czech Republic, e-mail: yamen.shmait@gmail.com

Adoption of Sustainable Rice Farming Technologies: Perceptions of "One Must Do, Five Reductions" Technologies

Helena Wehmeyer[1], Melanie Connor[2]

[1]*University of Basel, Dept. of Environmental Sciences, Philippines*
[2]*International Rice Research Institute (IRRI), Sustainable Impact Platform, Philippines*

Vietnam's rice sector has been growing rapidly over the past decades due to the fast intensification of rice production. Nevertheless, this development has come with increased environmental degradation. Today, Vietnamese rice farmers face several constraints, especially regarding climate change adaptation. Therefore, the national programme 'One Must Do, Five Reductions' (1M5R) was introduced to mitigate the negative effects of intensive rice farming by introducing farmers to sustainable rice farming practices and technologies. The objective of this study was to determine the adoption and diffusion rate of the recommended 1M5R technologies and examine farmers' perceptions of the technologies. Furthermore, perceived input savings and 12 dimensions of livelihood change were analysed to evaluate the different elements of economic, social, and environmental change. Data were collected by means of a digital survey questionnaire application in An Giang and Can Tho Province in the Mekong River Delta. A farmers' perception survey was conducted in 2019 interviewing 465 farmers. Data were analysed using uni- and multivariate statistics. The findings show that most farmers adopted at least one recommended technology and that they were satisfied with it. The most highly rated benefits were "easy to apply" and "satisfies my preferences". Most farmers had adopted the 1M5R technologies for more than 8 years on average and perceived a reduction in input use as well as beneficial livelihood changes. Nevertheless, the results demonstrate that even with a broad diffusion of 1M5R technologies, it is difficult for farmers to reduce inputs to the recommended amounts. The seed rate and fertiliser application rate remain particularly high. As a result, farmers continue to overuse inputs which negatively affects their agricultural profitability and further deteriorates natural resources. Additionally, considering the devastating effects of climate change on the Mekong River Delta, the long-term adoption of such technologies in combination with attaining the 1M5R input requirements are crucial in helping farmers in their climate change adaptation process. Therefore, extension needs to be adapted to better support farmers in attaining the 1M5R requirements through technology use, and continued research is needed to determine the effects on farmers' socioeconomic situation and the rice agroecosystem.

Keywords: Adoption, input use, sustainable rice farming, technology perceptions, Vietnam

---

Contact Address: Helena Wehmeyer, University of Basel, Dept. of Environmental Sciences current address: International Rice Research Institute, 4031 Los Baños, Philippines, e-mail: h.wehmeyer@irri.org

Temporal Dynamics of Vegetation Cover and Agricultural Development in the High-Atlas and Anti-Atlas of Morocco from 1990 to 2020 Using Landsat 5-7 MSS and 8 OLI/TIRS Data

Nacer Aderdour[1,2], Thanh Thi Nguyen[1], Hassane Rhinane[2], Andreas Buerkert[1]

[1]*University of Kassel, Organic Plant Production and Agroecosystems Research in the Tropics and Subtropics, Morocco*
[2]*Hassan II of Casablanca, Geology, Morocco*

Since the 1990s Morocco's agriculture is characterised by the coexistence and the transformation of modern and smallholder traditional agriculture. In some oases of the Atlas Mountains, the effects of urbanization lead to intensified irrigated agriculture while others are abandoned. To better understand these effects, this study aimed at (1) analyzing the land use land cover (LULC) changes, (2) assessing the dynamics of vegetation and agricultural cover, (3) and determining the drivers of LULC changes at different scales. Based on Landsat data from the 1990s to the 2020s we used an automatic supervised classification of LULC. On-screen visualization and automatic object classification were combined to also analyze high-resolution Corona images of the 1970s and Google Earth images, and Spot 6 data of the 2020s. This approach allowed differentiating between open vegetation, bare land, forest, and water. In the High-Atlas Mountains, classification accuracy was 86% in 1990, 81 % in 2000, 80% in 2010, and 61 % in 2020. Results indicated the share of bare land to amount to 94 % in 1990, 88 % in 2000, 92 % in 2010, and 84 % in 2020, while forest and open vegetation accounted for less than 30% of the total area. During the same period, total forest area doubled from 3090 to 7362 km^2 associated with a reduction of bare land and open vegetation. Vegetation cover was between April and August whereby NDVI values > 0.3 accounted for 15% of the High Atlas Mountains. In the typical oasis of Targa N' Touchka, abandoned land increased by 0.57 km^2 while the agricultural area shrunk by 1.72 km^2 from 1970 to 2020. This transformation reflected the migration of the young population to the cities, making the livelihoods of the remaining oasis farmers more vulnerable.

Keywords: NDVI, remote sensing, rural-urban transformation, urbanisation

Contact Address: Nacer Aderdour, University of Kassel, Organic Plant Production and Agroecosystems Research in the Tropics and Subtropics, Hay Nassim IMM 478 ETG 1 APPT 2 CASA, Casablanca, Morocco, e-mail: nacer.ader1@gmail.com

Effects of Staggered Planting Periods and Potassium Fertilisation on the Performances of Cassava Cultivars in South-Kivu, Democratic Republic of Congo

Rutega Damas Birindwa[1,2], Wivine Munyahali[2], Pieterjan De Bauw[1], Gerd Dercon[3], Roel Merckx[1]

[1]*KU Leuven, Dept. of Earth and Environmental Sciences, Belgium*

[2]*Catholic University of Bukavu, Fac. of Agronomy, DR Congo*

[3]*International Atomic Energy Agency, Division of nuclear techniques in food and agriculture, Austria*

Cassava is the most important crop in DRC, where it is serving as a staple and a cash crop. Over the last years, cassava production seems increasingly constrained by disturbances due to climate change. To respond to these more frequent climate disturbances, farmers adjust the planting period in order to cope with erratic events that are most crucial during the first 100 days after planting. While this period coincides normally with the most regular rainfall, even during these 100 days some untimely interruptions for fairly long periods can occur which affects the performance of cassava.

A full-factorial field experiment was conducted in two contrasting sites in South-Kivu (DR-Congo) including the highland and forest region (i.e., Kalehe) versus the dry savannah at medium altitude (i.e., Uvira). Effects of planting period (3 different planting dates), two cultivars (landrace versus improved), and fertiliser (NP versus NPK) on yields of cassava were tested. Yield components were collected at 12 months after planting by determining the above-ground biomass (stems and leaves), fresh storage roots yield, and harvest index.

Cassava yield was influenced by location, in favour of the highland, the cultivar, being larger for the improved cultivar coinciding with the better harvest index. The yield was also influenced by the planting periods. The earlier the planting was done, the larger was the yield. As far as fertilisation was concerned, the presence of K led also to the highest yield levels. Interacting with planting time, the presence of K was able to mitigate the negative effect of late planting on yield.

Keywords: Biomass, cultivar, fertilisation, harvest index, *Manihot esculenta*, planting period, potassium

Contact Address: Rutega Damas Birindwa, KU Leuven, Dept. of Earth and Environmental Sciences, Leuven, Belgium, e-mail: damas.birindwarutega@kuleuven.be

Evaluating the Sustainability of Maize-Legume Strip Cropping Technology in the Context of Smallholder Farmers in Northern Ghana

Nurudeen Abdul Rahman[1], Asamoah Larbi[2], Bekele Hundie Kotu[1], Irmgard Hoeschle-Zeledon[3]

[1]*International Institute of Tropical Agriculture (IITA), Ghana*

[2]*Agriculture and Food Security, United States*

[3]*International Institute of Tropical Agriculture (IITA), Nigeria*

Agricultural intensification, producing more output per unit area of land through efficient use of resources has increased food production by smallholder farmers in West Africa including Ghana. Many agricultural technologies have been promoted to intensify the smallholder farming in Ghana, but there is a dearth of evidence on the sustainability of these technologies. A 3-year (2014–2016) on-farm study was conducted to evaluate the sustainability of maize-legume strip cropping technology in northern Ghana. The strip cropping technology involved the intercropping of maize with two widely grown legumes (cowpea and groundnut) in northern Ghana. Four strip cropping options: two rows of maize and two rows of cowpea or groundnut (2M:C and 2M:2G) and two rows of maize and four rows of cowpea or groundnut (2M:4C and 2M:4G) and sole cropping of each crop were laid-out in a randomised complete block design with five replications. We used Sustainable Intensification Assessment Framework (SIAF) to assess the sustainability of the above treatments. The SIAF measures the sustainability of agricultural technology based on five domains: productivity, economic, environment, human and social domains. We conducted the assessment in three steps: 1) Measured selected indicators from the five SIAF domains which were useful to answering research question, 2) Converted the actual values of the indicators into scores using the Min-Max scaling (scores lie within 0–1 range) and 3) Aggregated the scores of the indicators under each of the domain and Calculated sustainability index using geometric rules considering each SIAF domain as an edge of a pentagon. Specific indicators used in this study were: grain yield and land equivalent ratio (productivity), net income and variability of net income (economic), biological nitrogen fixation and intercepted photoactive radiation (environment), calorie and protein production (human) and technology rating by gender (social). The results showed that the 2M:2G strip cropping option recorded sustainability index of greater than one, indicating better sustainability than the other treatments. This suggests that 2M:2G strip cropping option can enhance the sustainability of smallholder farming systems in northern Ghana and similar agro-ecologies in West Africa and better their wellbeing through its effects on income, nutrition, and gender equity.

Keywords: Cowpea, groundnut, maize, northern Ghana, SIAF, sustainability

---

Contact Address: Nurudeen Abdul Rahman, International Institute of Tropical Agriculture (IITA), Education Ridge, 0233 Tamale, Ghana, e-mail: a.nurudeen@cgiar.org

Compost in Growing Vegetables: An Effort in Reducing Organic Waste Disposal into the River

Tedi Yunanto, Adang Saputra, Suparno, Wahid Sugiman, Farisatul Amanah, Jasin Nugraha, Ratu Alamanda, Firza Farid

Ministry of Energy and Mineral Resources, Indonesia

Citarum is the longest river in West Java Province, Indonesia, heavily polluted by waste from nearby activities, such as household waste. The community dumps organic and inorganic waste into the Citarum River every day. Based on the data from the local government in Karawang Regency, West Java Province, the total waste disposed of was about 900 tons per day with the estimation of 60 % of organic waste from nearby activities. This condition is one of the causes of the overflowing of the Citarum River during the rainy season and causing a flood. Many actions had been taken due to reducing organic waste. Organic waste compost has been developed for the surrounding community along the river bank. This project aims to: train people to reduce the organic waste dumping to the river; transform the waste into a valuable product such as compost; test the productivity of the compost as a growing media for vegetables. Twenty-five participants have joined the compost training which only eight of them continued to the planting experiment. The aerobic composting was conducted using microbes. After composting organic waste from the settlement area, the compost was put into a polybag with a diameter of 30 cm. The vegetable seed was planted inside the polybag (compost media) and placed in the participant's yard. The result showed that Ipomea reptans Poir (kangkong), Brassica juncea subsp. integrifolia var. crispifolia (green mustard) and Brassica rapa subsp. Chinensis (bok choi) crop was harvested in only a month after first planting and repeated monthly. In the early 2 two months, total productivity was 34 kgs of kangkong, 16.5 kgs of green mustard, and 18.2 kgs of bok choi. In general, the products were used for daily consumption and local direct selling. The estimated prices were: IDR 50,000 kg^{-1} for kangkong; IDR 15,000 kg^{-1} for green mustard; and IDR 30,000 kg^{-1} for bok choi. The revenue from the sale was used to buy seeds for the following plants. First, form the evaluation; this project requires a community organisation in charge of management on a large scale. The organisation also needs capacity building to empower the sustainability of the project.

Keywords: Organic waste, polybag plantation, waste reduction

Contact Address: Tedi Yunanto, Ministry of Energy and Mineral Resources, Bandung Polytechnic of Energy and Mining, Jl. Jenderal Sudirman No. 623, 40211 Bandung, Indonesia, e-mail: genom.tedi@gmail.com

Nitrogen Source Affecting the Competitiveness between Lowland Rice and Weeds under Low and High Vapour Pressure Deficit

Duy Hoang Vu[1], Sabine Stürz[2], Folkard Asch[2]

[1]*Vietnam National University of Agriculture (VNUA), Dept. of Cultivation Sciences, Vietnam*

[2]*University of Hohenheim, Inst. of Agric. Sci. in the Tropics (Hans-Ruthenberg-Institute), Germany*

Implementation of water-saving irrigation practices in lowland rice results in increased availability of nitrate (NO_3^-) in the soil and favours germination of upland weeds. Since plant species show a specific preference for either ammonium (NH_4^+) or NO_3^- as nitrogen (N) source, changes in both soil NO_3^- concentration and weed flora may affect the competition between rice and weeds. In addition, vapour pressure deficit (VPD) affects leaf gas exchange and plant growth, which may alter the N uptake of plants. The study was conducted to evaluate the effects of N source on the competition between two rice varieties (NU838 and KD18) and two weed species (*Echinochloa crus-galli* and *Solanum nigrum*) at low and high VPD. Rice and weeds were grown hydroponically as monoculture or mixed culture with different N sources (75%/25 % or 25%/75 % NH_4^+/NO_3^-). Independent of N source, rice and *E. crus-galli* took up a larger share of NH_4^+, whereas *S. nigrum* took up a larger share of NO_3. A high correlation between water uptake rate and total N uptake rate was found in *S. nigrum* and *E. crus-galli* but not in rice. Moreover, in contrast to the other species, growth of *S. nigrum* was not reduced at high VPD, resulting in increased competitiveness and total N uptake of the weed. In competition, high NO_3^- increased the competitiveness of NU838 against *E. crus-galli*, but decreased the competitiveness of NU838 against *S. nigrum*. Our results suggest that increased availability of NO_3^- in aerobic rice soils may be disadvantageous for rice in competition with upland weeds, especially at high VPD, whereas, it may reduce pressure of common lowland weeds.

Keywords: Alternative wetting and drying, nitrogen source, plant competitiveness, water-saving irrigation, weeds in lowland rice

Contact Address: Duy Hoang Vu, Vietnam National University of Agriculture (VNUA), Dept. of Cultivation Sciences, Ngo Xuan Quang - Gialam, 131000 Hanoi, Vietnam, e-mail: vdhoang87@gmail.com

Impact of the Increase in Aridity Levels on the Value of Insect-Pollination in Drylands: Farmers' Perspective

Soukaina Anougmar[1], Jean-Michel Salles[1], Nicola Gallai[2], Stefanie Christmann[3]

[1]*Institut Agro Montpellier (INRAE), CEE-M, France*

[2]*ENSFEA Toulouse, France*

[3]*Intern. Center for Agricultural Research in the Dry Areas (ICARDA), BCI, Morocco*

Pollinating insects are facing worrying declines in many parts of the world. Farmers in drylands in low and middle income countries are particularly vulnerable to pollinator decline because of the growing dependency on insect-pollination. The increase in the levels of aridity, as a result of climate change, and agricultural intensification are two main factors endangering pollinators in dry areas. Actions for the protection of pollinators depend on farmers' awareness of the importance of insect mediated pollination, the impact of the level of aridity on their perceptions of pollinating insects and how the increase in aridity levels will affect their perceptions in the future. In this study, we conduct a discrete choice experiment with Moroccan farmers to a) explore their preferences for insect mediated pollination and, b) examine the impact of aridity on their preferences and the value they attribute to its benefits. The survey took place in 5 different climatic ranges within the Moroccan territory, dry sub-humid, semi-arid (irrigation-based), semi-arid (rainfall-dependent), arid and hyper-arid, with a total of 492 farmers. Our findings show that farmers have a high willingness to pay for the protection of insect mediated pollination benefits and that the contribution of pollinators to the quality of fruits and vegetables is the most valued benefit. The study demonstrates that the aridity has a strong effect on farmers' preferences and that farmers' willingness to pay increases at higher levels of aridity. It, also, shows that the value that farmers attribute to the benefits of pollinators will increase as a result of climate change.

Keywords: Climate change, discrete choice experiment, insect pollination, level of aridity

Contact Address: Soukaina Anougmar, Institut Agro Montpellier (INRAE), CEE-M, Montpellier, France, e-mail: soukaina.anougmar@gmail.com

Land Use Planning with Agro-Ecosystem Paradigm Approach in Rural Areas of Arid Regions of Iran

Naghmeh Mobarghaee[1], Houman Liaghati[2], Mostafa Keshtkar[1]
[1]*Shahid Beheshti University, Environmental Planning, Iran*
[2]*Shahid Beheshti University, Environmental Sciences Research Institute, Iran*

Increasing demand for different land uses and physical changes in the land, can lead to changes in ecosystem functions. This has created an increasing research focus on rural land use planning issues. In this case, it is very important to know the structure of the land, how to use it, and identifying the actual and potential appropriated areas for each land use. Most of planners have a relative understanding of importance of ecosystems and a limited use of methods that lead to sustainable exploitation. Lack of analysis in structure and function of the ecosystem leads to the formation of conflicts between land capabilities and consumers. Therefore, there is an urgent need to identify, quantify and mapping the ecosystem services in order to better understanding of benefit provides. Therefore, accurate assessment of ecological potential and its application in agro-ecosystem plan, can play an important role in rural stability and imbalance elimination. In this study, the evaluation of potential macro land use, was performed by considering the agro-ecosystem plan integration with land use planning, by using the optimal Multi-Objective Land Algorithm (MOLA) in rural areas of Isfahan province. The results show that more than 50 % of the area has ecological potential for arboriculture and extensive recreation. But contrary to the existing procedure, there is no possibility for agriculture, settlement development and industrial development. Also the region has a very good potential in solar energy that is a very important parameter in sustainable development. Based on socio-economic information, policies and opinions of experts and with the aim of developing the use of new energy and reducing the risk of natural and man-made disasters, land use planning was developed based on agro-ecosystem promotion. The results showed that by using the new spatial paradigm, the ability of performing land use planning to achieve sustainable development, has been provided in rural area. Also it can be concluded that by using this paradigm, the spatial organisation of landscape can create a reduction in environmental and human problems and moving towards sustainable development in these unique and valuable areas.

Keywords: Agro-ecosystem paradigm, arid regions of Iran, land use planning, rural area

---

Contact Address: Naghmeh Mobarghaee, Shahid Beheshti University, Environmental Planning, Velenjak, Tehran, Iran, e-mail: n_mobarghei@yahoo.com

Cultivation of Cowpea Challenges in West Africa for Food Security: Analysis of Factors Driving Yield Gap in Benin

Firmin Anago[1], Brice Oussou[2], Gustave Dagbenonbakin[2], Lucien Amadji[1]

[1]*University of Abomey-Calavi, Lab. of Soil Sciences of Crop Production School, Benin*

[2]*National Institute of Agricultural Research of Benin, Laboratory of Soil Science, Water and Environment, Research Agricultural Center of Agonkanmey, Benin*

Feeding the world in 2050 need to found the ways to boost yields of the main local crops. Among those crops, cowpea is one of the important grain legumes that is playing an important role in the livelihood of millions of people in West Africa, especially in Benin. Unfortunately, cowpea on-farm yields are very low. In order to understand the main factors explaining cowpea yield gaps, we collected and analysed detailed survey data from 298 cowpea fields in Benin during the 2017, 2018 and 2019 rainy seasons, respectively. Composite soil samples were collected from cowpea fields at flowering stage and analysed at laboratory. Data on-farm field management practices and field conditions were recorded through interviews with 606 farmers. Average cowpea grain yields were low and seldom surpassed 700 kg ha^{-1} on farmer's fields. Significant differences were observed between cowpea grain yields from northern to southern Benin ($p < 0.05$), and the lowest yields were observed in northern Benin. These low yields are related to crop management practices, soil nutrient contents, and the interaction ofboth. According to the model of regression tree from northern to southern Benin, mineral fertiliser used, insecticide sprays to control pests, and improvement of P, N, K, and cation sum content into top soil would increase cowpea grain yields. Insect pest, diseases, and soil fertility decline are the highest major constraints that limit cowpea cultivation in Benin. Future research should focus on formulating specific fertiliser for effective cowpea cultivation in Benin, as well as insect pest, and diseases control.

Keywords: Cowpea yield, crop management practices, food security, soil fertility

Contact Address: Firmin Anago, University of Abomey-Calavi, Lab. of Soil Sciences of Crop Production School, Abomey-Calavi, Benin, e-mail: firmin.anago@gmail.com

Molecular Characterisation of Apomixis in *Cenchrus ciliaris* and its Implication for Improvement

Alemayehu Teresssa Negawo[1], Ermias Habte[1], Meki Shehabu Muktar[1], Alieu Sartie[2], Chris S. Jones[3,1]

[1]*International Livestock Research Inst. (ILRI), Feed and Forage Development, Ethiopia*

[2]*Seed Systems Specialist, Pacific Community (SPC), Fiji*

[3]*International Livestock Research Institute (ILRI), Kenya*

Livestock production is a key element in addressing sustainable development goals, and forage grasses play an important role in enhancing the contribution of livestock. Despite playing this key role, the contribution of grasses to sustainable livestock production is limited by the lack of enough true-to-type seed to support forage production. Buffel grass is an important polymorphic forage grass grown worldwide for its forage quality and agronomic characteristics including high biomass yield, drought tolerance and adaptability to a wide range of soil conditions and agro-ecologies. In apomictic species like buffel grass, breeding efforts are also hindered by the lack of information on the reproductive behaviours of the existing germplasm. Thus, generating information of the reproductive behaviour of buffel grass in the collection held in the International Livestock Research Institute (ILRI) Genebank would be useful for improvement and utilisation of this forage. Here we report on the characterisation of the reproductive nature of the buffel grass collection using molecular markers linked to the apomixis sequence genomic region (ASGR). The results showed the amplification of genomic DNA sequence linked to ASGR in most of the accessions held in the collection. Amplification of genomic regions linked to both apomictic and sexual reproduction modes was observed in seven accessions in the collection. No accession was identified with a purely sexual mode of reproduction. Overall, the results show that most of the accessions in the collection are obligate apomictic lines while a few are facultative apomictic lines. The identified facultative lines could be used to develop sexual crosses which could be used to enhance the improvement of this forage and thereby its contribution to improved livelihoods for millions of households in low and middle income countries that are dependent on livestock production.

Keywords: Apomixis, *Cenchrus ciliaris*, characterisation, improvement

Contact Address: Chris S. Jones, International Livestock Research Institute (ILRI), Nairobi, Kenya, e-mail: c.jones@cgiar.org

Socio-Economic and Ecological Changes in Farming Systems of Targa N'Touchka (Anti-Atlas, Morocco)

Johanna Reger[1], Irene Holm Sørensen[2], Andreas Buerkert[3], Eva Schlecht[1]

[1]*University of Kassel / University of Göttingen, Animal Husbandry in the Tropics and Subtropics, Germany*

[2]*University of Kassel / University of Göttingen, Social-Ecological Interactions in Agricultural Systems, Germany*

[3]*University of Kassel, Organic Plant Production and Agroecosystems Research in the Tropics and Subtropics, Germany*

For many centuries Berber communities have settled in Moroccan mountains and practised oasis agriculture combined with livestock husbandry. Traditionally, livestock husbandry provided a savings and risk mitigation strategy, year-round income, and a high contribution to covering the nutritional needs of the rural population. Currently, these few remaining mountain oasis systems in Morocco experience rapid socio-economic and ecological changes, which affects the livelihoods of the communities depending on them.

To identify the ongoing transformations of the agricultural sector in a typical mountain oasis, 61 semi-structured interviews were conducted with farmers in Targa N' Touchka (29.886667, -9.204444). The survey allowed to characterise the current situation of agriculture and its contribution to household income, food security, and gender-specific task division. While traditional agricultural practices lose importance and only provide a supplementary income, migrants' remittances and other non-agricultural income sources are the main pillar of rural livelihoods, accounting for 81 % of the household income on average. Migration and off-farm work are male-dominated with a direct influence on the socio-economic conditions in the mountain oasis. Left-behind family members are mostly women, leading to an increasing importance of female labour in all agricultural tasks. The latter is reinforced by formal education of family members, of whom 80 % do not participate in farming activities (age group 0 to 20 years). In combination, the declining economic importance of agriculture and lack of agricultural labour lead to extensification of cereal crop farming or even an abandonment of fields. In contrast, cultivation of alfalfa as livestock feed has gained importance at the expense of other crops. While only 39 % of households are also herding their small ruminants, remaining animal dietary requirements are covered to a high extent by green fodder and by external fodder resources, purchased at the local market. The shift from cereal to green fodder cultivation points to the dwindling importance of food self-sufficiency. The targeted rearing of animals for consumption or sale at times of high demand during religious holidays seems to be a labour-saving adaptation to the new circumstances.

Keywords: Morocco, oasis agriculture, transformations

Contact Address: Eva Schlecht, University of Kassel / University of Goettingen, Animal Husbandry in the Tropics and Subtropics, Steinstraße 19, 37213 Witzenhausen, Germany, e-mail: tropanimals@uni-kassel.de

Response of Dual-Purpose Sorghum Varieties to Fertiliser and Sowing Dates in Mali's Sudanian Zone

Madina Diancoumba[1], Omonlola Nadine Worou[1], Baloua Nebie[1], Abdoulaye Tangara[1], Mamourou Sidibe[1], Aly Togo[2], Mbairassem Bendigngar[1], Birhanu Zemadim[1]

[1]*International Crops Research Institute for the Semi-arid Tropics, Crop Improvement, Mali*

[2]*DA-TA Flaq's Technologies, SIDNET-Agronomy, Mali*

Sorghum is an important multipurpose' traditional cereal crop widely grown and consumed by smallholder farmers in the drylands of Africa. Low soil fertility, limited use of fertiliser and erratic rainfall distribution induced by climate change and land degradation, underline the need to ascertain an appropriate sowing time and an effective fertiliser strategy to reduce risks and increase sorghum productivity. The objective of this study is to analyse the interaction between sowing date (early, medium and late) and fertiliser types (mineral, organic, the combined and the control) to assess the synergies and trade-offs between these agricultural practices and to evaluate their impact on the agronomic performance of two dual-purpose sorghum varieties (Soubatimi and Peke) in the Sudanian region of Mali. Experiments were carried out at ICRISAT-Mali research station, during 2019 and 2020 rainy seasons and laid out in a split-split plot design. Destructive and non-destructive measurements on crop phenology parameters, leaf number and size, leaf area index and crop biomass at three stages and crop yield at harvest were assessed as well as changes in soil volumetric water content during crop growth. Results show that the highest average grain and stover yields were obtained with Soubatimi followed by Peke both sown early (DS1) under combined fertiliser treatment (mineral+organic). In general, a negative effect of late sowing was observed on grain yields in both varieties. In 2019, the interactions of sowing date × variety × fertiliser types were observed to impact grain and stover yields in the two varieties. This study revealed the potentials and constraints of production, of two improved dual-purpose varieties that are important sources of feed for livestock and food for human consumption.

Keywords: Dual purpose sorghum, fertiliser, Mali, Sudanian zone, sowing date

Contact Address: Omonlola Nadine Worou, International Crops Research Institute for the Semi-arid Tropics, Crop Improvement, Route de Guinee, 320 Bamako, Mali, e-mail: n.worou@cgiar.org

Salinity Constraints and their Implications for Smallholder Farming in North-aceh, Indonesia

Elvira Sari Dewi, Issaka Abdulai, Gennady Bracho-Mujica, Reimund P. Rötter

University of Goettingen, Dept. of Crop Sciences: Tropical Plant Production and Agricultural Systems Modelling (TROPAGS), Germany

Smallholder farmers in Indonesia's coastal lowlands continue to face climate-related challenges threatening their livelihoods such as increased salinity on agricultural land. We investigated the effects of salinity on farming practices, income, and challenges perceived for crop production in the Blang Nibong village in North Aceh. Indonesia. We conducted a survey of 120 smallholder farmers chosen in consultation with local leaders, considering their agricultural activities and susceptibility to salinity. Open and closed questions were formulated to assess farmers' perception of major crop production constraints (e.g. salinity, intra- and inter-annual variability of rainfall and temperatures), and potential adaptation strategies to better cope with these constraints. The study reveals that for decades, farmers in the study region have primarily grown rain-fed rice using traditional monoculture techniques. Results also indicate that all farmers (respondents) perceived salinity as the primary crop production constraint leading to plant mortality, decreased soil health and water quality, limited plant growth, and low yields. Farmers have also indicated that the high interannual variability in the duration and amount of growing season rainfall remains a major constraint for crop production in the region. Additionally, an area that can be cultivated is also relatively limited (> 0.5 ha), resulting in low total production. Thus, results from the survey also show the implications reflected in farmers' income. In fact, farming activities are not contributing positively to farmers' income, and off-farm activities generate the greatest proportion of income. Based on farmer's efforts to overcome the salinity problem on their farms, the surveyed households were divided into adaptive and non-adaptive farmers. The option preferred by the non-adaptive group is to convert their land to pasture (81 %) while the adaptive group would rather prefer the strategy of improving the irrigation system (77 %) to be implemented. Other strategies mentioned, but with low interest by farmers, include integration of crops and livestock, grow salt-tolerant rice cultivars, diversification of crop rotations, improvement of soil quality, and maintenance of land cover.

Keywords: Adaptation strategies, farmer's perception, lowland coastal farming, salinity constraints, salinity risks

Contact Address: Elvira Sari Dewi, University of Goettingen, Dept. of Crop Sciences: Tropical Plant Production and Agricultural Systems Modelling (TROPAGS), Grisebachstraße 6, 37077 Göttingen, Germany, e-mail: elvira@unimal.ac.id

Maize-Jackbean Intercrop as Influenced by Different Mixture Proportions and Planting Patterns

Ademola Adebiyi[1], Philip Adetiloye[2], Olusegun Adeyemi[2], Olusegun Oduwaye[2]

[1]*International Institute of Tropical Agriculture (IITA), Root & Tuber Cropping Systems Agronomy, Nigeria*
[2]*Federal University of Agriculture, Abeokuta, Department of Plant Physiology and Crop Production, Nigeria*

Intercropping is attractive to small-scale farmers for efficient growth resource utilisation and risk reduction caused by crop failure in sub-Saharan Africa. Despite the adherence to intercropping, there is a lack of sufficient knowledge of the best planting patterns and intercrop proportions for jackbean in the existing cereal-legume based intercropping. Therefore, this study evaluated the performance of maize-jackbean intercrop as influenced by different mixture proportions and planting patterns. The study was a 4 × 2 factorial experiment fitted into a randomised Ccmplete block design, with three replicate conducted at the Federal University of Agriculture, Abeokuta in 2015 and 2016. Treatments were mixture proportions MP (½:½, ½:1, 1:½ and 1:1) of maize: jackbean respectively and planting patterns PP (alternate single row (ASR) and alternate double row (ADR)). Data collected on growth parameters, yield and yield components of both crops were subjected to Analysis of Variance at 5 % probability and means were separated using the LSD. MP (1:½) significantly increased plant height, leaf area, LAI, the number of leaf and leaf area duration of maize producing a maize grain yield of (3.91 $t\,ha^{-1}$) and (5.50 $t\,ha^{-1}$) in 2015 and 2016 respectively. Similarly, growth and seed yield of jackbean were significantly increased with MP (1:1) producing the highest jackbean seed yield (0.46 tha^{-1}) and (0.65 $t\,ha^{-1}$) in both years respectively and MP (1 :½) significantly had the highest LER of 1.42 and 1.02 in both years respectively. ASR increased the growth of maize and had a higher yield (2.48 $t\,ha^{-1}$) and (4.06 $t\,ha^{-1}$) in both years compared to ADR. Contrarily, ADR produced higher growth and yield of Jackbean (0.37 tha^{-1} and 0.52 tha^{-1}) in both years respectively. Conclusively, maize-jackbean MP 1: ½ (53,333 plants/ha: 15,015 plants/ha) was appropriate for increased productivity of maize, and the use of the ASR pattern improved the growth and yield of maize in the maize-jackbean intercropping.

Keywords: Intercropping, planting-patterns, sub-Saharan Africa

Contact Address: Ademola Adebiyi, International Institute of Tropical Agriculture (IITA), Root & Tuber Cropping Systems Agronomy, Oyo Road, 200001 Ibadan, Nigeria, e-mail: a.adebiyi@cgiar.org

Shift in Cropping System: Is a Sustainable Means of Improving Crop Production and Maintaining Soil Health?

Md. Amirul Islam[1], Shyam Pariyar[1], Timothy J. Krupnik[2], Mathias Becker[1]

[1]*University of Bonn, Inst. Crop Sci. and Res. Conserv. (INRES) - Plant Nutrition, Germany*

[2]*International Maize and Wheat Improvement Center (CIMMYT), Sustainable Intensification Program, Bangladesh*

Sustainable intensification has been getting popular in terms of crop production, enhancing resilience to climate induced stresses. It comprised with various integrated approaches to maximise crop yields while maintaining soil quality. Thus, changes in cropping pattern, i.e. shifting in cropping systems might responsible for producing higher yield whilst arable land is limiting. In recent decades, there is increase in freshwater demand for agricultural purposes. Due to higher demand of water, there is a possibility of reduction in traditional rice or rice-wheat system in South Asia including Bangladesh which can be altered by maize, legumes, vegetables. Previous study conceptualised the most accepted concept of land use transition stages, where there is tendency of replacing dry season rice by high value crops that termed as "diversification phase". The replacement in cropping systems is mostly occurring because of water shortage in a region including other factors such as economy, climate change and globalisation. An experiment conducted in Africa revealed that shifting from pastoral land to crop cultivation occurred due to unavailability of economic alternatives. Such system shifts are important to generate more income by the marginal farmers and fulfiling the demand by urban consumers. It was reported that shifting from rice to maize improves the yield of dry season crop along with affecting organic matter, C and N present in soil as well as interfering availability of P and other micronutrients. In addition, the minimum water conserving crop cultivation cause improvement in the declining trend of soil fertility status e.g. soil aeration improves soil C mineralisation and augmented N loss. In this review paper, we highlight the potential effects and trade-off of system shifts on the availability of nutrient fluxes, carbon mineralisation and water use efficiency. These influence the supply-demand synchrony of nutrients and water management strategy in the newly emerged cropping systems, especially focussing on south Asian agriculture systems.

Keywords: Crop production, resource use efficiency, supply–demand synchrony, sustainable intensification, system shift

Contact Address: Md. Amirul Islam, University of Bonn, Inst. Crop Sci. and Res. Conserv. (INRES) - Plant Nutrition, Karlrobert-kreiten Strasse 13, 53115 Bonn, Germany, e-mail: m.a.islam@cgiar.org

Resilience and Economic Benefits of Water Harvesting and Changing Planting Date in Maize Systems of Semi-Arid Tanzania

Abiud Gamba[1], Kelvin Mtei[2], Todd Rosenstock[3], Anthony Kimaro[2]

[1]*Nelson Mandela African Institution of Science and Technology, Dept. of Sustainable Agriculture, Biodiversity and Ecosystems Management (SABEM), Tanzania*
[2]*Nelson Mandela African Institution of Science and Technology, Dept. of Water and Environmental Science and Engineering (WESE), Tanzania*
[3]*World Agroforestry (ICRAF), Tanzania*

In sub-Saharan Africa (SSA), climate variability and poorly adapted farming practices threaten food security, economic growth and development among rural households. Unpredictable rainfall challenge smallholder farmers to determine the appropriate timing for planting and result in crop failure and declining crop yields. There is inadequate information on the resilience and economic benefits of improved technologies to inform farmers in semi-arid conditions. This study adopted a split plot experimental design in 2017/18 and 2018/19 cropping seasons to test the resilience and economic benefits of select management practices (ox-cultivation/control, tied ridges, Chololo pits and intercropping) and planting dates (early, normal and late) with three replications. Long-term annual rainfall recorded was below average by 40 % which lead to declined yield at 6 % in 2019 cropping season due to droughts. Chololo pits at early and tied-ridges at late planting dates significantly ($p = 0.047$ and $p = 0.001$) increased maize grain yield (3.1 and 3.6 t ha^{-1}) respectively in both cropping seasons as compared with ox-cultivation at normal planting dates. Similar practices had higher biomass yield (ranging from 4.6 to 4.7 t ha^{-1}) and economic benefits of average margin rate of return of 89.82 % and 83.64 % also marginal net return of 426.05 USD ha^{-1} and 460.92 USD ha^{-1} in two cropping seasons. In both cropping seasons, Chololo pits and tied ridges demonstrated the highest resilience on soil moisture retention at 10.8 % and 13 % that influenced maize growth and yield. Chololo pits and tied ridges at early and late planting dates can be recommended as climate-smart because they yield resilience and economic productivity benefits that smallholder farmers in semi-arid areas of central Tanzania can capitalize on.

Keywords: Climate change, climate-smart agriculture, climate change, economics, planting date, resilience, water harvesting

---

Contact Address: Anthony Kimaro, World Agroforestry (ICRAF), Tanzania Country Programme, P.O. Box 6226, Dar-es-Salaam, Tanzania, e-mail: a.kimaro@cgiar.org

Agro-ecology

Oral Presentations

Posters

Two Enabling Factors for Farmer-Driven Pollinator Protection in Low- and Middle-income Countries

Stefanie Christmann[1], Aden Aw-Hassan[2], Yasemin Güler[3], Hasan Cumhur Sarisu[4], Marc Bernard[5], Moulay Chrif Smaili[6], Athanasios Tsivelikas[1]

[1]*Intern. Center for Agricultural Research in the Dry Areas (ICARDA), BCI, Morocco*

[2]*Independent Consultant, Canada*

[3]*Plant Protection Central Research Institute (CRIPP), Turkey*

[4]*Fruit Research Institute (MAREM), Turkey*

[5]*Federal Agency for Agriculture and Food (BLE), Germany*

[6]*Institut National de la Recherche Agronomique (INRA), Morocco*

Reward-based wildflower strips are the most common approach for pollinator protection in high-income countries. However, farmers greatly dislike or even reject them even if a payment is provided. Low- and middle-income countries cannot afford this practice. A promising pilot study in Uzbekistan introduced an alternative approach, Farming with Alternative Pollinators, focusing on farmers as target group, marketable habitat enhancement plants and a method-inherent incentive: higher income per surface achieved already in the first year. We hypothesised that higher income would be a replicable enabling factor across continents, but a knowledge-raising campaign would be necessary in many low and middle-income countries. We assessed the replicability of the incentive with a small number of farmers 2015–2016 in Morocco but focused on assessing if farmers have sufficient knowledge to recognise wild pollinators and use this approach. We conducted 766 interviews using a standardised questionnaire with randomly selected smallholder farmers in three culturally different farming societies of low- and middle-income countries (Morocco, Turkey and Benin). Farming with Alternative Pollinators induced higher income (75 % (2015), 177 % (2016)) also in Morocco. The trial and the survey show the indispensability of a knowledge-raising campaign as second enabling factor. Farmers often don't know the value of wild pollinators, they cannot recognise beneficial insects and disregard them as pests, they destroy nests unwillingly, they are not aware of the habitat requirements of wild pollinators working in their fields. Many farmers have only vague knowledge concerning pollinator-dependency of their crops and even less knowledge concerning the reasons of crop failure. This bears the risk of decisions, which do not solve the problem, but are counterproductive for pollinators. However, based on capacity building, Farming with Alternative Pollinators could have indeed high potential to promote pollinator protection in low- and middle-income countries.

Full article: DOI: 10.1080/14735903.2021.1916254

Keywords: Farmers' knowledge, farming with alternative pollinators, incentive, motivation, scalable

Contact Address: Stefanie Christmann, Intern. Center for Agricultural Research in the Dry Areas (ICARDA), BCI, POB 6299, 10112 Rabat, Morocco, e-mail: s.christmann@cgiar.org

Host- Plant, Insect- Pest Compensations, and Microclimate as Drivers for Healthy and a Sustainable Agro-ecosystem

ANTHONY IJALA[1], SAMUEL KYAMANYWA[1], JENINAH KARUNGI[1], THOMAS HILGER[2], PAUL OKULLO[3]

[1]*Makerere University, College of Agriculture and Environmental Sciences, Dept. of Agricultural Production, Uganda*

[2]*University of Hohenheim, Inst. of Agric. Sci. in the Tropics, Germany*

[3]*Nabuin-ZARDI, National Agricultural Research Organisation, Crops and Natural resources Program, Uganda*

Host-plants and insect-pests' compensational relationships are known to enable plants and insects to survive and adopt to changing environmental conditions. In the mount Elgon region of Uganda there exists a mosaical pattern of different coffee farming systems with increasing altitudes, and their combinations create differing microclimates, which influence the host-plant and pest behaviours. The objective of this study was to determine the host-plant and Toxoptera aurantii (Hemiptera: Aphididae) compensations with microclimate in Arabica coffee under conditions of different altitudes and farming systems. A two-year study on the coffee leaf biomass, T. aurantii numbers on the leaf surface, and damage intensity of T. aurantii, was conducted using 72 Arabica coffee farms with mixed coffee polycultures (farming systems). Two independent factors were considered: altitude as a major factor and the farming system as the second factor. There was evidence of significant host-plant and insect-pest compensations; host-plant/microclimates; and insect-pest /microclimates. Linear regression analysis revealed a - relationship (number of leaves /branch / T. aurantii numbers). A + relationship (number of leaves / branch infested by T. aurantii/T. aurantii abundance). Also T. aurantii abundance had a + relationship / RH or/ambient temperature). The Arabica coffee leaves/ branch had a – relationship (ambient temperature or/ RH). While the T. aurantii infested leaves /branch only had a + relationship with RH. Regarding the soil variables it was only soil temperature which had a + relationship with the number of leaves /branch. The T. aurantii infested leaves /branch had a + relationship (soil temperature or/soil moisture). An understanding of the key relationships of host-plant and insect-pest compensations; host-plant/microclimates; and insect-pest /microclimates enables control decisions for a healthy and sustainable future in Agriculture.

Keywords: Leaf-biomass, microclimate, relative-humidity, temperature

Contact Address: Anthony Ijala, Makerere University, College of Agriculture and Environmental Sciences, Dept. of Agricultural Production, Kampala, Uganda, e-mail: tonyijala209@gmail.com

Can the Push–Pull Technology Reduce Stem Borer Damage and Increase Sorghum Yield in Western Somaliland?

Nasir Mohamed Osman, Christoph Studer

Bern University of Applied Sciences (BFH), School of Agricultural, Forest and Food Sciences (HAFL), Switzerland

Sorghum is a key crop for the livelihood of millions of smallholders in Africa. In Somaliland the cereal serves as a main staple crop, it provides fodder for livestock and additional income. However, lepidopterous stem borers regularly cause major yield losses.

The "push-pull" technology is an ecological approach to minimise stem borer damage and increases cereal and fodder yields while providing additional benefits. The intercropping strategy involves pullplants (grasses as *Brachiaria* sp.) that attract stem borer moths and, on the other hand, the repellent intercrop *Desmodium* sp., which drives them away from the cereal crop ("push").

The performance of the push-pull technology in terms of stem borer control, sorghum grain yield and fodder production was compared with the farmers' practice of sorghum production (broadcast sole crop) in a field trial with block design in Gabiley Region, Somaliland. Stem borer infestation (number of stem borers inside and outside the plants) and damage (leaf damage, number of damaged plants, attacked and dead growing points, and stem borer entry holes) were assessed, and sorghum grain and stover yield as well as Brachiaria production measured.

Practically no Desmodium (*Desmodium intortum*) germinated. Nevertheless, the push-pull technology still provided an effective control of stem borers by significantly reducing infestation and leaf damage ($p < 0.001$). Sorghum grain yield was significantly higher in the push-pull than in farmers' practice plots (+49 % and +146 % in terms of fresh and dry weight). Also sorghum stover production was significantly higher in the push-pull system than under farmers' practice (+10 % and +42 % in terms of fresh and dry weight). The additional fodder produced by Brachiaria (4434 kg ha^{-1}) in the push-pull system is particularly attractive to farmers in Somaliland, where livestock is the backbone of farming and economy. To satisfy the farmers' interest in the technology and prepare for dissemination, future research should focus on ways to assure availability of Desmodium and Brachiara planting material and methods for better establishment of Desmodium under field conditions. Overall, this trial has shown that push-pull offers a valuable option for sustainable intensification of Somaliland's agriculture, may contribute to food and fodder security and enhance resilience and livelihoods.

Keywords: Push-pull, somaliland, sorghum, stem borer

Contact Address: Nasir Mohamed Osman, Bern University of Applied Sciences (BFH), School of Agricultural, Forest and Food Sciences (HAFL), Länggasse 85, 3052 Zollikofen, Switzerland, e-mail: nqaylo@gmail.com

Integrated Pest and Pollinator Management (IPPM) as a Novel Tool to Merge Ecosystem Services: Lessons Learnt from Avocado in Kenya

Thomas Dubois[1], Nadia Toukem[1], H. Michael G. Lattorff[2], Rose N. Sagwe[2], Abdullahi A. Yusuf[3], Elfatih M. Abdel-Rahman[4], Marian Salim Adan[4], Beatrice Muriithi[5]

[1]*International Centre of Insect Physiology and Ecology (icipe), Plant Health, Kenya*
[2]*International Centre of Insect Physiology and Ecology (icipe), Environmental Health, Kenya*
[3]*University of Pretoria, Department of Zoology and Entomology, South Africa*
[4]*International Centre of Insect Physiology and Ecology (icipe), Data Management, Modelling and Geo-Information Unit, Kenya*
[5]*International Centre of Insect Physiology and Ecology (icipe), Social Science and Impact Assessment, Kenya*

Ecosystem-based services such as pollination and integrated pest management (IPM) are key drivers of sustainable productivity and resilience of African agricultural production systems. Although pollination and IPM could interact in multifarious ways to ensure healthier agricultural ecosystems and improve food security, they have mainly been promoted in isolation with minimal efforts to demonstrate their synergies. We present a study that tested the combined effect of beehive supplementation and IPM treatments through an innovative integrated pest and pollinator management (IPPM) approach against the oriental fruit fly Bactorcera dorsalis and the false codling moth Thaumatotibia leucotreta among smallholder avocado (*Persea americana*) farmers in Muranga, Kenya. Production systems were characterised for the role of pollinators and insect pests at the landscape level using freely available Sentinel^{-2} remotely sensed data to calculate normalised difference vegetation index (NDVI) as a proxy for vegetation productivity. Based on socio-economic surveys of 529 farmers, ex-ante studies revealed that farmers were willing to pay USD 56, 70 and 195/farm for IPM, pollination supplementation and IPPM, respectively. Subsequently, one of four treatments (IPPM, IPM, *Apis mellifera* beehive supplementation or control) was implemented on avocado orchards with 36 participating farmers. Pollination deficits, amounting to 27 %, was eliminated completely in farms in high and medium vegetation productivity areas following beehive supplementation, while in low vegetation production areas, pollination deficit remained at 4 %. Results equally showed the benefit of IPPM on pest populations, especially for low vegeta-

Contact Address: Thomas Dubois, International Centre of Insect Physiology and Ecology (icipe), Plant Health, icipe road, kasarani, 30772-00100 Nairobi, Kenya, e-mail: tdubois@icipe.org

tion productivity areas. IPPM is a novel concept, which may be highly beneficial in pollination-dependent crops in East Africa. Awareness was created through training of 1,411 farmers and local government representatives to help increase the diffusion rate of IPPM among avocado-growing communities.

Keywords: Avocado, integrated pest and pollinator management, Kenya, landscape, normalised difference vegetation index, pollinator

Rearing *Eiphosoma laphygmae*, a Potential Biocontrol Agent of the Fall Armyworm

Tabea Allen[1], Marc Kenis[2], Lindsey Norgrove[1]
[1]*Bern University of Applied Sciences (BFH), School of Agricultural, Forest and Food Sciences (HAFL), Switzerland*
[2]*CABI, Switzerland*

Fall armyworm (FAW), *Spodoptera frugiperda*, is an invasive pest, native to the Americas. Since detection in Nigeria in 2016, it has established in most of Africa, Asia and has reached Australia, jeopardising the food security of millions in its wake. The parasitoid wasp, *Eiphosoma laphygmae*, is a potential candidate for classical biocontrol. A detailed risk assessment in quarantine is necessary before any candidate can be recommended. Both testing biocontrol agents and ultimately releasing them relies on rearing and this can be problematic and a bottleneck given their specificity.
Our objective was to improve rearing of *E. laphygmae* by assessing how the diet of the host FAW and their age at exposure impacted parasitoid sex-ratio, emergence rate and FAW mortality. We exposed FAW larvae reared on a modified McMorran-Grisdale diet versus those reared on maize plants. We also exposed one-, two-, three- and four-day old FAW larvae. Work was conducted in a quarantine facility at 23°C, 60 % relative humidity and a photoperiod of 16h light and 8h dark.
E. laphygmae did develop successfully in one-, two-, three and four-day old FAW larvae, with female progeny in all tested host ages. On average, parasitism rate was 18 % and mortality of FAW larvae was 12 %. Of those parasitised FAW, the parasitoid emergence rate was 82 % of which 77 % were male. Parasitism rates were numerically lower for four-day old FAW larvae than for the younger larvae. Parasitism rates were numerically lower on FAW reared on the modified McMorran-Grisdale diet than on those reared on maize plants. We did not detect any other effects of either FAW larval age or diet on parasitoid sex ratio, mortality of host-larvae or emergence rate of the parasitoid. We recommend using maize plants to rear FAW and then exposing them at one- to three-days old. Future work should verify these results in larger-scale studies and consider whether manipulating other nutritional or environmental variables can increase parasitism rates and reduce the male progeny bias.

Keywords: Biocontrol, maize, parasitoid wasps, rearing, *Spodoptera frugiperda*

Contact Address: Tabea Allen, Bern University of Applied Sciences (BFH), School of Agricultural, Forest and Food Sciences (HAFL), Bern, Switzerland, e-mail: tabea.allen@bfh.ch

Effect of Plant Pathology and its Impact on Agriculture, Beekeeping and Food Security in Kwara State, Nigeria

Ajao Mufutau[1], Yusuf Oladimeji[2]

[1]*Kwara State University, Bioscience and Biotechnology, Nigeria*

[2]*Ahmadu Bello University, Zaria, Dept. of Agricultural Economics and Rural Sociology,, Nigeria*

Honey bee (*Apis mellifera*), the most domesticated western honeybee is faced with multiple factors and stressors affecting its health, productivity and survival leading to bee population decline and subsequently low agricultural production and food security. This work was therefore designed to identify, assess pest, predators and their effects on bees,agricultural production and suggest management strategies for pollinator conservation for sustainable food security.The method adopted involved a simple random sample of fifty-seven (57) beekeepers was chosen for questionnaire administration. This consisted of six (6) institution based,twenty-one (21) non-governmental and thirty (30) private/individuals bee farms from 10 Local Government Areas (LGAs) of Kwara State, North-Central, Nigeria. Furthermore, a field survey and sample collection of honey bee pests and predators was also conducted at the six apiaries at the study area. The results of views of respondents, field collection and laboratory analysis revealed ants as dominant pest and predators, 63.5 % in Asa; small hive beetle, 95.2 % Irepodun; termite 68.1 % in Moro; and wax moth 78.9 % in Ilorin South as arthropod of beekeeping practices, pollinators, hampering agricultural production and food security. The identified pests and predators manifest in low honey production, irritation and disease condition leading to hive abandonment, forced swarming, reduced pollination efficiency, colony population decline and death, hence food insecurity. Suggested coping strategies among others include providing the beekeeper with adequate training on bee health maintenance, pollination service and skilled management expertise needed to identify and combat these problems.This is imperative if its impact on agriculture, beekeeping and food security would be achieved.

Keywords: Bee survival, coping strategy, food security, pests

Contact Address: Ajao Mufutau, Kwara State University, Bioscience and Biotechnology, PMB 1530, ILORIN, Department of Bioscience and Biotechnology, PMB 1530 KWASU Malete, Kwara state, +234 ILORIN, Nigeria, e-mail: drajaoadeyemi@gmail.com

Effects of Time and Level of *Striga* Infection on Pearl Millet Varieties in North Darfur

Yahia Eldie, Hamdi Ahmed
Al Fashir University, Biological Sciences, Sudan

This study conducted to revise the interaction between the parasitic weed *Striga hermonthica* (Del.) and pearl millet (*Pennisetum glaucum*). The main objective of the study was to investigate the effects of time and level of *Striga* infection on the interaction between the host plant and parasite. (Dimbie) and the Ashana pearl millet varieties were grown in pots with and without seed infestation with Striga. Both pearl millet genotypes responded to infection by the *Striga* parasite, but it was evident that Dimbie was more strongly affected than Ashana pearl millet in plant height; final leaf number, green leaf area, and total dry weight which were significantly reduced by infection. The Ashana landrace showed significantly lower and delayed attachments of *Striga hermonthica* than the Dimbie cultivar, and this could be explained by a delay in the onset of attachments. *Striga hermonthica* infection had a stronger effect on the sensitive cultivar, although the parasite affected growth and dry matter allocation in both cultivars. The reduction in biomass production was accompanied by a relatively increased allocation of dry matter to the roots. It was observed that the pearl millet genotypes have different sensitivities to *Striga* infection. The tolerant millet variety Ashana is highly resistant to *Striga* infection while sensitive variety Dimbie is slightly resistant to the weed *Striga* infection and therefore the hypothesis is rejected. It is concluded that differences in root manner and the resulting early infection and higher *S. hermonthica* numbers are partly responsible for the stronger effects of the parasite on the Dimbie cultivar.

Keywords: Attachment, infection time, pearl millet tolerance, *Striga hermonthica*

Contact Address: Yahia Eldie, Al Fashir University, Biological Sciences, PO Box 125, El Fasher, Sudan, e-mail: yahiaeldie0@gmail.com

Evaluating the Various Soilless Potting Media for Healthy Mango Nursery Production

Sami Ullah[1], Ishtiaq Ahmad Rajwana[1], Kashif Razzaq[1], Gulzar Akhtar[1], Hafiz Nazar Faried[1], Muhammad Amin[2], Atif Iqbal[3], M. Shahid Fareedi[1]

[1]*MNS-University of Agriculture, Dept. of Horticulture, Pakistan*

[2]*The Islamia University of Bahawalpur, Dept. of Horticultural Sciences, Pakistan*

[3]*Mango Research Institute, Multan, Dept. of Agriculture, Pakistan*

In Pakistan, traditionally mango nurseries are grown under the canopy of mango trees in soil, being highly susceptible to diseases present on mature trees. However, for healthy seedling production, different soilless potting media combinations used in the world. A study was conducted to evaluate different soilless media combinations including, sugarcane bagasse, coconut fiber, peat moss and organic compost in different ratios for healthy mango nursery production. A total of six different media combinations were tested replicated thrice, each replication contained 10 numbers of polythene containing one mango seedling. Randomised Complete Block Design was applied and various parameters including, new flushes emergence (%), plant height (cm), stem girth (mm) and numbers of leaves was recorded. Maximum percentage of flush emergence, highest plant height (52.66 cm), maximum numbers of leaves (17.33) and stem girth (9.48 mm) were recorded in media combination (Bagasse + coconut fiber + peat moss having ratio 65 % + 5 % + 30 %). The physical and chemical characteristics of potting media also determined, media combination (Bagasse + coconut fiber + peat moss having ratio 65 % + 5 % + 30 %) exhibited ideal preferred range; pH (7.0), electrical conductivity [(EC): 922.5 μS/cm], water holding capacity [(WHC): (44 %)] and air filled porosity [(AFP); 12.5%] as compared to other combinations. It is concluded that the soilless potting media combination of Bagasse + coconut fiber + peat moss having ratio 65 % + 5 % + 30 % found to be most effective for vegetative parameters of seedlings and can be used to produce as soilless potting media for healthy mango nursery production.

Keywords: *Mangifera indica*, Mango nursery, peat moss

Contact Address: Sami Ullah, MNS-University of Agriculture, Dept. of Horticulture, 60000 Multan, Pakistan, e-mail: sami.ullah1@mnsuam.edu.pk

Allelopathic Potential of Rhizobacteria against *Leptochloa chinensis* (L.) Nees in Rice under Gnotobiotic Conditions

Iram Afzal, Zahir Ahmad Zahir, Ayyub Muhaimen, Ali Qasim
University of Agriculture, Faisalabad, Inst. of Soil & and Environmental Sciences, Pakistan

Prevailing water crisis, labour shortage, and environmental issues have necessitated the use of alternative rice cultivation methods; however, dealing with associated risks such as weed infestation has become more challenging. *Leptochloa chinensis* (L.) Nees, that emerges simultaneously with rice seedlings has threatened rice productivity under direct-seeded conditions. Therefore, developing a sustainable weed control strategy has become necessary to control production losses by weeds. This study aimed to evaluate the potential of rhizobacteria in the suppression of *L. chinensis*. Isolated rhizobacteria were screened through cyanide production and lettuce seedling *in vitro* bioassay. Eight isolates (IR4–10, IR7–6, IR7–7, IR7–11, IR3–1, IR3–4, IR8–1, IR5–13) were cyanogenic and caused mortality of lettuce seedling up to 14 folds over uninoculated control. Moreover, biochemical, and microbiological characterisation revealed the presence of multiple characteristics in these isolates. Eight cyanogenic isolates were selected for weed suppression trial and growth promotion trial on rice under gnotobiotic conditions. Statistical analysis of data indicated that all isolates were effective in weed suppression; however, four isolates (IR4–10, IR7–6, IR7–7, and IR7–11) suppressed *L. chinensis* most effectively by reducing root length, shoot length, and fresh biomass. The most effective four isolates were not suppressive to rice growth. A few of these promoted growth by improving root length, root diameter, number of tips, shoot length, fresh biomass, dry biomass, and chlorophyll contents. The results may lead to the formulation of bioherbicides for successful suppression of *L. chinensis* under direct-seeded conditions if the capability of selected four allelopathic bacterial isolates is confirmed under field conditions.

Keywords: Allelopathic bacteria, bioherbicide, direct seeded rice, *Leptochloa chinensis*

Contact Address: Ali Qasim, University of Agriculture, Institute of Soil and Environmental Sciences, 38040 Faisalabad, Pakistan, e-mail: qasim.ali@uaf.edu.pk

Genetic Diversity and Population Structure of *Moniliophthora roreri* in Cocoa Producing Areas in Guatemala

José Alejandro Ruiz-Chután[1], Julio Ernesto Berdúo-Sandoval[2], Luis Montes[2], Marie Kalousová[1], Bohdan Lojka[1], Carlos Villanueva-González[3], Amílcar Sánchez-Pérez [2]

[1]*Czech University of Life Sciences Prague, Fac. of Tropical AgriSciences, Dept. of Crop Sciences and Agroforestry, Czech Republic*

[2]*Universidad de San Carlos de Guatemala, Facultad de Agronomía, Guatemala*

[3]*Universidad Rafael Landívar, Facultad de Ciencias Ambientales y Agrícolas, Guatemala*

Moniliasis, caused by *Moniliophthora roreri*, is one of the most devastating cocoa diseases in the western hemisphere. From its centre of origin in the Magdalena Valley, Colombia, the pathogen has spread to eleven countries, including Guatemala, causing severe production losses. Despite reports of the dispersal of *M. roreri* to Central America from a single clone, the genetic diversity of the pathogen has not been studied in Guatemala, and the biological evolution of the pathogen is unknown. To clarify this aspect, 69 isolates of *M. roreri* obtained from four cocoa-producing departments were analysed, and genetic diversity was evaluated using the molecular marker AFLP. We identified a low level of genetic diversity by the Shannon index (0.0578) and the proportion of polymorphic loci (12.28 %). The overall genetic diversity (Ht) was 0.1289, and the mean genetic diversity within populations (Hw) was 0.1310. Molecular analysis of variance showed a variance between regions, populations and within populations of 3 %, 6 % and 91 %, respectively, indicating a weak genetic structure. The Bayesian clustering analysis implemented in STRUCTURE revealed that the most likely number of genetic groups in the data was two, although We did not observe geographical differentiation. The low genetic diversity identified is congruent with reports of the colonisation process of M. roreri in Central America through a single haplotype. However, given the high mutation rate of *M. roreri*, We suggest constant monitoring of its evolution, quarantine practices that limit its dispersion. We also recommended evaluations of cocoa clones tolerant to the new *M. roreri* genotypes preventing increased losses to Guatemalan producers.

Keywords: AFLP, AMOVA, gene flow, moniliasis

Contact Address: José Alejandro Ruiz-Chután, Czech University of Life Sciences Prague, Fac. of Tropical AgriSciences, Dept. of Crop Sciences and Agroforestry, Prague, Czech Republic, e-mail: josealejandro.ruiz@icloud.com

In vitro Antagonism of Native Guatemalan Isolates of *Trichoderma harzianum* and *T. viride* for Biological Control of *Rhizoctonia solani*

José Alejandro Ruiz-Chután[1], Julio Ernesto Berdúo-Sandoval[2], Marie Kalousová[1], Bohdan Lojka[1], Amílcar Sánchez-Pérez [2]

[1]*Czech University of Life Sciences Prague, Fac. of Tropical AgriSciences, Dept. of Crop Sciences and Agroforestry, Czech Republic*

[2]*Universidad de San Carlos de Guatemala, Facultad de Agronomía, Guatemala*

Rhizoctonia solani is a basidiomycete present in the soil that currently affects the production of potato crops in Guatemala's western highlands, causing the disease known as black scurf. The current classification of *R. solani* is based on the ability of the hyphae to merge and form an anastomosis group (AG), 13 AG being recognised so far. Due to the negative impacts of chemical disease control, a viable alternative is presented in usage of fungi from the genus *Trichoderma* as a biological control strategy. The research objective was to identify the AG of *R. solani* present in the western highlands of Guatemala through specific PCR. The potential of native isolates of *T. harzianum* and *T. viride* as biological control agents characterised by the molecular marker AFLP was also evaluated. Of the 78 isolates of *R. solani*, the groups $AG^{-3}PT$, AG-4-HG-II and AG-5 were identified with an incidence of 46.16 %, 14.10 % and 17.96 %, respectively, while 17 isolates could not be assigned. Isolates of *T. harzianum* showed a higher average percentage inhibition of radial growth (PIRG) of *R. solani* (56.71 %) than *T. viride* (37.58 %) ($p < .001$). We determined that the phylogenetic groups of *T. harzianum* and *T. virens* are equally effective against the 3 AG identified, highlighting the benefit of using native isolates. We suggest using sequencing of the ITS region for the identification of unassigned isolates of *R. solani* through sequence comparison of NCBI genebank sequences and field evaluations to assess the efficiency of *T. harzianum* and *T. viride* native isolates.

Keywords: Anastomosis groups, genetic diversity, percentage inhibition of radial growth (PIRG)

Contact Address: José Alejandro Ruiz-Chután, Czech University of Life Sciences Prague, Fac. of Tropical AgriSciences, Dept. of Crop Sciences and Agroforestry, Prague, Czech Republic, e-mail: josealejandro.ruiz@icloud.com

Phytotoxicity Effect of Olive Mill Wastewater (OMWW) on Cress Germination

Sebastian Romuli, Jonatan Wuensch, Andrés Hernandez-Pridybailo, Joachim Müller

University of Hohenheim, Inst. of Agricultural Engineering, Tropics and Subtropics Group, Germany

The hot weather conditions in the Fès-Meknès region of Morocco lead to high water evaporation rates, nutrient loss from soil and the enrichment of dissolved salts on the soil surface. On the other hand, the region is one of the biggest olive oil producers in the country and generates considerable amounts of olive mill wastewater (OMWW). The OMWW is inadequately stored or discharged into the environment without any treatment. Although OMWW may present an environmental risk because of its phytotoxic effects due to the high polyphenol content, among other reasons, its utilisation as a fertiliser and a biopesticide is still under investigation. In this study, the phytotoxic effect of OMWW on the growth of cress (Lepidium sativum) was semi-automatically evaluated by analysing RGB images taken from germination tests according to DIN EN 16086–2. The cress seeds were sown under soil or perlite as a medium, which contained 0 % (control), 5 % (equivalent to a field application of 20 t/ha), 10 % (40 t/ha) and 15 % (60 t/ha) of OMWW, respectively. After 72 h, digital images of the cress were obtained, and the seedling growth parameters were determined by SmartRoot plugin from ImageJ software. When soil was used, it was observed that the higher the OMWW concentration, the shorter the root and hypocotyl length of cress. Nevertheless, the germination rate was not significantly affected ($p < 0.05$). On the other hand, no difference between the 5 % treatment and the control was found when perlite was used, while an OMWW concentration of 10 % and 15 % affected the root and hypocotyl length significantly at $p < 0.05$. Additionally, the results revealed a decrease in the cress seedling's fresh weight and dry matter content as the OMWW concentration was increased. This study shows the potential of using OMWW directly as a fertiliser and a soil amendment to improve soil properties and plantation yields.

Keywords: Natural resource, soil substrate, sustainable production, waste management

Contact Address: Sebastian Romuli, University of Hohenheim, Inst. of Agricultural Engineering, Tropics and Subtropics Group, Garbenstr. 9, 70599 Stuttgart, Germany, e-mail: sebastian_romuli@uni-hohenheim.de

Virulence and Diversity of *Fusarium oxysporum* f.sp. *cubense* Infecting Bananas in Tharaka Nithi County, Kenya

MALAKA MUSIME[1], MWONGERA THURANIRA[2], MAINA MWANGI[1], NCHORE SHEM[1], LUBABALI HUDSON[2], GATHAMBIRI CHARITY[2]

[1]*Kenyatta University, School of Agriculture and Enterprise Development, Kenya*

[2]*Kenya Agricultural and Livestock Research Organisation, Kenya*

Banana belongs to Musaceae family and is cultivated worldwide supporting the livelihoods of many people and the economies of many countries. One of the diseases that is significantly reducing banana yield in Kenya is the Panama disease caused by *Fusarium oxysporum* f.sp. *cubense* (*Foc*). This experiment was carried out in Tharaka Nithi County in Kenya to assess the pathogenicity of *Foc* isolated from Gros Michel banana varieties, Cavendish banana varieties and from plantains. Laboratory work and greenhouse experiments were done at the Horticulture Research Institute (HRI) in Kandara, Kenya. Conidia were harvested from 14 days old cultures on PDA media. The spore suspension was adjusted to 10–6 spores/ml by counting using a hemocytometer slide. To inoculate the plants, the soil was removed gently from the 2-month-old tissue cultured banana seedlings to expose the roots. The banana roots were wounded and immersed in *Foc* spore suspension for 30 minutes while shaking before planting into the sterile compost soil. A control was immersed in sterile water. All isolates were tested in four replicates of one plant per isolate. Disease severity was recorded every week for a period of 16 weeks following the disease scale of 1–5. For external symptoms, 1 indicated no symptom/healthy, 2 indicated initial yellowing mainly on the lower leaves, 3 indicated yellowing of all the lower leaves including some discolouration on the younger leaves, 4 indicated intense yellowing on all leaves while 5 indicated plant dead/complete wilting. Data was analysed using one way ANOVA. The isolates that revealed a scale of 3 included AP005 and AP003. Those that revealed a scale of 2 included AP001, AP002, AP004,AP006, AP007, AP008, AP009, AP010,AP011, AP012, CV001,CV002, CV004,CV005, CV006, KL001,KL003, KL004,EM001,UK001, FH001, and GN001. The results indicated that some of the Foc isolates used in the experiment were pathogenic. This therefore indicates that *Foc* is one of the diseases affecting bananas in Tharaka Nithi County in Kenya. The management strategies are therefore necessary for the control of *Foc* pathogen in Kenya. There is still a laboratory and greenhouse ongoing experiment for this work.

Keywords: Cavendish, Gros Michel, Panama disease, pathogenicity

Contact Address: Malaka Musime, Kenyatta University, School of Agriculture and Enterprise Development, Box 4, 00232 Ruiru, Kenya, e-mail: samuelmalaka540@gmail.com

Environmental Risk Assessment of Pesticide Pollution in Rice Fields in the Mekong Delta

Loan Vo Phuong Hong
University of Bonn, Fac. of Agriculture, Germany

The risk assessment is essential for any environmental evaluation, in respect of the impact of study results on human health and ecosystem which contributes not only to current situation but also to sustainable future. This potential risk assessment is associated with 10 pesticides in rice fields predicted by the RICEWQ model in the Mekong Delta in 2009. The first evaluation with individual pesticides in water is addressed through the comparison to known toxic thresholds suggested by the European Union, Japanese Ministry of Environment, Environmental Protection Agency, in Australia, the UK and Taiwan while the second consideration is with regard to combined multi substances. The pesticides post from low risks found in isoprothiolane, propiconzole, buprofezin, cypermethrin and fenobucarb, to highest risks of butachlor and pretilachlor while difenoconazole and fipronil are still be at acceptable safety margins. Although these high concentrations decrease to below the limit within a day, the exceeding residues potentially harm human health and/or water bodies during irrigation practices. Related to soil compartment, almost all residues are still under the critical values for "sediment dwelling organisms" and "soil macro-organisms". The adverse effects are determined as human health (buprofezin), fresh water (butachlor), inhalation (fenobucarb), invertebrate families (fipronil), fish species and alga (cypermethrin), soil organisms (buprofezin), or risk of bioaccumulation etc. Consequently, the requirements to reduce the loading and probably pesticide pathways through mitigation statements are necessary including proper irrigation practices, climate conditions related to pesticide usage, the implementation of the Integrated Pest Management, widely disseminated information on health concerns, safety using/treatment processes and storage, encouragement of pesticide surveys. Finally, the pesticide residues are higher than the estimated standards, these issues do not stand to inform that this area is under pollution status. The Vietnamese community needs to carefully establish the legislation in accordance with fully sustainable development. Regulation of registration processes and the improvement of the standard system on pesticides, offer a good opportunity to evaluate the impact of pesticide on human health and the environment.

Keywords: Human health, Mekong Delta, mitigation method, pesticide, rice field, RICEWQ, water

Contact Address: Loan Vo Phuong Hong, University of Bonn, Fac. of Agriculture, Bonn, Germany, e-mail: voloan_78@yahoo.com

Embelia schimperi Extracts Potential to Control *Meloidogyne incognita* and *Pratylenchus zeae*

Beth Wangui Waweru, Njira Njira Pili

Moi University, Biological Sciences, Kenya

Embelia schimperi, a plant belonging to Myrsinaceae family, occurs in the tropical and sub-tropical regions of Africa, eastern Asia, North & South America and Australia. The tree thrives best in zones with an altitude ranging from 1,700–2,800 m above sea level. The tree is one of the medicinal plants used traditionally for treatment of intestinal tape worm, dysmenorrheal, bacterial and fungal infections in humans. Previous research has revealed the presence of phenols, alkaloids, tannins and flavonoids in the tree extracts. In the present study, *Embelia schimperi* stem extracts were prepared using hexane, dichloromethyl and ethanol as solvents. 1 g of the extract was dissolved in 10 ml of 10 % dimethyl sulphoxide (DMSO) and treated as the stock solution and later screened for egg hatchability potential and nematicidal activity against *Meloidogyne incognita* eggs and *Pratylenchus zeae* J2 respectively. The nematode eggs and juveniles were exposed to different concentrations of the extract (25, 50, 75 and 100 %) for 24, 48, 72 and 96 hrs in 96 well plates. The sample was stored at room temperature (19) and covered with parafilm. Dimethyl sulphoxide was used for control experiment. Results showed gradual increase in P. zeae juvenile mortality with increase in extract concentration. Hexane solvent recorded 100 % mortality after 72 hours at 75 % concentration compared to dichloromethane which had 53 % mortality at the same extract concentration. Ethanol solvent reduced *M. incognita* egg hatchability to 23 % compared to dichloromethane (33 %) and hexane (42 %) after 96 hours at 100 % solvent concentration. The results give an insight of *E. schimperi* potential to control plant parasitic nematodes.

Keywords: *Embelia schimperi*, *Meloidogyne incognita*, nematicidal activity, *Pratylenchus zeae*

Contact Address: Beth Wangui Waweru, Moi University, Biological Sciences, P. O Box 3900, 30100 Kesses, Kenya, e-mail: bethwwr54@gmail.com

Legume Root-Exuded Phenolics Inhibit Development and Phytotoxin Biosynthesis in *Fusarium oxysporum* f. sp. *cubense* TR4

Evans Were[1], Jochen Schöne[2], Altus Viljoen[3], Frank Rasche[1]

[1]*University of Hohenheim, Inst. of Agric. Sci. in the Tropics (Hans-Ruthenberg-Institute), Germany*

[2]*University of Hohenheim, Inst. of Phytomedicine, Germany*

[3]*Stellenbosch University, Dept. of Plant Pathology, Germany*

Banana Fusarium wilt (FW) is a devastating disease caused by the root-infecting fungus *Fusarium oxysporum* f. sp. *cubense* (*Foc*). Suppression of *Foc* by intercropping banana with leguminous plants has been suggested as an alternative strategy for managing FW. However, the underlying mechanisms and mode of actions of the tripartite interaction of *Foc*, banana and legume remain uncertain. Hence, legume root-exuded metabolites that may influence host-pathogen and root-soil-microbiome interactions need to be discovered and their mode of actions unraveled. This study, through hydroponic culture and metabolite profiling, investigated the potential of root-exuded phenolic acids and flavonoids of *Desmodium uncinatum* and *Mucuna pruriens* to inhibit the growth and biosynthesis of virulence factors in *Foc* tropical race 4 (*Foc* TR4). Out of 12 metabolites, 4 phenolic acids (benzoic, *t*-cinnamic, *p*-coumaric, *p*-hydroxybenzoic) were common in root exudates of *D. uncinatum* and *M. pruriens*, while *p*-coumaric and vanillin were only detected in *M. pruriens*. The flavonoid quercetin was only detected in *M. pruriens*. Bioassays using synthetic benzoic-, *t*-cinnamic-, or *p*-hydroxybenzoic acid, or a combination thereof, showed a concentration-dependent suppressive effect on *Foc* TR4. Low concentrations (0.01, 0.1 mM) of phenolic acids inhibited chlamydospore germination, production of macro- and micro-conidia, and synthesis of fusaric acid, whereas radial mycelial growth and synthesis of beauvaricin was promoted in *Foc* TR4. Mycelial growth of *Foc* TR4 was only inhibited at high concentrations (1 mM) of benzoic acid, *t*-cinnamic acid, and their combination. Our results highlight a mechanism by which root-exuded metabolites may directly suppress Foc during the earliest stages of pathogen development.

Keywords: Benzoic, *Desmodium uncinatum*, Foc TR4, *Mucuna pruriens*, *p*-hydroxybenzoic, *t*-cinnamic

---

Contact Address: Frank Rasche, University of Hohenheim, Inst. of Agric. Sci. in the Tropics (Hans-Ruthenberg-Institute), Stuttgart, Germany, e-mail: frank.rasche@uni-hohenheim.de

Agro-ecosociology - implications for sustainability

Oral Presentations

Posters

Agroecology Strengthens Social Reproduction: Implications for Food Security and Nutrition

Chukwuma Ume, Stephanie Domptail, Ernst-August Nuppenau
Justus-Liebig University Giessen, Inst. of Agric. Policy and Market Res., Germany

Agroecology is increasingly discussed and promoted to achieve secure food and nutrition in various parts of the world. However, critics suggest that agroecology may not be able to address the sub-Sahara African nutrition and development challenges in the long term because it constitutes a barrier to modernisation, locking farmers in a non-productive traditional agriculture and poverty trap. This shows that the pathways through which agroecology leads to improved food and nutrition security (FNS) among rural populations are still unclear. In fact, social and power dimensions intrinsically linked to an agroecology-based - or in fact any - intensification strategy appear to be ignored in the discussion and research on agroecology in sub-Sahara Africa. We claim that transitioning to agroecology, even at the farm level also transforms farming households' social and political characteristics, thereby affecting their overall handling of food and nutrition. Using primary survey data from rural Nigeria, 2021, this paper uncovers which pathways of causalities exist between agroecology and FNS and quantitatively assesses the moderating effect of physical and social reproduction on the relationship between agroecology and FNS. We first assume that farming households are socio-ecological systems, nested in larger socio-ecological systems (landscape) at the territory level. Second, we anchor our analysis within feminist economics, more specifically making use of the concepts of physical, social reproduction, and agency in our analysis. We consider FNS as a productive goal of the household, linked to several other reproductive dimensions. The concept of reproduction, found in feminist economics, is a useful innovative lens to empirically address the dimension of agency for FNS among farming households. Given the strong constraints for the participation of smallholders in the formal production economy, investigating how else they achieve to maintain themselves with alternative systems is crucial for food security in rural areas in Africa.

Keywords: Africa, agency, agroecology, physical reproduction, smallholders, social reproduction, sub-Saharan

---

Contact Address: Chukwuma Ume, Justus-Liebig-University Giessen, Inst. of Agricultural Policy and Market Research, Giessen, Germany, e-mail: chukwuma.ume@agrar.uni-giessen.de

Determinants of Income Diversification and its Effect on Food Security of Small Holder Farmers in Ethiopia

Mezgebu Aynalem Mengstie
Debre Markos University, Burie Campus, Ethiopia

In Ethiopia 83 percent of small-holder farmers participated in farming activities and only 27 percent in non-farm economic enterprises. This paper examines the determinants of income diversification and its effect on food security in Ethiopia. The study used two stages sampling in combination with stratified and simple random sampling procedures to select kebeles and households. Fractional response model were employed to analyse the data collected from a sample of 450 rural households. While the Simpson index of diversity were used to measure the extent of income diversification. Income diversification level has positive and significant effect on food security status of the rural farming households in Ethiopia. The level and type of income diversification depends on the accessibility and availability of different income sources. The mean results of degree of income diversification revealed that Simpson Index of Diversity (SID = 0.24) by rural households in the study area. Based fractional response model educational status, credit utilisation, distance from market and access to electric power affect at $p < 0.01$ percent probability level, sex of the household head affect at $p < 0.05$ percent probability level and, annual household income, special skill and family size significantly affecting degree of income diversification at $p < 0.1$ percent probability level. Finally, this thesis indicates the important policy implications suggesting that programs, projects and/or any interventions designed targeting to engage people in other income generating activities would augment their income sources which are made to increase the food security status at household level in Ethiopia. To reduce food insecurity, government policies should better aim at increasing access to non-farm activities for all rural households, particularly for households with little human resources, land and monetary assets (opportunities) by decreasing the constraints that hinders these households from participating in non-farm activities.

Keywords: Food security index, fractional response regression, income diversification

Contact Address: Mezgebu Aynalem Mengstie, Debre Markos University, Burie Campus, Debre Markos, Ethiopia, e-mail: mezgebu12aynalem@gmail.com

Sustainability Certification, Social Cohesion, and Prospects of Agroecological Change in Ghana's Cocoa Producing Areas

Franziska Ollendorf, Katharina Löhr, Stefan Sieber

Leibniz Centre for Agric. Landscape Res. (ZALF), Sustainable Land Use in Developing Countries (SusLAND), Germany

Departing from the persistence of major sustainability challenges in cocoa production such as severe poverty among cocoa producers and associated high rates of child labour and deforestation, we provide an empirical-based discussion of how sustainability certification shapes the prospects of a broader paradigm shift in Ghana's cocoa production towards a more holistic approach to sustainability, which we capture combining the concepts of social cohesion and agroecology. One of the core ideas driving the currently biggest cocoa certification scheme, Rainforest Alliance, is to overcome sustainability challenges with a sustainable intensification of production. In Ghana, most of the cocoa sustainability certification schemes are implemented by transnational corporations (TNCs) from the cocoa and chocolate industry, mainly the processing segment. While the share of certified beans has grown drastically over the past decade, researchers find sustainability effects in terms of farmers' livelihoods improvement and biodiversity conservation to be only marginal. Moreover, studies depict unexpected side-effects of TNC-led sustainability certification schemes, such as new patterns of inequalities among cocoa producers, negative effects on upgrading opportunities of local firms, and an increased control of TNCs over smallholder production. All together sustainability certification in its current form fails to mitigate sustainability challenges and even counteracts with the goal of a healthy and sustainable future of cocoa production. Our results indicate that sustainability certification endangers social cohesion in cocoa producing communities by fragmenting them into beneficiaries and non-beneficiaries and creating new disparities. Also, by privileging external top-down interventions over existing local knowledge and innovation capacities, and by prioritising intensification over diversification strategies, sustainability certification establishes new structures in the local cocoa sector which go opposite to agroecological approaches to agricultural change. Based on qualitative data we collected in the Ghanaian cocoa sector in 2021, our analysis offers an empirical insight in how different stakeholders perceive such hidden dynamics and how those shape their visions about alternative future pathways of cocoa production. The discussion aims to contribute to the understanding of institutional and governmental barriers to sustainable agricultural change.

Keywords: Agroecology, certification, cocoa, Ghana, social cohesion, sustainability governance

Contact Address: Franziska Ollendorf, Leibniz Centre for Agric. Landscape Res. (ZALF), Sustainable Land Use in Developing Countries (SusLAND), Eberswalder Straße 84, 15374 Müncheberg, Germany, e-mail: f.ollendorf@posteo.net

Divergences in Defining 'Sustainable Palm Oil' for Smallholders and How to Achieve It

ZOË OGAHARA[1], KRISTJAN JESPERSEN[2]

[1]*University of Copenhagen, Food and Resource Economics, Denmark*
[2]*Copenhagen Business School, Dept. of Management, Society and Communication, Denmark*

A growing literature casts a critical eye on the claims for sustainability upgrading within global value chains for tropical agricultural commodities. Palm oil is associated with deforestation, greenhouse gas emissions, land conflicts and labour abuses, but is also a highly efficient crop that is a potential vehicle for rural development in tropical countries. Thus, it is important that claims for the sustainable production of palm oil are scrutinised for positive and equitable outcomes where it is produced. This paper identifies the main tenets of 'sustainable palm oil' for smallholders from an extensive review of the academic literature and assesses sustainability policies from environmental, social and economic perspectives. A multi-step literature search yielded 102 relevant journal articles that address smallholders, oil palm and issues related to sustainability. It was found that the academic research agenda has largely focused on sustainability verification (certification) and environmental issues such as land use change, which are the concerns of downstream actors. Issues of more immediate concern to smallholders such as land tenure, access to inputs and contractual relations with large companies feature less prominently but are also crucial to long-term sustainability. This paper also examines key policies with the aim of achieving sustainability in light of findings from the literature, namely certification; intensification; organising farmers into cooperatives; contract farming; and training. These initiatives offer only partial solutions towards the sustainable production of palm oil by smallholders with varying strengths and weaknesses. In order to disrupt existing trends in academic research, the authors conclude that more research is needed in strategies that blend different approaches to promoting sustainable palm oil, such as policy mix analysis or evaluation of jurisdictional approaches to certification.

Keywords: Certification, global value chains, palm oil, smallholders, sustainability

Contact Address: Zoë Ogahara, University of Copenhagen, Food and Resource Economics, Rolighedvej 23, 1958 Copenhagen, Denmark, e-mail: zoe.sakura@ifro.ku.dk

Decoding Farmers and Stakeholders' Discourses on Conservation Agriculture's Usefulness in Zambia

Godfrey Omulo, Thomas Daum, Karlheinz Köller, Regina Birner

University of Hohenheim, Inst. of Agric. Sci. in the Tropics (Hans-Ruthenberg-Institute), Germany

Advocative discourses have credited Conservation Agriculture (CA) as a sustainable practice fit for curbing food insecurity and climate change impacts worldwide. Yet, heretic views citing CA's limitation in solving the escalating food demands and environmental degradation footprints are on the rise. Despite CA's wide research in SSA, few studies have sought to analyse how these contesting discourses affect the perception of CA usefulness in different contexts. Thus, this study was aimed at investigating the new insights of CA discourses as revealed by Zambian farmers and other stakeholders. Data was collected through in-depth interviews, focus group discussions, and a review of selected media publications within Lusaka and Central provinces. Using Deductive Qualitative Analysis (DQA) and Discourse Analysis (DA) techniques, farmers' and stakeholders' perspectives of CA were explored via MAXQDA 2020 software to generate categories, codes, and sets. Results revealed unreported discourses as adequate adoption levels, shorter years to CA's yield stability, CA as the future of farming, and CA as an ancient African farming method. Perceived CA benefits are driven by the awareness, training, and positive socio-economic evaluation of the practice. Hindrances to CA's utilisation are espoused by conflicting cultural norms, gender gaps, financial and mechanisation constraints. Negative mindset, inadequate agronomic management, overdependence on donor-aid disincentivize CA's usefulness among Zambian farmers. This implies that, despite the critics and the perceived low adoption, farmers and stakeholders underscored CA's irrefutable role in yield and soil quality improvement, time-saving and cost-effectiveness, and climate resilience amidst the current erratic rainfall patterns. In order to harness the potential of sustainable agricultural practices like CA, favourable government policies that seek to bridge the gap between theory and practice are indispensable.

Keywords: Climate change, conservation agriculture, discourse analysis, sustainable practices, Zambia

Contact Address: Godfrey Omulo, University of Hohenheim, Institute of Agricultural Sciences in the Tropics, Wollgrassweg 43, 70599 Stuttgart, Germany, e-mail: omuloh@gmail.com

Sustainable Finances for Green Investments: Findings from Experimental Adoption of Innovative Index Insurance in Uzbekistan

Laura Moritz, Lena Kuhn, Ihtiyor Bobojonov, Thomas Glauben

Leibniz Institute of Agricultural Development in Transition Economies (IAMO), Agricultural Markets, Germany

In times of climate change farmers in the Global South need an incentive to invest in a healthy and sustainable agriculture. Innovative index insurance is an often discussed climate change adaptation that improves farmers' climate resilience. By smoothing agricultural incomes over time, index insurance delivers an economic incentive to think sustainably and thus plays a fundamental role in green agriculture debates. However, index insurance still lacks global and voluntary demand, in particularly among the most climate-vulnerable smallholders that often lack the financial reserves to compensate yield losses. Yet, previous research has mainly focused on economic adoption determinants (product attributes as well as farmers' income, land size and risk attitude) but these alone can barely explain farmers' insufficient risk management behaviour. Our study is the first to also explore the influence of provider trust, product understanding, distinct peer effects, and precautionary savings as an alternative and more familiar *ex-ante* risk instrument. We present results from experiments with 199 Uzbek farmers in 2019. Located in the insurance pilot region, they approximate local farm conditions and offer marketable index insurance as well as a realistic savings option prior to real product launch. Applying multinomial logit models, results indicate significant and stronger peer group imitation effects (bigger surrounding) compared to peer neighbour imitation (closer surrounding). While this signals superior trust in other's perceptive skills, it requires a critical mass to establish credible innovation strategies. Further, credit uptake, provider trust and practical product understanding boost index insurance adoption. Although farmers' preference for precautionary savings seems inferior to the one of index insurance, our findings assume a slightly complementary relationship between the two. In addition, results suggest community-based extension interventions that can fully exploit peer interactions, and offering credit-bundled products to increase widescale and resilience-enhancing innovation diffusion. This is one prerequisite for financial stability vital to think about (green) future investments in times of uncertain climate change impacts in the first place.

Keywords: Agricultural index insurance, climate adaptations, precautionary Savings, rural development, Uzbekistan

Contact Address: Laura Moritz, Leibniz Institute of Agricultural Development in Transition Economies (IAMO), Agricultural Markets, Theodor-Lieser-Straße 2, 06120 Halle (Saale), Germany, e-mail: moritz@iamo.de

An Analysis of the Brazilian Agroecological Commitment: the Effectiveness of Public Policies between 2015 and 2021

Hayanne Rodrigues Carniel Cavalcante[1], Dedierre Gonçalves Silva[1], Jaqueline da Costa Paixão[1], Lucas Soares Fraga[1], Nara Rabia Rodrigues do Nascimento-Silva[2], Luciana Ramos Jordão[3], Thiago Henrique Costa Silva[1]

[1]*University Center Alves Faria, Law College, Brazil*
[2]*Federal University of Goiás (UFG), Nutrition, Brazil*
[3]*State University of Goiás, Law College (Southeast Campus: Morrinhos), Brazil*

Agroecology aims for quality production and a balanced ecosystem through the integrative participation of society through the reduced use of harmful products and strategies for handling and improving the soil. This practice of producing target to promote environmental preservation, social justice, cultural respect and economic viability. In 2018 Brazil received the silver award in the Future Policy Award, which rewards the best laws and policies in the world aimed at promoting agroecology, by the National Policy on Agroecology and Organic Production (PNAPO), instituted in 2012 (Decree nº 7.794/2012). The main instrument for this policy is the National Plan for Agroecology and Organic Production (Planapo) that was elaborated in the years 2016 and 2019. Through this promising record, it is questioned about the current Brazilian agroecological commitment. Thus, this research aimed to analyse the period 2015–2021, with a focus on the analysis of public policies related to agroecological production and the use of pesticides. Using the historical and dialogical method, based on the analysis of Planapo and its effectiveness, documentary research and indirect study of data were carried out to verify the performance of agroecological public policies. The legal and political initiatives that debate the release, use and impacts of pesticides, are opposed to the way of producing agroecology, especially in the last three years, with President Bolsonaro. It was inferred that there are constant budgetary reductions for the promotion of agroecology over the years, hampering (impeding) the execution of this sustainable productive model and increasing the vulnerability of farmers. In addition to a significant increase in the release of pesticides, according to the Ministry of Agriculture, Livestock and Supply, increased from 139, in 2015, to 474, in 2020. Besides, the Brazilian State has a policy of tax incentives for pesticide use, since 1997 and remains, and currently is being confronted in the Federal Supreme Court through Direct Action of Unconstitutionality No. 5553, discussed at a slow pace. However, despite the Brazilian constitutional forecast of the ecologically balanced environment as a fundamental right, the dismantling of agroecological policies in Brazil is ongoing, therefore, research is still under development.

Keywords: Agroecology, pesticides, public policies, sustainable production

Contact Address: Thiago Henrique Costa Silva, Federal University of Goiás (UFG), Law College, Avendida Virgilio Joaquim Ferreira, 74860615 Goias, Brazil, e-mail: thiagocostasilva.jur@gmail.com

Land Prices, Concentration and Sustainability: A Relational Analysis from Morrinhos, Goiás Scenary between the Years of 2019 and 2021

Victória Cardoso Carrijo, Thiago Henrique Costa Silva, Luciana Ramos Jordão, Rebeca Barbosa Moura, Igor Gomes de Araújo, Barbara Luiza Rodrigues, Thays Dias Silva

State University of Goiás, Law College (Southeast Campus: Morrinhos), Brazil

Land, whatever the society, has always been associated with immanent, material and immaterial values. In ancient Mesopotamia, land was associated to life and fertility, a source of food. As cultivation practices have been improved, allowing correction in a given region, allowing human population to stay permanently at some area. Initially collectively and then individually, a land came to have owners. Today, land cannot be limited to a single function, as it is synonymous with food on the table, sacred territory for traditional populations, place of life and, also, of economic power. In this last perspective, the owners started to have rights over their lands and the conflicts over their ownership or property are constant. In Brazil, it was no different, conflicts started with a sixteenth-century pose system, initiating an uncontrolled land concentration in the country. In 1850, the Agrarian Law presented the first land reform plan, however, to this day, it has not yet been carried out. Turning to looking at the reality of the municipality of Morrinhos, in midwestern Brazil, around the 18th century, there was a disorderly occupation amid a lack of land legislation, which was regularized with each approved land law, consolidating the latifundios and monoculture in the region. Due to the characteristics of geographic relief and soil, which makes the land conducive to the cultivation of soy and livestock, the economic value of the land is increasing in the municipality. Given this context, this research seeks to assess whether the rise in land prices, in the last 3 years, has accentuated inequality and land concentration in Morrinhos, going against sustainable development. It starts from the hypothesis that the land concentration in Brazil favours the intensive and unsustainable use of land and natural resources, increasing social inequality. In a qualitative approach, supported by the inductive method, bibliographic and documentary research and indirect analysis of data on market values and land concentration in Morrinhos were carried out, correlating them. Data is being collected at the local registry office and at the city hall. It is also intended to compare such data with sustainability indicators, demonstrating the relationships graphically.

Keywords: Environment, rural development, sustainability

Contact Address: Victória Cardoso Carrijo, State University of Goiás, Law College (Southeast Campus: Morrinhos), Av. 101A quadra 57 lote 13A, 75650000 Morrinhos, Brazil, e-mail: victoriacarrijo@hotmail.com

Environmental Conduct Adjustment Agreements Performed by the Public Prossecutors's Office in Morrinhos, Goiás, Brazil: Mechanisms for Promoting Sustainbility

Rebeca Barbosa Moura, Luciana Ramos Jordão, Thiago Henrique Costa Silva, Victória Cardoso Carrijo, Barbara Luiza Rodrigues, Niury Ohan Pereira Magno, Igor Gomes de Araújo, Thays Dias Silva, Letícia Versiane Arantes Dantas

State University of Goias, Law College (Southeast Campus: Morrinhos), Brazil

The Environmental Conduct Adjustment Agreements are instruments regulated in Brazil that aim to allow the adaptation of enterprises to the legislation and to avoid civil liability lawsuits. In theory, these extrajudicial agreements aim to recover the degraded environment and to protect diffuse rights by those who have harmed them. These agreements enable entrepreneurs to overcome legal problems, promoting sustainable development and the continuity of economic activities. Thus, it is possible for the entrepreneurs to maintain a environmental licenses, without subjecting themselves to fines, since by carrying out a repairing environmental project, it is possible to obtain longer terms to fulfil environmental obligations. Considering normative intention and the aim to achieve sustainability, this research, with a geographical profile of Morrinhos, a medium-sized municipality in midwestern Brazil, which has its economy based on rural activities, intends to verify whether these agreements are effective or not. Through a deductive method, in a qualitative approach, the research discusses the agreements made between 2017 and 2020 and, through an interview with the municipality's public prosecutor, it assesses the perception of public agents about the effectiveness of these agreements. Although the agreements aim to reduce the number of lawsuits and accelerate the response to society, it appears that public prosecutors in Goiás do not sign many agreements, especially considering the great amount of registered environmental crimes. Still, in the few cases that were raised, using a deductive method, it is possible to assesses the perception of public agents about the effectiveness of these agreements and to verify their contents and results. Considering the perception of the responsible public agents, it concludes that the Conduct Adjustment Agreements are able to guarantee the responsibility of those who violate environmental norms, guaranteeing the continuity of agricultural and agro-industrial production, however, in addition to their celebration, they need follow-up. Thus, it is inferred that improving monitoring methods is needed. Also, trained personnel and the use of technologies, such as drones and satellite images, would allow the identification of compliance with agreements and judicial decisions.

Keywords: Environment, human right to food, rural development, sustainability

Contact Address: Rebeca Barbosa Moura, State University of Goias, Law College (Southeast Campus: Morrinhos), Avenue A, block A, lot C-13. Neighborhood Garden America, Morrinhos, Brazil, e-mail: rebecabbmoura@gmail.com

Between the Economic, the Social and the Environmental: The Border of Diversity in Regularisation of Land in Brazilian Traditional Territories

Marucia Pereira Dos Santos Silva[1], Renata Priscila Benevides de Sousa[2], Rogério Fernandes Rocha[3], Maria Izabel de Melo Oliveira dos Santos[1], Luciana Ramos Jordão[4], Thiago Henrique Costa Silva[3]

[1]*Alves Faria University Center (UNIALFA), Law College; Public Policy and Agrarian Study and Research Group (GEPPA), Brazil*
[2]*Amazônia University (UNAMA), Law College (Santarém, Pará); Public Policy and Agrarian Study and Research Group (GEPPA), Brazil*
[3]*Federal University of Goiás (UFG), Rural Development Sector; Public Policy and Agrarian Study and Research Group (GEPPA), Brazil*
[4]*State University of Goiás, Law College (Southeast Campus: Morrinhos), Brazil*

Brazilian policies for rural land regularisation aim to protect collective rights, establish criteria that guarantee the effectiveness of the constitutional right to housing and encourage sustainable development by mitigating the deficit of land. Considering the importance of this process, this study analyses the projects that propose changes in the land regularisation laws in Brazil and the risks that the changes can cause to the environment and to the traditional peoples that relate to the land as sacred, source of life and place of social reproduction. Advances in the democratisation of the land titling process of traditional populations are discussed, with mechanisms that protect collective rights in a participatory manner. A land policy without consultation with those involved can mean insecurity for the exercise of culture and the way of doing and living in the territories of the communities involved. The research procedures will be the bibliographic review, the documentary analysis of legislative projects in progress in the Brazilian National Congress whose themes are land tenure regularisation, traditional communities and sustainable rural development. An example is bill no. 510/2021 which provides for land regularisation in areas of the Union's domain, extending the regularized area to up to 2,500 hectares, with the elimination of the need for prior inspection of the area to be regularized, which can be replaced by a declaration by the occupier himself. Norms like these encourage the process of land grabbing, threatening traditional territories, and the process of deforestation for land use for grazing and production of commodities. Therefore, in order to discuss this reality, the historical-dialectical method is adopted, with the aim of rethinking the paradigms for Brazilian rural development. For this, an indi-

Contact Address: Marucia Pereira Dos Santos Silva, Alves Faria University Center (UNIALFA), Law College; Public Policy and Agrarian Study and Research Group (GEPPA), Rua 24 Qd 60 Lt 04 Setor Santos Dumont, 74463680 Goiania, Brazil, e-mail: arucias@gmail.com

rect analysis of data on the current Brazilian land tenure issue is made, relating it to the disruptive thoughts of critical researchers of rural development in Brazil, such as Antônio Carlos Wolkmer , Maria Cristina Vidotte Blanco Tárrega e Alfredo Wagner Berno de Almeida. It can be concluded that Brazilian norms disregard the diversity of subjects and ways of life of traditional peoples, stimulate land concentration and, consequently, increase Brazilian social inequality.

Keywords: Land Regularisation, Public Policies, Traditional Communities

The Behavioural Receptiveness and Non-Receptiveness of Farmers towards Organic Cultivation System

Nitika Thakur
Shoolini University Solan, Biotech, India

A farmer generally possesses small, marginal and large-scale farm production systems. According to the general perception, the maintenance of such farms generally demands chemical cultivation system for increased quantity and yields. Increased use of chemicals in the agriculture sector has disturbed the harmony existing among plants, soil, animals and human health. The excessive use of chemicals is an inorganic food production method which made the health-conscious people explore and follow organic farming. The present review deals with the acceptance and perception diversions of farmers and consumers from farm level to marketing of the final product regarding goals to assure the sustainability of their farms for the future, motivation attributes to make their choices judiciously and by their personal view of preferences for farming. Furthermore, for the farmers, it is essential to stand individually with self-interest and a zeal to become a ßuccessful farmerfollowing the pattern, integration and certification of organic farming. On the other phase, social, cultural, psychological, economic and personal factors adversely affect the consumer market, and the incline has been observed positively for organic food with enhanced quality and nutrition. However, some flaws and barriers have been identified to switch to a state of organic farming from other modes of cultivation systems which are generally faced in the inter-conversion phase by the farmers. However, these barriers can be handled with the right strategies which rejuvenate soil health, food quality and sustainability. So, the present review presents the scenario and prospect of organic farming, highlighting the barriers which hinder a farmer from adopting and investing sustainably.

Keywords: Food quality and sustainability, inorganic farming, motivation attributes, organic, perception and preferences

Contact Address: Nitika Thakur, Shoolini University Solan, Biotech, Vill-Basal, 01792 Solan, India, e-mail: nitikathakur45@gmail.com

Influencing Factors on Consumers' Perceptions and Awareness of Food Health and Organic Food in Iran

Arezou Babajani[1], Hossein Shabanali Fami[2], Ali Asadi[2]
[1]*University of Hohenheim, Agricultural and Food Policy Department, Germany*
[2]*University of Tehran, Dept. of Agricultural Management & Development, Iran*

Food health concerns and environmental matters are the most important motives for consumers to learn more about healthy foods and especially organic foods and being aware of them. In Iran due to the lack of permanent organic markets, fewer consumers have been informed about organic foods. The objective of this study was to present some insights on Iranian consumers' attitudes toward food health and organic food and the factors which have some effects on their awareness about these products. Thus, this study has focused on consumer's awareness and perceptions of organically produced foods in Iran regarding their state of knowledge about organic products. Demographic characteristics, knowledge about organic production, and resources of their information about this kind of foods have been analysed to show differences between consumer's awareness on organic food products and finally a regression model has been used to ascertain the factors affecting consumer's awareness and differences between people who are aware and not aware about organic foods. The method of this study was survey method, by applying quantitative approach and data gathering has been done among Iranian consumers of city markets. In the present research, 454 respondents from six provinces of Iran have been asked by a questionnaire. Results indicate that people who are aware of organic food products are more educated, have more incomes and live-in big cities and mass media is the most important channel of informing people about organic products. According to the results, in response to the question whether the authorities and institutions in Iran are careful about proper using of chemical fertilisers and pesticides, the opinion of two group respondents (aware and not aware of organic products) were significantly different. Furthermore, the high prices of organic certified products and too few accesses to them in city markets are the main obstacles to purchase them in big cities in Iran.

Keywords: Food health concerns, organic food consumers, organic food products

---

Contact Address: Arezou Babajani, University of Hohenheim, Agricultural and Food Policy Department, Schwerzstraße 46 , 70593 Stuttgart, Germany, e-mail: arezou.babajani@uni-hohenheim.de

Linking Agricultural Sustainability with Climate Change Mitigation and Peacebuilding: A Framework for Assessing Interventions

Lisset Perez Marulanda[1], Augusto Castro-Nunez[2], Martin Rudbeck Jepsen[1]

[1]*University of Copenhagen, Dept. of Geosciences and Natural Resource Management, Denmark*

[2]*Alliance Bioversity-ciat, Climate Action,*

To achieve a sustainable future in rural conflict-affected areas, interventions to improve agricultural livelihoods that reinforce climate change mitigation and peacebuilding are needed. Therefore, a conceptual framework that allows an evaluation of the effects of these interventions is important. However, the current agricultural sustainability frameworks lack elements that integrate either peacebuilding or climate change mitigation, or both. An integrated framework that assesses simultaneously social, economic, environmental, climate change mitigation and peacebuilding dimensions would improve the evaluation of agricultural interventions. In this research, we develop such a framework. The proposed structure is based on the existing frameworks and takes into account different elements of them. Especially, the environmental, economic and social dimensions of sustainability that are frequently used. While peacebuilding and climate change mitigation are assessed transversally. The framework is adapted to the local context and allows obtaining results at various scales ranging from farm to value chain level. We tested the framework in a case study in Caquetá- Colombia, an area that has emerged from armed conflict. Furthermore, various agricultural interventions have been implemented as a means to achieve peace meanwhile reducing deforestation and forest degradation. To validate the framework, we interviewed 475 households and applied participatory approaches. For data analysis, we calculated an index that combines the dimensions through the Principal Component Analysis method. In the qualitative approach, we develop a barometer that reflects the effects of the human-environment interaction and visualises Caquetá's progress toward sustainability. Finally, we discuss how landscape-level climate change and peacebuilding outcomes vary according to the number of farmers who comply with the interventions in the agricultural sector.

Keywords: Agriculture, climate change mitigation, peacebuilding, rural development, sustainability

Contact Address: Lisset Perez Marulanda, University of Copenhagen, Dept. of Geosciences and Natural Resource Management, Copenhagen, Denmark, e-mail: lisset.perez@cgiar.org

Long-Term Impact of Fairtrade Coffee Certification on Household Income in India

Kedar Nepal[1], Ulrike Grote[1], Pradyot Jena[2]

[1]*Leibniz Universität Hannover, Institute of Environmental Economics and World Trade, Germany*

[2]*National Institute of Technology Karnataka, Surathkal, School of Management, India*

Coffee certification is increasing rapidly in global value chains in lieu of the price premiums received by coffee farmers. This is also true to fairtrade certification of small-scale coffee farmers. However, several studies on coffee certifications reveal mixed results on welfare impacts. One of the promising objectives of Fairtrade coffee is to improve the livelihoods of smallholder farmers. Against this background, we analyse whether Fairtrade certification will make discernible changes in livelihoods for small-scale coffee farmers of Araku valley in India. We use cross-sectional data of 386 households from 2011 and 2018. Propensity score matching and endogenous switching regression methods are applied to the impact evaluation in order to reduce self-selection bias. We find that certified members show significantly 8 % higher coffee income than non-certified members. The benefits being modest and certification systems gaining slower in momentum, challenges persist in escalating capacity of cooperatives and gearing up awareness of fairtrade in communities. Income gains have been significant among certified households indicate a pavement to sustainable livelihoods reducing income risk and improving poverty prevalent in the survey area. Higher farm gate prices and assured minimum prices of coffee have boosted the economic performance of coffee producers. Maturity of farmers i.e., age, participation in coffee training and years of cropping experience play a crucial role for households to be certified. Fairtrade certified farmers directly benefit from price premiums and indirectly from communal infrastructures. Fairtrade certification should thus be promoted to reduce poverty and improve smallholder livelihoods of coffee growing households in rural India.

Keywords: Coffee, endogenous switching regression, fair trade certification, India, propensity score matching

Contact Address: Kedar Nepal, Leibniz Universität Hannover, Institute of Environmental Economics and World Trade, Karl Wiechert Allee 15, 30625 Hannover, Germany, e-mail: knepal1991@gmail.com

Participatory Guarantee Systems: Agroecological Certification to Enhance Small-Scale Family Farmers' Adaptive Capacity to Climate Change

Pablo Urbina, Jürgen Pretzsch

Technische Universität Dresden, Inst. of International Forestry and Forest Products: Tropical Forestry, Germany

Food production is behind the success of developed nations. Once the food issue is solved, human and financial resources can be allocated to address other areas of development. But since the advent of the green revolution and its rapid spread over the world, the food production system has proven to be a major driving force of environmental change by destabilising the phosphorus and nitrogen cycle, biodiversity, freshwater usage, land use, and forest coverage. Additionally, climate change poses serious challenges to an already vulnerable food production system, with more frequent extreme weather events reducing crop yields. Therefore, the food production system needs to undergo a transformation that can guarantee foodstuff for nations constantly and effectively whilst in harmony with the earth system, resilient enough to withstand unforeseen human or natural driven shocks. Participatory Guarantee Systems (PGS) are alternative mechanisms for regulating food production standards. They rely upon the active participation of multiple stakeholders to provide an affordable quality assurance certification for local producers. Using a case study approach comparing two PGS from Peru, this article highlights how PGS contributes to strengthening farmers' adaptive capacity to climate change by empowering farmers, enhancing their food security and market access, and stimulate the implementation of agroecological practices. Using a non-probabilistic purposive critical case sampling technique secondary data from the period 2017–2020 of 54 farmers were collected to analyse the trend on agroecological practices, this was complemented with interviews to stakeholders of the organisational structure of both PGS to qualitatively assess farmers empowerments, enhanced food security and enhanced market access. The analysis shows that year after year farmers keep on implementing agroecological practices to get a certification, improving their practices based on inspectors' recommendations. The results from the interviews show farmers have access to more information on agricultural practices, make better decisions about their farm management, have a sustained food production through the year, and sell their agricultural products easier with a certification.

Keywords: Adaptive capacity, agroecological practices, certification, climate change

Contact Address: Pablo Urbina, Technische Universität Dresden, Inst. of International Forestry and Forest Products: Tropical Forestry, Zellescher Weg 41c C01061, 01217 Dresden, Germany, e-mail: pablo-urbina@outlook.com

Measuring Environmental and Economic Sustainability of Vegetable Production in Karnataka, India

Afrin Zainab Bi, K.B. Umesh

University of Agricultural Sciences, Dept. of Agricultural Economics, India

India has witnessed tremendous growth in vegetable production, especially during the post-green revolution. Increased demand, high value, and excellent response to external inputs have resulted in vegetable cultivation's intensification. It is imperative to understand how sustainable this intensification of vegetable production is? Sustainability assessment is complex and multidisciplinary in nature, yet a key to plan and achieve sustainability in agriculture production. Environmental and economic dimensions are the most crucial for assessing sustainable agriculture development at the farm level. Against this backdrop, economic and environmental sustainability was studied for the two major vegetables, viz., onion and tomato in Karnataka state, India. Indicators for assessing sustainability were systematically chosen. First, through an extensive review of literature candidate indicators for both the dimensions were selected and then they were finalized by the multidisciplinary expert panel. Mean composite indicator scores of economic sustainability were 0.57 and 0.62 for onion and tomato, respectively. Cultivating onion in rainfed conditions had healthier economic sustainability than irrigated conditions for all its principals, except productivity. Apart from profitability, drip cultivation was found more economically sustainable than flood irrigation for all its principals. The two crops had peripheral differences in composite environmental indicators, with average scores of 0.47 and 0.49 for onion and tomato. The environmental performance of flood and drip-irrigated cultivation of tomato farms was similar, although drip cultivation marginally outperformed the flood system in relation to space organisation and energy consumption. In total, though both the composite economic and environmental sustainability scores were low, the economic dimension had better accomplishment than the environmental dimension. Production practices that have environmental impact were far below the sustainable path in vegetable cultivation. Thus, imparting environmental consciousness directly or controlling farming practices indirectly through taxation and subsidy policies on inputs is required to maintain and enhance environmental health.

Keywords: Composite indicator, economic sustainability, environmental sustainability, farm-level, vegetables

Contact Address: Afrin Zainab Bi, University of Agricultural Sciences, Dept. of Agricultural Economics, 560065 Bengaluru, India, e-mail: afrinzainab22@gmail.com

Sustainable Development Goals in Agriculture and Responsible Investment: A Comparative Study from the Czech Republic and Ukraine

Alex Plastun[1], Inna Makarenko[2], Ricardo Chandra Situmeang[3], Tetiana Grabovska [4], Serhii Bashlai[5]

[1]*Sumy State University Ukraine, Chair of the International Economic Relations, Ukraine*

[2]*Sumy State University Ukraine, Chair of Accounting and Taxation, Ukraine*

[3]*Czech University of Life Sciences Prague, Fac. of Tropical AgriSciences - Dept. of Sustainable Technologies, Czech Republic*

[4]*Bila Tserkva National Agrarian University, General Ecology and Ecotrophology, Ukraine*

[5]*Sumy National Agrarian University Ukraine, Chair of Economics and Entrepreneurship, Ukraine*

This paper explores Sustainable Development Goals (SDGs - 2 and 12) in agriculture for the cases of the Czech Republic and Ukraine. The idea is to find best practices in SDGs 2 and 12 implementations within a responsible investment framework. For these purposes benchmarking (comparative analysis) is used. Using data from the Czech Republic and Ukraine over the period 2017–2020 a general comparative review of global and national targets of SDGs 2 and 12 in Ukraine and the Czech Republic is provided. The basis for the comparative analysis is information and statistical sources (The Czech Republic's Voluntary National Review (2017), Sustainable Development Goals Ukraine. Voluntary National Review. (2020), abovementioned normative sources on SDGs and the database of Sachs et al. (2020). A key method of comparing progress in achieving SDG 2 and 12 in the agricultural sector of Ukraine and the Czech Republic, as well as the role of responsible investment in this progress, is benchmarking (comparative analysis)
Results justify a merely incorporation and compliance of these targets on the national and global levels. Identified problems in SDG 2 and SDG 12 progress are common for Ukraine and the Czech Republic and concern unequal access to investment and financial resources. Recommendations and solutions to the most important problems based on responsible investment instruments are proposed in this paper. The results of the research can be useful for regulators (both in agriculture and financial market), society, and a wide group of other stakeholders in promoting responsible investment towards more comprehensive SDG 2 and 12 progress in the Czech Republic and Ukraine to 2030.

Keywords: Responsible investment, sustainable consumption

Contact Address: Ricardo Chandra Situmeang, Czech University of Life Sciences Prague, Fac. of Tropical AgriSciences - Dept. of Sustainable Technologies, Kamýcká 129, 16500 Prague, Czech Republic, e-mail: ricardositumeang@gmail.com

Credit Access and its Impacts on Small Coffee Farmers in Climate Change Adaptation in Vietnam

Phuong Anh Nguyen, Miroslava Bavorova
Czech University of Life Sciences Prague, Fac. of Tropical AgriSciences, Czech Republic

Coffee is the second most important agricultural commodity in Vietnam after rice in terms of export value. However, the impacts of climate change and conventional farming systems have had severe repercussions for the sustainable growth of this sector. Although adaptation strategies, such as practising water management, intercropping farming, and livelihood diversification, showed many benefits, their application has remained low in small-scale Vietnamese coffee farmers. Part of the reason involves the low accessibility of coffee farmers to agricultural credit. A closer look reveals constraints from the demand side of coffee farmers, especially smallholders, have substantially contributed to such low credit access. One of the main constraints comes from the difficulties of small households in fulfiling bank requirements in which a lack of collateral due to tiny farm size or unclear land ownership concerns is a common reason. Other constraints depend on farmers' attitude towards credit risk alongside their perception of loan repayment and procedures. The study aims to analyse the influences of credit access on coping strategies, namely water-saving techniques and mixed/multi-cropping practices during climate stimuli in Vietnam's Central Highlands. Both quantitative and qualitative primary data will be collected in early 2022. The quantitative data using a structured questionnaires survey includes about three hundred randomly selected respondents, while qualitative data will consist of about 15–20 expert interviews. Logit model will be used to identify the relationships between applying adaptation strategies (dependent variable) and selected farmers' socio-economic characteristics and various types of credit schemes. The effect of credit scheme characteristics, including guarantor, collateral, and interest rate, on adopting adaptation strategies will also be estimated. The proposed study will contribute to the existing literature on credit access, precisely its effect on smallholder coffee farmers who have highly relied on this financial package to engage in adaptation strategies in Vietnam. The results will help policymakers and non-governmental organisations to adjust their policies to improve credit accessibility and thus increase the level of adopting climate change adaptation strategies among small-scale coffee farmers in Vietnam.

Keywords: Adaptation strategy, climate change, coffee, credit access, intercropping farming, small farmers, Vietnam, water management

Contact Address: Phuong Anh Nguyen, Czech University of Life Sciences Prague, Fac. of Tropical AgriSciences, Kamýcká 129, Prague, Czech Republic, e-mail: nguyenp@ftz.czu.cz

An Agroecological Turn? Assessment of Kenya's Policy and Legal Framework

MARTIN OULU[1], VALENTINE OPANGA[2]
[1]*University of Nairobi, Dept. of Agricultural Economics, Kenya*
[2]*University of Bonn, Dept. of Geography, Germany*

Agroecology is emerging as one of the approaches to sustainably transform agricultural systems. Fanned by global crises such as climate change and Covid-19, agroecology is however more than just a set of principles and practices. It aims to create a people-led movement that is supported by rather than led by science and policy. This paper analyses Kenya's policies and laws to determine whether an agroecological turn is happening. Given the cross-cutting nature of agroecology, Kenya's policies and legislations are fragmented across different sectors but with little synergy. A sectoral approach still dominates, with many policies focused on attaining rapid economic growth. Even though the agricultural policies and strategies do not use the terms agroecology per se, they offer a good foundation for the promotion of and implementation of agroecological principles and practices (particularly levels 1 and 2 of Gliessman's five levels). The environment, forestry and wildlife related policies and legislations address conservation of biodiversity and genetic resources and broader environmental stewardship. However, most are focused on protected areas rather than agricultural landscapes even though most of Kenya's biodiversity are found in the agricultural landscapes. The trade policies provide little support to niche products such as organic products, even though export-led agriculture can stimulate adoption and upscaling of ecological agriculture to meet the often-higher international market standards. Some counties are showing leadership by developing agroecology and biodiversity policies even without similar policies or guidance from the national government. Overall, the policies provide a good framework for the promotion and enhancement of agroecology. However, poor implementation, little participation and influence of farmers, and poorly designed incentive structures remain. Making the transition from conventional to agroecological agriculture and food system requires a portfolio of incentives, but also the empowerment and participation of small-scale farmers in decision-making. These are still largely lacking in Kenya. The agroecological turn is thus yet to fully take root in Kenya, but elements of such a turn are evident. More research on the barriers and opportunities for the establishment and up-scaling of agroecology in Kenya and other sub-Saharan African countries is necessary.

Keywords: Agroecology, Kenya, law, policies, political ecology, sustainable food systems

Contact Address: Martin Oulu, University of Nairobi, Dept. of Agricultural Economics, Nairobi, Kenya, e-mail: ochiengmoulu@gmail.com

The Cost of Climate Change Adaptation and the Effects on Revenue in Oyo State, Nigeria

Khadija Atinuke Ibrahim, Oluwafunmiso Adeola Olajide

University of Ibadan, Dept. of Agricultural Economics, Nigeria

The effects of climate change can be perceived in basic terms of the increase in global temperature that has resulted in variations in weather conditions. These variations have resulted in weather-related disasters that have had negative effect on the environment and agriculture as a whole, especially in Africa where we practice rain-fed agriculture. This study focuses on the effect of climate change and the possible adaptation strategies that have been adopted over time. This study aims at not only identifying the climate adaptation strategies adopted by cassava farmers but the monetary cost incurred by farmers who adopt these strategies and how it affects their income and yield. We can therefore identify this cost as the 'cost of adapting to climate change'. The study was carried out in Ibarapa East Local Government Area of Ibadan, Oyo state where 110 cassava farmers were randomly selected from three towns. Data was collected through a well-structured questionnaire. The socioeconomic characteristics of the farmers were computed using statistical tools such as frequency, mean, andd standard deviation. The average age of the respondents was 49 years while 32.7 % of the farmers were females and 67.3 % were males. A 5-point scale Likert analysis was used to check the level of Climate Change perception of farmers and it was found out that farmers are actually observant of the changes that occur on their farms due to Climate change. Multiple regression analysis was carried out to check for the factors that actually affect Income and yield. The cost of adapting to climate change has a significant negative effect on the annual production and does not have any significant effect on the annual income of farmers. From the findings of the study, it was recommended that more training on climate change sensitisation should be focused on by the extension workers, and also subsidised inputs should be made available for farmers.

Keywords: Adaptation strategies, cassava, climate variability, cost

Contact Address: Oluwafunmiso Adeola Olajide, University of Ibadan, Dept. of Agricultural Economics, Ibadan, Nigeria, e-mail: funso.olajide@ui.edu.ng

The Role of Social Capital on Food Security: Empirical Evidence from Rural Tanzania

JUKWANG YOON, RASADHIKA SHARMA
Leibniz Universitat Hannover, Institute for Environmental Economics and World Trade, Germany

Over 80 percent of the population in rural Tanzania suffers from food insecurity. In this regard, social capital can be considered an important resource for the rural poor, who possess little or no other forms of capital, to attain food security. Against this background, this paper explores the role of household social capital – social network, trust, norms, and reciprocity; as well as knowledge sharing and collective action - on food security in rural Tanzania. We employ unique data on 900 households in villages of Morogoro and Dodoma obtained under the "Trans-SEC" project. The social capital index is constructed with standardisation and exploratory factor analysis, and, is cross-validated by comparison with alternative indices. We capture household food security using various food security indicators such as Food Consumption Scores (FCS), Coping Strategy Index (CSI) and Months of Adequate Household Food Provisioing (MAHFP). These represent the four pillars of food security - availability, access, utilisation, and stability. Structural equation modelling is used to examine the effect of social capital and the mediation effect of knowledge sharing and collective action on food security. The results show a strong positive correlation between social capital and almost all food security indicators. Both, knowledge sharing, and collective action demonstrate a positive correlation with caloric and protein intake as well as diet diversity. Additionally, collective action is negatively correlated with Coping Strategy Index and stability of food provision. However, knowledge sharing and collective action only partially explain the effect of social capital on food security. The significant direct effect of social capital on food security indicates that other effects of social capital such as direct transfer of food or money through social connections have strong impacts on food security. The paper highlights that increased household social capital is key to improved food security in rural contexts. It recommends strengthening social capital both horizontally and vertically. For instance, diffusion of knowledge can be enhanced by promotion of agricultural cooperatives and associations among farmers (horizontal ties) and encouraging extension services and cooperation with research institutions (vertical ties).

Keywords: Collective action, food security, knowledge sharing, social capital

Contact Address: Jukwang Yoon, Leibniz Universitat Hannover, Institute for Environmental Economics and World Trade, Hannover, Germany, e-mail: jyoon06@gmail.com

Is Organic Farming Financially Competitive at the Farm Gate Level in the Tropics?

Amritbir Riar, Eva Goldmann, Laura Armengot, David Bautze, Gurbir Bhullar, Johanna Rüegg, Noah Adamtey, Monika Schneider, Beate Huber

Research Institute of Organic Agriculture (FiBL), Dept. of International Cooperation, Switzerland

Climate change, coupled with land degradation and growing population, has severely threatened food security. The effects of the aforesaid challenges are more profound in the tropics; home for half of the global human population by 2050 and accounts for 18.7 % of the global economic activities. Decisions regarding more sustainable farming practices, like organic farming, are more complex in tropical contexts as many smallholder farmers in tropics operate at the interface of sustainability, agroecology and economic viability of farms often tangled in socio-economic and biophysical challenges. The economic performance of organic in comparison to conventional is often the decisive argument for producers. The present study draws on twelve years of economic data from four long-term experiments located in Bolivia, India, and Kenya. The trials are based on three different main crops (cocoa, maize, and cotton) and compare the performance of organic and conventional farming systems. We found that the economic performance of organic farming systems is site-specific and is not necessarily dependent on main crop performances. Associated crops also provide a similar contribution to the income in conventional and organic farming; however, it was also evident that relative contribution of the associated crops would increase in the organic systems if fair prices are paid for the associated crops, it will reduce the dependence on one crop (export crop) and make farmers more resilience to market fluctuations. Diverse and balanced crop production is an integral element of organic agriculture, hence effort need to be taken to focus research and development of organic farming systems beyond often export-orientated main crops towards associated crops.

Keywords: Cacao, cotton, farming systems, long term experiments, maize, organic farming

Contact Address: Amritbir Riar, Research Institute of Organic Agriculture (FiBL), Dept. of International Cooperation, Frick, Switzerland, e-mail: amritbir.riar@fibl.org

Increasing Farmers' Involvement in Participatory Breeding of Organic Cotton in India

STEFAN GRAF[1], NATHALIE OBERSON[1], CHRISTOPH STUDER[1], MONIKA MESSMER[2], TANAY JOSHI[2], AMRITBIR RIAR[2]

[1]*Bern University of Applied Sciences (BFH), School of Agricultural, Forest and Food Sciences (HAFL), Switzerland*

[2]*Research Institute of Organic Agriculture (FIBL), Frick, Switzerland*

Seeds of cotton are increasingly difficult to source for organic farmers in India. The reason for this being the wide use of Genetically Modified (GM) cotton, on which breeding and seed production focusses. GM is banned in organic production. There is therefore a need to breed varieties adapted to the organic sector. Participatory Plant Breeding (PPB), meaning breeding which involves farmers in the breeding process, is applied all around the world where market conditions do not provide adequate varieties to farmers. The Seeding the Green Future (SGF) project aims of securing GM-free cotton seeds, safeguarding and improving the germplasm for organic farming systems since 2011 in India through PPB, as PPB is well adapted to the context.

To increase the knowledge base on farmers' preferred cultivar traits as well as identifying opportunities and challenges in participation to improve the PPB methodology, trainings of farmers in seed multiplication, cultivar evaluation and hybridisation of the SGF project were observed. Furthermore, 53 open ended interviews and 126 structured interviews were conducted with farmers.

Farmers have a broad range of criteria in selecting cotton cultivars, differing between the considered species *Gossypium hirsutum* and *G. arboreum*, between male and female farmers, as well as between different groups of farmers. All quality traits important to the industry (fibre length, strength, and finesse) were mentioned by farmers. That different farmers seek different types of varieties shows the importance of using in-situ, decentralised selections by the target groups of farmers. To do so, farmers need access to segregating material with suitable characteristics based on which they can select their own local varieties. The farmers need more hands-on experience in breeding during the trainings.

To avoid the "sword of Damocles" of GM cotton, local seed production with hand gins is recommended, as farmers complain they cannot get their seeds back from the large-scale ginneries. Storage of seeds is also preferably done locally, either by individual farmers or by farmers' groups. Furthermore, using *G. arboreum*, which is not yet available as GM, can be used, or alternatively perceptually distinct *G. hirsutum* which are visually distinct from available GM cultivars.

Keywords: Farmer participation, *Gossypium*, organic cotton, participatory plant breeding, plant breeding, PPB

Contact Address: Stefan Graf, Bern University of Applied Sciences (BFH), School of Agricultural, Forest and Food Sciences (HAFL), Laenggasse 85, 3052 Zollikofen, Switzerland, e-mail: stefan.graf@bfh.ch

Role of Multi-Stakeholder-Partnerships for Achieving a Higher Degree of Sustainability in Cross-Border Agri-Food Supply Chains

Julia Reimers[1], Christine Wieck[2]

[1]*University of Hohenheim, Dept. Agricultural and Food Policy, Germany*
[2]*University of Hohenheim, Dept. Agricultural and Food Policy (420a), Germany*

The UN Global Compact Network Germany defines sustainability in supply chains as the management of ecological, social and economic effects and the promotion of good corporate governance over the entire life cycle of products and services. However, how sustainability is to be implemented and interpreted in agricultural supply chains in practice is controversial, not least against the background of the currently discussed German "Supply Chain Act". The promotion of multi-stakeholder partnerships (MSP) to promote cooperation between all actors in a supply chain from production, trade and industry as well as representatives from politics, civil society and research seems reasonable and is increasingly being promoted, also within the framework of the Sustainable Development Goals, SDG. Nevertheless, the effects that can be achieved through MSP for different supply chains have so far hardly been scientifically researched.
With the help of a political network analysis, this paper analyses if, how and through which channels MSP and / or individual actors within the food system negatively and / or positively influence political decision-making processes in Germany in favour of more sustainability. Based on the "German Initiative for Sustainable Cocoa" (GISCO), the "Forum for Sustainable Palm Oil" (FONAP) and the "Partnership for Sustainable Orange Juice" (PANAO), standardised interviews with several hundred of representatives from politics, the private sector, civil society and science are analyzed, using the snowball procedure. Results are presented and interpreted via sociograms and network measures, allowing to draw conclusions on whether and under which conditions MSP should be further promoted in future.

Keywords: Agri-food policy, MSP, network analysis, standards, sustainability

---

Contact Address: Julia Reimers, University of Hohenheim, Dept. Agricultural and Food Policy, Stuttgart, Germany, e-mail: j.reimers@uni-hohenheim.de

Monitoring Processes in Carbon Credit Projects and their Potentials for Farmer Organisations - A Case Study from Kenya

ANNKATRIN SCHOBER, REGINA BIRNER, CHRISTINE BOSCH
University of Hohenheim, Inst. of Agric. Sci. in the Tropics (Hans-Ruthenberg-Institute), Germany

Carbon Credits and Voluntary Carbon Markets are experiencing an upswing and growing interest from different actors involved in global emission trading as well as development projects linked to climate change. However, the accounting of carbon sequestration which underlies the certificate of a carbon credits requires sophisticated data sets. Depending on the certification standard a specific monitoring process has to be developed and implemented in order to assure the collection of reliable data. This leads to various challenges, especially in the African context where for instance input documentation in agriculture is rather rare if not non-existent. Nevertheless, the establishment of these monitoring processes and resulting data might have beneficial effects that exceed the generation of carbon credits. Besides increasing transparency of the certification, reliable data on agricultural practices might be used to optimise extension service on sustainable agricultural practices that increase soil fertility, climate resilience, and improve yields. The Kenya Agricultural Carbon Project, implemented 2009 in Western Kenya, set an example for activity-based monitoring of the implementation of so called Sustainable Agricultural Land Management (SALM) practices. Yet, knowledge gaps on the potential benefits of the collected data for local actors like farmer groups and extension service, and challenges along the monitoring process remain. This study has the objective to systematically map the governance challenges, and to explore the potential role of farmers organisations. Expert interviews have been conducted to identify the most influencing aspects linked to these topics. Preliminary findings show challenges specifically on the local level. Trust and expectations of farmers towards data collecting institutions, project timeframe and pressure, the role of data cross checks, and data ownership are identified as critical aspects regarding monitoring processes. Additionally, projects often do not use existing organisational structures such as farmer organisations. This, however, could be adavtageous regarding the critical aspects. The potential role of existing farmer organisations should therefore be examined further. From the perspective of such an organisation, participation in these projects might lead to institutional strengthening or even the development of a business model in case carbon revenues are used to fund extension service.

Keywords: Carbon credits, farmer organisations, Kenya, monitoring processes

Contact Address: Christine Bosch, University of Hohenheim, Inst. of Agric. Sci. in the Tropics (Hans-Ruthenberg-Institute), Wollgrasweg 43, 70599 Stuttgart, Germany, e-mail: christine.bosch@uni-hohenheim.de

Social-Ecological Systems Analysis of a Bolivian Artisanal Fishery: Community-Based Governance and Rural Economies towards Agrobiodiversity

Carla Rene Baldivieso Soruco[1,2], Michelle Bonatti[2], Stefan Sieber[1,2]

[1]*Humboldt-Universität zu Berlin, Germany*

[2]*Leibniz Centre for Agric. Landscape Res. (ZALF), Sustainable Land Use in Developing Countries (SusLAND), Germany*

Studying the organisational ways on how agrobiodiversity is created in different social-ecological systems could provide insights to develop initiatives based on an adequate understanding of rural economics ways of life. In this regard, the present paper elaborates on the governance of a rural space in the Gran Chaco region in Bolivia. The main goal of the research was to study which mechanisms of governance are developed by the Weenhayek riverine population, as people highly dependent on the riverine system of the Pilcomayo stream for their livelihood. In a regional context of major changes with respect to their economies and their ways of life, given the threats to agrobiodiversity in the life zones of the Pilcomayo river.

The study is build based on a qualitative case study, through an in-depth exploration from multiple perspectives of the social – ecological system. The research developed a triangulation of methods as a strategy to combine various practices of research, linking participant observation and interviews developed in exploratory field work, with document analysis and media analysis. The categorisation of data into defined concepts and its systematisation was developed with support of the NVivo 12 software. In total, thirty-seven third-tier variables were unpacked for the study, using the Socio-Ecological Systems (SES) analytical framework developed by Ostrom and colleagues (2020) and the Agrobiodiversity Knowledge Framework developed by Zimmerer et al. (2019).

As results, it is proposed that Weenhayek community-based governance organisations are dependent on socioeconomic and ecological factors. Regarding social variables, their greater or lesser integration into the modern market economy plays an important role. Considering ecological factors, the ecological characteristics of the river determine the form of organisation for fishing. The flows affecting the SES generate social and ecological distributive conflicts that determine the effectiveness of community governance. The emergence of community-based institutions to overcome social - distributive issues and "social dilemmas" is observed as a main feature. The understanding that people have of their SES determines their societal representations, as exemplified in their cultural norms and traditional ecological knowledge, based on which they manage the agrobiodiversity of their life zones.

Keywords: Agrobiodiversity, governance, rural economies, social ecological systems

---

Contact Address: Carla Rene Baldivieso Soruco, Humboldt-Universität zu Berlin, 10559 Berlin, Germany, e-mail: baldivic@hu-berlin.de

Ecovillages as a Model of Sustainable Settlements in Rural Development Context: Principles and Main Elements

Arezou Babajani

University of Hohenheim, Agricultural and Food Policy Department, Germany

Sustainable rural development has many definitions and approaches especially in production regard, but this kind of development has some other aspects such as spatial and social issues. Ecovillage design is created as a response to the environmental and social problems of modern way of life and could be described as a new field of knowledge which is defined by Global Ecovillage Network (2005) as an intentional or traditional community using local participatory processes to holistically integrate ecological, economic, social, and cultural dimensions of sustainability to regenerate social and natural environments. The main objective of this paper is to identify the principles and features of ecovillages and their relations to sustainable rural development. Results of this study shows that ecovillages have some main principles that based on them sustainable rural development could take place. These principles are classified in some topics like sustainable living, environment improvement, optimal water, and soil management, improving health, cultural, social, and spiritual development, optimal management of energy resources, sewage management, and sustainable farming managements. On this basis an ecovillage has three dimensions, which related to each other in a settlement space, including ecologic dimension, socio-economic dimension, cultural/ spiritual dimension. Literature of ecovillages reveals some basic features or elements. These features which are essential for a successful development of such settlements include using new technologies e.g., solar energy, having no standard dimensions, requiring a planned architecture, and landscaping, offering onsite employment opportunities to the village residents or visitors, having local governments, having specific social management plans, and being the reflection of the consciousness of the nature. In fact, ecovillages are formed to fulfil the socio-ecological needs of the people in a sustainable way.

Keywords: Ecovillage, rural development, sustainable development

Contact Address: Arezou Babajani, University of Hohenheim, Agricultural and Food Policy Department, Schwerzstraße 46 , 70593 Stuttgart, Germany, e-mail: arezou.babajani@uni-hohenheim.de

Ethnomedicinnal Evaluation of Herbaceous Plants in the Guineo-Congolaise Zone of the Benin

Eudes Florent Sobakin, Serge Adomou, Carlos Ahoyo, Thierry Dèhouégnon Houehanou, Gérard Nounagnon Gouwakinnou, Armand Natta

University of Parakou, Lab. of Ecology, Botany and Plant Biology, Benin

In underdeveloped countries, plants occupy a central place in human life in society with regard to their multiple functions in the fields of food, medicine, cultural, agro-forestry, fuel and technology. However, knowledge about herbaceous medicinal plants used in the treatment of various diseases or ailments is not as popular as woody species are. The present study proposes (i) to identify the main herbaceous medicinal plants in the Guinean-Congolese zone of Benin and (ii) to assess the effect of socio-environmental factors such as age, sex, religion. , ethnicity, level of education and phytogeographic district, determining the variation in knowledge related to the use of herbaceous medicinal plants. A study is carried out through individual and semi-structured interviews with 310 people on medicinal herbaceous species in the Guinean-Congolese zone of Benin. The medicinal use values (UV) and relative citation frequencies (FC) for each species were determined. The Kruskal-Wallis and Mann Whitney inference tests are performed on socio-environmental factors in order to identify factors having a significant effect. Principal Component Analysis was applied to the medicinal use value (UV) on the variables having a significant effect. This multivariate analysis allow us to assess the use pattern according to factors. A total of 150 herbaceous species have been identified in the treatment of various diseases and which are grouped into 75 families and 135 botanical genera with a significant effect of gender, ethnicity, religion and educational level. Among the 150 species identified, 27 species are used more in the treatment of diseases than the others. This study could contribute to the preservation and sustainable use and conservation of natural biological resources and to the strengthening of knowledge about medicinal herbaceous plants in the Guinean-Congolese zone of Benin.

Keywords: Benin, diseases, Guineo-congolaise, herbaceous, medicinal use

Contact Address: Thierry Dèhouégnon Houehanou, University of Parakou, Laboratoire of Ecology, Botanical and Plant Biology, Parakou, Benin, e-mail: houehanout@gmail.com

Thanakha Biotrade Market Potential through the Implementation of Access and Benefit Sharing in Myanmar

José Tomás Undurraga[1], Alessandra Giuliani[2], Saw Min Aung[3]

[1]*University of Freiburg, Faculty of Environment and Natural Resources, Chair of Silviculture, Germany*

[2]*Bern University of Applied Sciences (BFH), School of Agricultural, Forest and Food Sciences (HAFL), Switzerland*

[3]*Helvetas Myanmar, Regional Biotrade Project Southeast Asia, Myanmar*

The world's demand for biological resources as natural ingredients is growing rapidly, creating new opportunities for investment in natural-based products, but also threats over local biodiversity. The Nagoya Protocol (NP) on Access and Benefit-Sharing (ABS) seeks to safeguards user's access to biological resources and traditional knowledge (TK), in exchange for sharing the benefits derived from their use with the provider country. Myanmar, one of the most rich-biodiversity countries, signed the NP in 2014, but the implementation is still at early stages. Thanakha is an NTFP traditionally grown and used because of its skincare and cosmetic properties in Myanmar. Due to the current interest of global corporates and the cosmetic industry, there is an increasing BioTrade export potential associated with Thanakha.

Through a mixed research design, this study analysed the potential of implementing ABS or other measures to support the development of a sustainable Thanakha Market System (MS) and reduce threats to local biodiversity and TK. In 2018, we collected qualitative data through semi-structured interviews with key informants involved in the implementation of ABS measures in Myanmar. Quantitative data was collected through structured interviews with Thanakha farmers from Myanmar's Dry Zone.

Results showed a weak ABS regulatory framework in Myanmar. Current R&D activities on Thanakha can trigger ABS obligations, but current deficiencies with benefit-sharing practices raise uncertainties about the potential of ABS implementation (low level of traceability, low farmers' access to relevant information and fair prices, a low commitment of companies with local development, and lack of agreements when sourcing Thanakha for R&D or commercial purpose). Constraints for the development of a sustainable Thanakha MS relates to a lack of farmers' access to finance, low processing capacity at the local level, and deficient traders' fulfilment of international market standards. This study showed the relevance of implementing BioTrade principles and ethical sourcing practices for promoting a sustainable Thanakha MS, as well as for supporting the implementation of ABS requirements in Myanmar. Companies willing to implement BioTrade principles must tackle several challenges in their sourcing practices. Other benefit-sharing possibilities involve rural infrastructure investments, traditional varieties breeder's rights or geographical indications.

Keywords: Access and Benefit Sharing, biodiversity, BioTrade, Myanmar, Thanakha

Contact Address: José Tomás Undurraga, University of Freiburg, Faculty of Environment and Natural Resources, Chair of Silviculture, Stühlingerstrasse 15, 79106 Freiburg Im Breisgau, Germany, e-mail: jose.undurraga.q@gmail.com

From Farm to Fork: Is the European Green Deal a Green Deal also for Tropical Countries?

Ana Cecília Kreter[1], Philip-Marcel Cantos-Siemers[1], Pastre Rafael[2], Guilherme Soria Bastos Filho[3], Dietrich Darr[1]

[1]*Rhine-Waal University of Applied Sciences, Fac. of Life Sciences, Germany*
[2]*Inst. for Applied Economic Research, Dept. of Macroeconomic Policies and Studies, Brazil*
[3]*Ministry of Agriculture, Livestock and Food Supply, Ministry Office, Brazil*

The European Commission established the European Green Deal in December 2019 to address the climate and environmental challenges of its member states. The aim of this policy is for the EU to become the first climate-neutral continent by 2050. For the agricultural sector, the Farm to Fork strategy aims to promote sustainable practices, food security and healthy nutrition through adoption of stringent measures and quality standards in food production, processing, consumption and food loss and waste prevention. In addition, the policy aims to improve the status of the EU's ecosystems, both in quality and quantity, through the reduced use of agrochemicals, biodiversity conservation targets, and the recovery of degraded areas. At the same time, the EU is one of the world's largest importers of agricultural commodities. According to EUROSTAT, the total value of agricultural products imported into the EU in 2019 reached US$ 135 billion. While the Green Deal is seen by some as a vehicle to enforce sustainability standards in the EU, it is also feared that the agricultural productivity decrease associated with the farm to fork strategy will increase the EU's dependence on food imports and lead to a considerable displacement of land use and deforestation, especially in tropical countries, in the coming decades. Based on current trade agreements and international trade data, this paper investigates the possible bottlenecks of this policy for the agribusiness commodity suppliers, using Brazil as a case study. This paper is divided into three sections. First, the goals established by the EU Commission regarding agricultural production and rural areas will be presented. Then we will detail the EU's agricultural commodity trade with tropical countries, indicating the main origins and the products. This section also includes some forecasts for both population and production growth in order to establish a consumption prognosis for the next ten years. Finally, this paper presents the Brazilian Forest Code and the bottlenecks that the Green Deal may generate there in the future.

Keywords: Agricultural commodities, Brazil, trade market

Contact Address: Ana Cecília Kreter, Rhine-waal University of Applied Sciences, Faculty of Life Sciences, Marie-Curie-Straße 1, 47533 Kleve, Germany, e-mail: ana.kreter@hsrw.org

Trees for people

Squaring the Circle: Is Agroforestry Compatible with Modern Capitalism?

Patrick Worms

World Agroforestry (ICRAF), Belgium

The evidence that the inclusion of trees in agricultural production systems brings, on average, more benefits than disadvantages is overwhelming and comes from almost all biomes on our planet. Further, the benefits agroforestry trees provide are likely to increase as the heterogeneity of weather is boosted by worsening climate change conditions. Lastly, many classes of natural catastrophes can be mitigated to some extent by trees: mangroves dampen tsunamis, deep-rooted trees prevent some landslides, hedges and tree alleys reduce waterborne soil erosion, etc. As a dual Belgian and German citizen, I am particularly struck by the July floods that killed over 200 of my countrymen and were worsened by monoculture land use patterns in both forestry and farming.
However, despite this mountain of evidence, agroforestry seems to be honoured far more in speeches than practice. The world seems increasingly set on intensifying petrochemical agricultural production systems. These are of course optimised on an ongoing basis, but nevertheless are unlikely to provide the required adaptation, mitigation, biodiversity, resilience and profitability targets that should be the basis of a holistic assessment of any farming system. This talk will explore whence this paradox arises, review a number of options to increase the adoption rate of agroforestry and other agroecology-based interventions, and discuss with the market based mechanisms alone are likely to bring the agricultural transformation we need.

Contact Address: Patrick Worms, World Agroforestry (ICRAF), Brussels, Belgium, e-mail: p.worms@cgiar.org

The Role of Shade Tree Pruning in Cocoa Agroforestry Systems: Agronomic and Economic Benefits

LAURA ESCHE[1], MONIKA SCHNEIDER[1], JOHANNA RÜEGG[1], JOACHIM MILZ[2], ULF SCHNEIDEWIND[2], LAURA ARMENGOT[1]

[1]*Research Inst. of Organic Agriculture (FiBL), International Cooperation, Switzerland*

[2]*ECOTOP, Bolivia*

Cocoa (*Theobroma cacao*) is commonly produced in full-sun monoculture cropping systems to increase yields in the short term. Nevertheless, cocoa is a suitable crop for production under shaded conditions and is traditionally cultivated in agroforestry systems in Latin America. To ensure productivity and profitability, however, the development of best practices for shade management is crucial, but shade tree pruning is not commonly practised. This study investigates the influence of pruning shade trees in cocoa-based organic agroforestry systems in Bolivia on agronomic and economic performance. Four organic agroforestry farms were selected, where shade trees were not pruned for at least 10 years. At each site, half of the plot was kept unpruned and the other half was pruned, while all other management practices were kept equal. Data on yield formation were collected subsequently for two harvesting seasons. The trial results show a significant increase in cocoa yield under pruning conditions ranging from 28 % to 82 % compared to unpruned plots. This is attributed to an increase in flowering and fruit set in pruned plots. No differences in the incidence of pests and diseases in the pods were found between both treatments. To evaluate, whether pruning shows an economic benefit for farmers, different scenarios of yield increase based on the minimum, average, and maximum yield of local cocoa producers were used. Other sources of income, such as by-crops, were not considered in the calculations. For the average yield level of 287.4 kg ha^{-1} (dry), an increase of 51 % in two consecutive years will cover the pruning costs. Despite the promising results and indication, that the yield increase will last for more than two years, the initial costs for pruning might still discourage farmers. Therefore, financing programs for farmers that support agroforestry tree pruning are necessary to increase both cocoa production and farmer's income.

More information on the project can be found here: https://systems-comparison.fibl.org

Keywords: Agroforestry, Bolivia, cocoa, pruning

Contact Address: Laura Esche, Research Inst. of Organic Agriculture (FiBL), International Cooperation, Ackerstrasse 113, 5070 Frick, Switzerland, e-mail: l.esche95@gmail.com

Wild Tree Species Diversity and Analysis of their Cultural Importance Value: Using a Community-Based Protected Area in Northwestern Ghana

John-Baptist S. N. Naah

University of Cologne, Institute of Geography, Germany

Use of wild tree species by smallholder farmers for various purposes is crucially important for their daily livelihoods. However, the growing demand for these natural resources could lead to their overexploitation and environmental change. The aims of this study were to i) document wild tree species, uses and analyse their cultural importance, ii) investigate socio-demographic variables of smallholder farmers influencing their traditional knowledge on wild tree species and uses, and iii) examine smallholder farmers' perceptions about the establishment of the Wechiau Community-based Hippo Sanctuary (WCHS). 135 smallholder farmers were interviewed in nine villages belonging to the Waala and Birfor ethnic groups in the WCHS. The primary data were subjected to rigorous statistical analysis e.g. using Cognitive Salience index reflecting cultural importance and univariate analysis. Given the results of this study, the WCHS is enriched with 43 ethnoecologically important wild tree species belonging to 22 families and 41 genera. Also, eight topmost wild tree species in descending order of cultural importance included *Vitellaria paradoxa, Burkea africana, Diospyrous mespiliformis, Bombax costatum, Parkia biglobosa, Pterocarpus erinaceus, Terminalia avicennioides* and *Acacia gourmaensis*. The family cultural importance for Fabaceae and Sapotaceae is predominantly high as reflected in the frequency and ranking of citations of wild tree species under these families by local informants. The 43 wild tree species cited by local informants were categorised into nine different uses including food (9 species), forage (30), firewood (40), medicine (6), construction (9), soil improvement (3), social use (2), gardening (5) and fiber/ropes (2). Among these use categories, firewood, forage, food and construction topped the list as the most culturally important to the smallholder farmers. The traditional knowledge on varied wild tree species and their uses was significantly affected by age of smallholder farmers ($p < 0.05$), but not ethnicity and other socio-demographic factors. This study thus suggests the need for community-based conservation measures for sustainable management of natural resources for rural livelihood improvement in the tropics and sub-tropics.

Keywords: Community-based conservation, cultural importance, ecotourism, hippo sanctuary, smallholder farmers, wild tree species

Contact Address: John-Baptist S. N. Naah, University of Cologne, Institute of Geography, Albertus Magnus Platz, D-50923 Cologne, Germany, e-mail: jeanlebaptist@yahoo.co.uk

Modelling On-Farm Above-Ground Woody Biomass Production for Smallholder Agroforestry Systems in Semi-Arid Tanzania

Johannes Michael Hafner[1,4], Götz Uckert[1], Jonathan Steinke[1], Harry Hoffmann[2], Stefan Sieber[1], Anthony Kimaro[3]

[1] *Leibniz Centre for Agric. Landscape Res. (ZALF), Sustainable Land Use in Developing Countries (SusLAND), Germany*

[2],

[3] *World Agroforestry (ICRAF), Tanzania Country Programme, Tanzania*

Fuelwood is the main source of cooking energy in rural Tanzania; the dependency on fuelwood from off-farm sites is high. Ongoing forest degradation and deforestation negatively affects women and children who are mainly responsible for collecting fuelwood in rural areas.

On-farm produced fuel can reduce or off-set households' need for off-farm fuel without compromising food production at the same time. In semi-arid Tanzania, on-farm above-ground woody biomass production (AGB) of *Gliricidia sepium* (*G. sepium*) and pigeon pea have not been quantified yet. Our aim was to enable farmers to estimate their on-farm fuel production based on dendrometric variables such as root collar diameter assessed 20 cm above the collar (RCD20) and stem height in order to optimise their farm production plans. Using a destructive sampling approach, we assessed 112 *G. sepium* and 80 pigeon pea plants with an age of less than 12 months. We fitted arithmetic (non-linear) and log-transformed (linear) models to estimate AGB from the legumes *G. sepium* and pigeon pea intercropped with maize.

Log-transformed linear models met the assumptions of linear regression using the ordinary least square (OLS) estimator. Findings suggest that RCD20 is the most efficient predictor for AGB production for *G. sepium* and pigeon pea in semi-arid Tanzania. While stem height alone was not a good predictor to estimate ABG fuel biomass production of *G. sepium* (R^2 = 40.6 %) and pigeon pea (R^2 = 83.2 %), RCD20 as a single predictor explained 89.6 % (R^2) of the variation of AGB production of *G. sepium* and 91.4 % (R^2) of pigeon pea. A combination of RCD20 and stem height to predict AGB production was neglected due to multicollinearity of the variables and due to the fact that additional effort is required to assess both RCD20 and stem height.

Using the developed models, smallholder farmers can estimate there on-farm AGB by assessing dendrometric indicators such as RCD20 (and stem height) of *G. sepium* and pigeon pea. On-farm fuel production reduces households' dependency on external fuelwood which might have major socio-economic knock-on effects on rural livelihoods especially with regard to food security.

Keywords: Agroforestry, allometric modelling, *Gliricidia sepium*, intercropping, on-farm fuel production, pigeon pea, semi-arid Tanzania

Contact Address: Johannes Michael Hafner, Leibniz Centre for Agric. Landscape Res. (ZALF), Sustainable Land Use in Developing Countries (SusLAND), Müncheberg, Germany, e-mail: johanneshafner@gmx.net

The Effect of Species Identity, Litter Diversity and Habitat Quality on Litter Decomposition Rate in Gerese District Southwest Ethiopia

Seyoum Getaneh Aydagnehum[1,3], Olivier Honnay[2], Ellen Desie[1], Simon Shibru[3], Bart Muys[1]

[1]*KU Leuven, Dept. of Earth and Environmental Sciences, Belgium*
[2]*KU Leuven, Plant Conservation and Population Biology, Belgium*
[3]*Arba Minch University, Dept. of Biology, College of Natural Sciences, Ethiopia*

Attempts to rehabilitate degraded highlands are made by the Ethiopian government, mainly by tree plantation. But so far, less attention was given to species identity and stand composition. In this study, three indigenous and two exotic trees species were selected for litter decomposition study. The objective was to identify a better combination of tree leaf litters for the restoration of degraded land. Litter bags were incubated into potential restoration sites (disturbed natural forest and plantation) in comparison to intact natural forest. Tested litters includes three natives, two exotics and their mixtures (five monospecific litters, ten 3 species litter mixtures and one 5 species litter mixture) in comparison to standard green tea and rooibos tea. A total of 1033 litter bags were retrieved for weight loss analysis after one, three, six and twelve months of incubation. Beside the linear mixed model both Student t-test and Spearman rank correlation were employed for data analysis. The results reflect the significant impact of litter quality and diversity. *Millettia ferruginea* was shown to have a comparable fast decomposition rate as green tea. Both *Cupressus lusitanica* and *Syzygium guineense* were shown to have a lower decomposition rate than the slowly decomposing rooibos tea. The decomposition rate of *Croton macrostachyus* and *Eucalyptus globulus* were in between the above two groups. Student t-test was confirming the existence of non-additive effects of litter mixture. Significant correlation was also observed between litter mass loss and initial leaf litter chemical composition. The findings suggests that litter diversity and admixture of native species play a significant role in restoring degraded land.

Keywords: Antagonistic effect, forest types, litter bag, litter mixture, litter quality, non-additive effect, tea bag

Contact Address: Seyoum Getaneh Aydagnehum, KU Leuven, Dept. of Earth and Environmental Sciences, 3000 Leuven, Belgium, e-mail: seyoumgetaneh.aydagnehum@kuleuven.be

(Un)Sustainable Amazon: Deforestation Burn Out Illegal Land Grabbing Legitimation in Brazil

Igor Gomes de Araújo, Luciana Ramos Jordão, Thiago Henrique Costa Silva, Rebeca Barbosa Moura, Victória Cardoso Carrijo, Barbara Luiza Rodrigues, Niury Ohan Pereira Magno, Thays Dias Silva, Letícia Versiane Arantes Dantas

State University of Goiás, Law College (Southeast Campus: Morrinhos), Brazil

The Brazilian environmental issue is always at the centre of environmental agendas whether at the national or global level, especially when it comes to the Legal Amazon. However, since 2019, with the tenure of President Bolsonaro, the sustainability measures already consolidated in the country remain threatened. To portray these threats, Provisional Measure 910 and Bill 2963/2019 are being studied, which aim at restructuring the means of territorial occupation, encouraging the process of land grabbing, burning and deforestation in the Legal Amazon. In this context, it is asked: do the rules implemented in the Bolsonaro government act as precursors and maintainers of environmental problems in the Amazon? To answer, in a qualitative perspective, guided by historical and dialectical methods, bibliographic, documentary research and indirect data analysis were carried out. According to the Amazon Environmental Research Institute (IPAM), there was a 50 % increase in deforestation in 2019, compared to 2018. Between August 2019 and July 2020, according to the National Institute for Space Research (INPE), there were 11,088 km^2 of lost area of the biome, three times more than agreed in the 2009 Climate Convention. As for the fires, INPE registered 103,161 fires in the forest in 2020, a value 15.7 % higher than in 2019. In turn, land grabbing is also a constant and growing phenomenon, with 23 % of its public lands not designated, equivalent to 11.6 million hectares, being irregularly declared as private property, according to research carried out by IPAM, together with researchers from the Federal University of Pará. In 2020, for the simple dissemination of data, INPE managers and researchers were criticised by the government and removed from their positions. The interference system in environmental agencies by the federal government, such as the Brazilian Institute for the Environment and Renewable Natural Resources and the Chico Mendes Institute for Biodiversity Conservation, has been the object of denunciation by researchers and public servants since 2019. Therefore, it is concluded by the list of causality between the implementations of these new norms and the political agenda of weakening environmental guidelines with the increase of environmental degradation, especially in the Amazon.

Keywords: Amazonian territory, ambiental degradation, sustainability, territorial occupation

Contact Address: Igor Gomes de Araújo, State University of Goiás, Law College (Southeast Campus: Morrinhos), AV. A QD. A LT C-13, MORRINHOS, Brazil, e-mail: igor.gomesa@gmail.com

How Far Are Mangrove Ecosystems Conserved by the Ramsar Convention in Benin (West Africa)?

Antoine Elie Padonou[1], Innocent Gbaï[2], Moustapha Kolawolé[2]

[1]*Nationale University of Agriculture, School of Tropical Forestry, Benin*
[2]*University of Abomey-Calavi, Fac. of Humanities and Social Sciences, Benin*

Mangroves provide humanity with a variety of ecosystem services worldwide. Mangrove sites in Benin are embedded in the internationally recognised wetlands of Ramsar sites 1017 also known as West Complex. These mangrove ecosystems are characterised by high biological productivity, which translates into significant biodiversity that benefits many animal and plant species. It offers abundant timber and fishery resources for various agricultural, aquaculture and other activities. In addition, it serves as a refuge for many endangered species and is an essential link in the course of migratory birdlife. However, the rising demography coupled with human activities jeopardises the sustainable management of these ecosystems. In addition, climate change is also expected to have a severe impact on the mangrove ecosystems especially in Benin. Several initiatives were set with the Ramsar Convention on Wetlands for mangroves conservation since 2000. Land use/land cover changes (LULC) were used at Ramsar Site 1017 in Benin for the periods 1995, 2005 and 2015 to assess the impact of the Ramsar Convention on mangrove ecosystems conservation. The observed changes during 1995–2005 and 2005–2015 were considered to predict LULC change towards 2070 using Markovian chain. From 1995 to 2005, a total area of 3.43 ha of mangroves was degraded while 2.65 ha were restored during the 2005–2015 period. Future scenarios predicted that the mangroves area was expected to decrease more than half of the area of 1995 by 2070 assuming the dynamic of 1995–2005 and increase by 1.1 % of the area of 2005 by 2070 with the dynamic of 2005–2015. Implementation of conservation policies, projects and awareness raising activities could contribute to effective restoration of the mangrove ecosystems.

Keywords: Land use/land cover, mangroves, policy, Ramsar site 1017, West Africa

Contact Address: Antoine Elie Padonou, Nationale University of Agriculture, School of Tropical Forestry, Ketou, Benin, e-mail: padonouelie@gmail.com

NTFP Certification Schemes in the Walnut Forests of Kyrgyzstan and Lessons Learned from Other Countries

Klara Dzhakypbekova
University of Bonn, Center for Development Research (ZEF), Germany

The walnut-fruit forests on the Ferghana and Chatkal mountain ranges in Kyrgyzstan are considered as a global biodiversity hotspot, they also have the soil-protection and water regulating functions, and are the major source of income to the local households. Collection of walnuts and other NTFPs in Kyrgyzstan constitute an important basis for the local livelihoods. Previous studies recommended that the logistical and infrastructural improvement of the existing value-chains in accordance with the international sustainability standards might positively change the socio-economic welfare of the local livelihoods, and promote sustainable forest management to reduce deforestation. The NGOs and donor programs introduced the NTFP certification projects of the Fair Trade certification (in 2008) and Forest Stewardship Council (FSC) NTFP certification (in 2018) within the selected pilot communities of the walnut forests in Kyrgyzstan both aiming to improve the socio-economic welfare, sustainable forest utilisation and conservation of the walnut forests. However there is a limited understanding about the possible effects and impacts of these certification programs on the local livelihoods and the socio-ecological situation in the local forests. The global experience has also shown that the collection, utilisation and trade of the NTFP products offers significant advantages to the national economies and the subsistence of the local livelihoods. NTFP certification was promoted since early 2000-s the NGOs and donor programs as a market-oriented tool to encourage sustainable extraction and commercialisation of non-timber forest products. It is important to analyse the current state of the art of the existing NTFP certification cases along with their success or failure factors, as well as other important contextual conditions. The given study investigates the state of the art for NTFP certification in Kyrgyzstan based on the expert interviews, and compares the local situation with the existing experiences of similar certification schemes in other countries using the systematic literature review methods. The challenges and opportunities for the future of NTFP certification experiences in the country are described and further recommendations for policy- and decision-makers are provided.

Keywords: Kyrgyzstan, non-timber forest products, walnut forests

Contact Address: Klara Dzhakypbekova, University of Bonn, Center for Development Research (ZEF), Genscherallee 3, 53113 Bonn, Germany, e-mail: klara.dzhakypbekova@gmail.com

Mapping Tree-Based Systems in Tropical Landscapes: Fostering Sustainability at Agricultural Frontiers (Zambia, Ecuador, Philippines)

RUBEN WEBER[1,2], MELVIN LIPPE[1], SVEN GÜNTER[2,1]

[1]*Johann Heinrich von Thünen Institute, Institute of International Forestry and Forest Economics, Germany*

[2]*Technical University of Munich: Center of Life and Food Sciences Weihenstephan, Inst. of Ecosystem Dynamics and Forest Management in Mountain Landscapes, Germany*

Global forest maps, such as Hansen's Global Forest Change (GFC) dataset, are inestimable tools to monitor land cover and land use effectively at large-scale. This is a precondition to support international environmental objectives such as defined by the SDGs or Forest Landscape Restoration (FLR) and to design policies that support the future life on Earth. However, the applicability of large-scale remote sensing products at local levels still faces operational challenges. This is particularly critical in tropical landscapes, due to the lack of reference data or the permanent cloud cover presence. Finding consistent and valid forest definitions, which target the vast diversity of tropical vegetation can also be challenging. For instance, defining adequate tree cover thresholds of the GFC product that match specific forest characteristics is strongly dependent on the region of application. Additionally, other tree-based systems such as agroforestry and fast-growing vegetation cause further classification uncertainty. The accurate differentiation of forests, agroforestry and other tree-based systems can help to support natural resource management in tropical landscape mosaics, especially those located at the agricultural frontiers. We collected 16,000 ground control points/photographs and digitised 18,000 ha (with details on land use and disturbance history) across thirty-six landscapes, located in nine regions of Zambia, Ecuador and Philippines, which constitute a gradient of pantropical deforestation contexts. We generated forest masks covering our landscapes (0.36 Mha) and regions (15 Mha), by combining remote sensing data from different satellites (Landsat-8, Sentinel-1, SRTM). Furthermore, we validated the quality of our maps across our deforestation gradient and compared them to seven national and global forest datasets (including GFC), which are commonly used in international reporting such as FAO or UNFCCC. We observed recurrent difficulties in distinguishing forest from other vegetation, especially from mixed tree-based systems (e.g. perennial crops, palms and other agroforestry arrangements). Our results highlight that the importance of ground truthing as accompanying method to establish efficient land cover and land use monitoring systems. This is especially relevant in tropical regions of advanced stages of deforestation and early stages of reforestation; precisely, where current FLR initiatives (Agenda 2030, Bonn Challenge), are likely to occur.

Keywords: Deforestation, forest monitoring, land cover, land use, remote sensing

---

Contact Address: Ruben Weber, Johann Heinrich von Thünen Institute, Institute of International Forestry and Forest Economics, Leuschnerstraße 91, 21031 Bergedorf, Germany, e-mail: ruben.weber@thuenen.de

Which Nature Do Nature-Enhanced Plantations Create: A Workers' Perspective

Adriana Gomez, Stephanie Domptail

Justus Liebig University Giessen, Inst. of Agricultural Policy and Market Research, Germany

Oil palm plantations are industrial agricultural landscapes. Such landscapes can be homogeneous and oversimplifying biodiversity conveying little or no human well-being. Yet, integrating nature on plantations can support biodiversity conservation and generate social landscape values. Possibly, the integration of nature on plantations can support their naturalisation in the view of local actors, legitimized through the benefits and values the integrated nature brings in. As plantation workers are the key local population concerned, affected by and interacting with the plantation, our analysis of the plantation landscape is done through their eyes. Little research has been done on how plantation workers relate to and perceive nature (e.g., gallery and riparian forests) and non-natural areas (e.g., crop such as oil palm) in their plantation. We address this gap by analysing workers' perception of nature in the Macondo oil palm plantation. The Macondo oil palm plantation, in eastern Colombia, integrates simultaneously land sharing and land sparing approaches to create a nature-enhanced industrial landscape. We claim that the nature-enhanced design generates human-wellbeing that creates a better working place and legitimize the existence of the plantation in the landscape. The paper addresses the following empirical questions: 1) what facets of nature do the social landscape values perceived by workers reflect? 2) How does nature on the plantation affect workers and their activities in the plantation? 3) How does the workers' perception of nature in the oil palm plantation reflect the naturalisation of the plantation? Using content analysis based on focus group and individual interview material of workers depicting their perception of nature on Macondo collected in October 2018 – April 2019, we show that workers derive a great variety of values from the integration of nature on the farm. They also experience an ambivalent relationship to nature on the plantation, typically for interactions with "nature in the outside world", that is, both benefits and fascination and disservices and fear. Through the perceived benefits to humans and other species, a process of legitimisation can be traced which points to the possible naturalisation of such nature-enhanced plantation.

Keywords: Human well-being, integrating nature, landscape social values, natural areas, naturalisation, non-natural areas, plantation workers

---

Contact Address: Adriana Gomez, Justus Liebig University Giessen, Inst. of Agricultural Policy and Market Research, 35396 Giessen, Germany, e-mail: adriana.gomez@agrar.uni-giessen.de

Assessment of Fiber Characteristics and Suitability of Ten Hardwood Species Grown in Sudan for Paper Production

Hanadi Mohamed Shawgi Gamal[1], Majdaldin Rahamtallah Abualgasim Mohammed[2], Ashraf Mohamed Ahmed Abdalla[1]

[1]*University of Khartoum, Fac. of Forestry; Forest Products and Industries, Sudan*
[2]*National Center for Research, Natural Resources and Desertification Inst., Sudan*

Sudan is considered one of the rich African counties with great diversity of tree species; it encompasses about 3156 species belonging to 1137 genera and 170 families. However, Sudan is almost entirely dependent on imports to satisfy its needs for pulp and paper despite its richness in different hardwood species which could be good sources of pulp production. There is an urgent need to evaluate the locally available raw materials as potential sources for pulp and paper industry. The present study was carried out to investigate the fiber characteristics of ten hardwood species growing in Sudan as well as to assess their fiber variation and suitability for pulp and paper making. The study species include: *Ailanthus excelsa, Albizia amara, Balanites aegyptiaca, Boswellia papyrifera, Diospyros mespiliformis, Eucalyptus camaldulensis, Euphorbia tirucalli, Ficus sycomorus, Sterculia setigera, Tamarix aphylla* and *Ziziphus spina-christi*. The wood materials were collected from the low rainfall woodland savannah in Sudan, from two states namely: South Kordofan State and Sennar State. Fibers dimensions and their derived values were investigated and used to consider the suitability of the selected species for pulp and paper making. Significant variations have been detected in fiber characteristics among the ten studied species. Depending on the study results, eight species out of ten species have been ranked as good source for pulp and paper making. The remaining two species (including *Balanites aegyptiaca* and *Diospyros mespiliformis*) have been ranked as poor source for pulp and paper making; however, mixing their wood with soft-wood is suggested to improve their properties.

Keywords: Fibers characteristics, hardwood species, paper production, Sudan

Contact Address: Hanadi Mohamed Shawgi Gamal, University of Khartoum, Fac. of Forestry; Forest Products and Industries, Alemarat Street 61, Khartoum, Sudan, e-mail: hanadishawgi1979@yahoo.com

Spring Restoration through Sustainable Land Management in the Mid-Hills of the Indian Himalaya

Jaclyn Bandy[1], Hanspeter Liniger[2], Rajesh Joshi[3], Ranbeer Rawal[3], Sanjeev Bhuchar[4]

[1] *WOCAT (World Overview of Conservation Approaches and Technologies); University of Bonn; University of Hohenheim , Germany*

[2] *University of Bern, Centre for Development and Environment, Switzerland*

[3] *G.B. Pant National Inst. for Himalayan Environment and Sustainable Dev., Sikkim Regional Center, India*

[4] *International Centre for Integrated Mountain Development (ICIMOD), Watershed Managment, Nepal*

Of the estimated 3 million springs in the Indian Himalayan region, roughly 60 % have dried up or become seasonal. This phenomenon is occurring across the entire Himalaya, threating water security, biodiversity and livelihoods in both mountain communities and downstream areas. Rapid population growth has brought about extensive land-use changes, mainly through large-scale deforestation, overexploitation of forest resources, development and poor cultivation practices. These land use changes–coupled with climate change and rainfall variability–are impacting spring flows. The demand to restore springs has especially increased in the mid-hills of the Himalaya, and over half of the springs in the north Indian state of Uttarakhand have reportedly dried up. However, a lack of coherent perspectives and knowledge on the interrelations of springs, land use and land cover, surface runoff and groundwater regimes has hindered action to address spring restoration. The case study examines the micro-watersheds, or "springsheds" of two rural villages facing water scarcity in the mid-hills of Uttarakhand. The unique contribution of this work lies in using ground-based knowledge to classify and analyse land cover/land use types, followed by applying a watershed tool to estimate runoff and a water-balance approach to estimate groundwater recharge and potential spring discharges. Built on basic, systematic field-mapping (i.e., land use and cover, soil type, slope and springshed delineation) the calculations for runoff, groundwater and spring discharge are deduced against available water flow data for mapped springsheds. The results demonstrate that spring recharge is dependent on monsoon rainfall to sustain year-round flow, emphasising the need to reduce runoff and increase groundwater recharge during monsoon. Sustainable forest management, water harvesting and regenerative agriculture practices indicate that springs flows can be preserved

Contact Address: Jaclyn Bandy, WOCAT (World Overview of Conservation Approaches and Technologies); University of Bonn; University of Hohenheim , 70594 Stuttgart, Germany, e-mail: jaclyn.bandy@uni-hohenheim.de

or increased through locally adapted management practices. This includes afforestation and natural regeneration of broadleaved tree species and native grasses, agroforestry (e.g., tree species for fodder, fuelwood, fruits, nuts and medicine) traditional cropping systems with reinforced terraces, groundwater recharge structures (e.g., infiltration ponds, recharge trenches, check walls/ dams) and protection of traditional spring water harvesting structures (Naula and Dhara). The solutions are outlined from the perspective of the land users who implemented them.

Keywords: Agroforesty, drought mitigation, forestry, land use mapping, mountain watershed management, natural resource conservation, spring restoration, surface water hydrology, sustainable land management, traditional practices

High Diversity Agroforestry Model for Coffee in Nicaragua

Matthias Baumann[1], Ingrid Fromm[2], Byron Reyes[3], Jürgen Blaser[4]

[1] *Research Institute of Organic Agriculture (FiBL), International Cooperation, Switzerland*

[2] *Bern University of Applied Sciences, School of Agricultural, Forest and Food Sciences (HAFL), Switzerland*

[3] *Alliance Bioversity-CIAT, Applied Economics, Nicaragua*

[4] *Bern University of Applied Sciences (BFH), School of Agricultural, Forest and Food Sciences (HAFL), Switzerland*

Arabica coffee cultivation in Nicaragua is becoming increasingly threatened by climate change effects such as rising temperature, droughts, heavy rainfalls, landslides and hurricanes. Thus, the suitability of coffee production is projected to decrease in the coming decades. The Nicaraguan NGO Aldea Global has the goal of improving coffee producers' situation by helping them build resilience through the implementation of highly diverse agroforestry systems, with the support of the International Center for Tropical Agriculture (CIAT). Data for this research project was collected by interviewing 309 coffee producers in the department of Jinotega. From this data, coffee production and profitability indicators such as yield, costs, income, net revenue and shading effect were analyzed. The results showed that the mean net revenue from coffee production was overall positive, but some had a net loss. Around a quarter of the coffee plots were already highly diverse and the most popular shade species is by far Musa spp. Neither the species nor the density of shade species had a significant influence on the coffee yield.

For some producers, especially on lower altitudes, incremental adaptation measures might not be enough anymore, and they might be forced to shift to alternative cash crops such as Robusta coffee or cocoa. To implement climate change adaptation strategies, producers need support from the government and NGOs, as well as financial support and extension services, for example in shade tree management to achieve high quality timber and fruits.

To lift producers out of the vicious circle of "low input – low income", sustainable intensification and diversification of the coffee agroforstry system – in accordance with effective climate change adaption strategies – could be a promising approach to increase producers productivity, efficency, profitability and climate change resilience.

Keywords: Adaptation strategy, agroforestry, arabica coffee, climate change, Nicaragua, producers survey, socioeconomics

Contact Address: Matthias Baumann, Research Institute of Organic Agriculture (FiBL), International Cooperation, Mittelholzerstr. 26, 3006 Bern, Switzerland, e-mail: matthias.baumann86@gmail.com

The Effect of two Different Rainfall Zones in Wood Properties of *Balanities aegyptiaca* Growing in Sudan

Hanadi Mohamed Shawgi Gamal[1], Majdaldin Rahamtallah Abualgasim Mohammed[2], Ashraf Mohamed Ahmed Abdalla[1]

[1]*University of Khartoum, Fac. of Forestry; Forest Products and Industries, Sudan*

[2]*National Center for Research, Natural Resources and Desertification Inst., Sudan*

Modern research on wood has substantiated that the climatic condition where the species grow has significant effect in wood properties. Understanding the extent of variability of wood is important because the uses for each kind of wood are related to its characteristics; furthermore, the suitability or quality of wood for a particular purpose is determined by the variability of one or more of these characteristics. With the great variation on the climatic zones of Sudan, great variations are expected in wood properties between and within species. This variation need to be fully explored in order to suggest best uses for the species.

The present study demonstrates the effect of rainfall zones in some physical and anatomical wood properties of *Balanities aegyptiaca* growing in Sudan. For this purpose, thirty healthy trees were collected randomly from four states located in two zones (zone one with 273mm rainfall and zone two with 701 mm rainfall annually). From each sampled tree, 2 stem discs of 3 cm thickness were cut at 10 % and 90 % from stem height. The investigated wood properties were: wood basic density, hardness strength, fibres and vessels dimensions. The study result reveals significant differences between zones in wood density, hardness strength as well as in vessels diameter, lumen diameter and wall thickness. From these results, Balanities aegyptiaca seems to be well adapted with the change in rainfall and may survive in any rainfall zone. However, the detected significant variation in wood properties lead to expected variation also on its suitability for industrial utilisation. The thing which needs to be fully explained to promote the optimal uses of wood resource in Sudan.

Keywords: *Balanities aegyptiaca*, rainfall zones, variation, wood properties

---

Contact Address: Hanadi Mohamed Shawgi Gamal, University of Khartoum, Fac. of Forestry; Forest Products and Industries, Alemarat Street 61, Khartoum, Sudan, e-mail: hanadishawgi1979@yahoo.com

Flowering and Fruiting Pattern in Various Ecotypes of the Neotropical Oilseed Palm *Acrocomia*

Catherine Meyer[1], Claudio E.M. Campos[2], Lucas Dos Santos Acácio[2], Thomas Hilger[1], Sérgio Motoike[3], Georg Cadisch[1]

[1]*University of Hohenheim, Inst. of Agric. Sci. in the Tropics (Hans-Ruthenberg-Institute), Germany*

[2]*Federal University of Viçosa, Department of Forestry, Brazil*

[3]*Federal Univerity of Viçosa, Department of Plant Science, Brazil*

Acrocomia oilseed palms are endemic to the sub-humid tropical regions of South and Central America. They are considered a promising economic and sustainable alternative to African oil palm, allowing to expand the growing areas of oilseed palms into semi-arid regions. *Acrocomia* has a specific period of flowering, normally in the first month of the rainy season, and it takes around 12–14 months for the fruits to maturate. *Acrocomia* is protogynous, as such cross pollination is predominant and the success of reproduction is higher, the more plants flower at the same time. Understanding the differences of flowering patterns between *Acrocomia* ecotypes is of importance as the time of flowering is crucial for the reproductive success and not the number of female flowers. Even though, flowering patterns were already studied in *Acrocomia*, this study is the first to assess the flowering pattern and fruit set of contrasting ecotypes of *A. totai* and *A. aculeata* grown in the same location. The study was conducted at the Acrocomia Active Germplasm Bank of the Universidade Federal de Viçosa in Araponga, MG, Brazil. In the flowering season of October 2019 to January 2020, the open inflorescence were counted once per week on 300 palms. In five ecotypes originating from various regions of Brazil, the exact opening data of all inflorescences were recorded. Bags were put around the infructescences in late 2020 to collect all fruits which were counted after the harvest in March 2021. The first inflorescences opened in early September 2019 whereas the last ones were seen in the beginning of February 2020. The peak of flowering was end November 2019 with more than 200 open inflorescences per week, even though a shift among ecotypes was observed. *A. totai* ecotypes showed a trend to flower scattered over the whole flowering period. In contrast, ecotypes of *A. aculeata* flowered over shorter periods of time. Fruit set was highest in the inflorescences flowering in November, so at the peak of the flowering season. Thus, we see differences in flowering patterns which need consideration in planning of breeding programs and, subsequently, plantations including several ecotypes.

Keywords: *Acrocomia aculeata*, *Acrocomia totai*, Brazil, fruit set, oilseed palms

Contact Address: Catherine Meyer, University of Hohenheim, Inst. of Agric. Sci. in the Tropics (Hans-Ruthenberg-Institute), Garbenstrasse 13, 70593 Stuttgart, Germany, e-mail: catherine.meyer@uni-hohenheim.de

Ecotypical Flower Biometry in the Neotropical Oilseed Palm *Acrocomia*

Catherine Meyer[1], Claudio E.M. Campos[2], Thomas Hilger[1], Sérgio Motoike[3], Georg Cadisch[1]

[1]*University of Hohenheim, Inst. of Agric. Sci. in the Tropics (Hans-Ruthenberg-Institute), Germany*
[2]*Federal University of Viçosa, Department of Forestry, Brazil*
[3]*Federal Univerity of Viçosa, Department of Plant Science, Brazil*

The neotropical oilseed palm *Acrocomia* is considered a sustainable alternative to African oil palm in sub-humid tropical regions due to its high oil yield and adaptation to many diverse environments. The inflorescences of *Acrocomia* are panicles with several hundred rachillae bearing female and male flowers. The female flowers are globose, sessile and spirally arranged at the basal end of the rachillae. The apical two thirds of the rachillae bear dense spirals of solitary male flowers. The inflorescences are surrounded by a bract leaf before anthesis. *Acrocomia* is protogynous and upon bract opening, female flowers are receptive for around 24–48 hours. Anthesis of male flowers starts 24 hours after bract opening and their abscission is initiated on the fourth day. General morpho-anatomic and biometric studies on *Acrocomia* flowers already provided relevant information on the reproductive biology of *Acrocomia*; however, differences between ecotypes were not assessed up to now. This study aimed to gain detailed knowledge on flower biometry in ecotypes of *A. totai* and *A. aculeata* originating from different climatic regions in Brazil. The study was conducted at the Acrocomia Active Germplasm Bank of the Universidade Federal de Viçosa in Araponga, MG, Brazil. Nine rachillae from each inflorescences of five ecotypes were sampled in the flowering season of 2019. The length of the rachillae, and the number of female and male flowers were recorded. Their fresh and dry weight were measured. The height and diameters of female flowers were determined. *Acrocomia totai* ecotypes showed higher numbers of small, light weighted female flowers per rachillae. *A. aculeata* tended to bigger and heavier female flowers but fewer in numbers per rachillae. In all ecotypes the number of female and male flowers decreased on the rachillae from the basal to the apical end of the inflorescence. However, only 17 % of the rachillae had 6–8 female flowers, considered optimal for a good yield formation. *Acrocomia* shows a requirement for genetic improvement but also a high capacity due to a wide genotypical and ecotypical variability important for the selection of superior breeding material.

Keywords: *Acrocomia aculeata*, *Acrocomia totai*, Brazil, flower biometry

Contact Address: Catherine Meyer, University of Hohenheim, Inst. of Agric. Sci. in the Tropics (Hans-Ruthenberg-Institute), Garbenstrasse 13, 70593 Stuttgart, Germany, e-mail: catherine.meyer@uni-hohenheim.de

Flowering in the Oilseed Palm *Acrocomia*: Rachillae Morphology in Various Ecotypes

Catherine Meyer[1], Thomas Hilger[1], Sérgio Motoike[2], Georg Cadisch[1]

[1]*University of Hohenheim, Inst. of Agric. Sci. in the Tropics (Hans-Ruthenberg-Institute), Germany*
[2]*Federal Univerity of Viçosa, Department of Plant Science, Brazil*

The oilseed palm *Acrocomia*, endemic to South-America, is a promising sustainable alternative to the African oil palm due to its drought tolerance and adaptability to a wide range of soils and climatic conditions. *Acrocomia* palms produce several inflorescences for the one-time flowering during the first months of the rainy season. The inflorescences are panicles with first order branching consisting of a main rachis with several hundred short rachillae. Those rachillae bear the female and male flowers at the basal and apical end, respectively. Even though the general morphological structure of the rachillae is well described, differences between various *Acrocomia* species and ecotypes from differing regions of origin are not accounted for up to now. This study aimed to assess the variability in the rachillae morphology of various ecotypes of the widespread species *A. totai* and *A. aculeata*. The study was conducted at the Acrocomia Active Germplasm Bank of the Universidade Federal de Viçosa in Araponga, MG, Brazil. During the flowering season 2019, each inflorescence of five *Acrocomia* ecotypes was assessed at its day of opening. Nine rachillae along the inflorescences were sampled. The length of the rachillae, as also of the basal and apical parts, were measured. The sequence of floral structures was recorded from the basal to the apical end of the rachillae. Most rachillae followed the general structure of solitary female flowers at the basal end, followed by pairs of male flowers and the male cluster at the apex. However, some uncommon structures were recorded. The most surprising one was the occurrence of branched rachillae, as it was not described up to now, and making the inflorescence a panicle with second order branching. Those second-order branches were found, in particular, in two ecotypes of *A. totai*. Female flowers were rarely found on those secondary branches. Furthermore, rachillae without female flowers were present in all ecotypes especially with increasing occurrence from the base to the apical end of the inflorescences. Inflorescences show a high morphological variability reflecting their ecotypes, useful for genetic improvement. Additionally new traits were discovered underlining the complexity of Acrocomia's inflorescences.

Keywords: *Acrocomia* spp„ Brazil, Oilseed palms, rachillae morphology

Contact Address: Catherine Meyer, University of Hohenheim, Inst. of Agric. Sci. in the Tropics (Hans-Ruthenberg-Institute), Garbenstrasse 13, 70593 Stuttgart, Germany, e-mail: catherine.meyer@uni-hohenheim.de

Priority Tree Species for Forest Restoration, Enhancing Livelihoods and Ecosystem Services in Ethiopia

Déglise Guillaume[1], Giuliani Alessandra[1], Vinceti Barbara[2]

[1]*Bern University of Applied Sciences (BFH), School of Agricultural, Forest and Food Sciences (HAFL), Switzerland*

[2]*Bioversity International, Tree Biodiversity for Resilient Landscapes , Italy*

The few remnants and highly degraded Ethiopian forests cannot adequately provide ecosystems services as well as livelihood security to rural Ethiopian households. This is why the Alliance of Bioversity and CIAT, in collaboration with local communities and institutional partners launched the project Trees for Needs, aiming at improving livelihoods of rural households and ecosystems services by restoring forests in Eastern Ethiopia. This study provides an extensive literature review of the current state of knowledge about reforestation practices in Ethiopia and a household survey targeting rural households in Oromia. The literature review looks at the reforestation practices used in Ethiopia over the last 30 years, focusing more particularly on aspects such as community involvement, land tenure, tree species selection in reforestation projects. The survey, performed in Eastern Ethiopia, Oromia, reaching 320 households in eight villages across four districts in the zones of Jimma and Illu Aba Bora, enabled to identify the tree species most preferred by smallholders and to understand which traits are most favoured. According to the literature, reforestation projects lack sufficient involvement of local communities. These tend to be included at a late stage in the decision-making processes, leading to a low willingness to participate in reforestation activities. Moreover, land tenure issues are a major constraint. The State owns the land, including reforested and restored areas. By not granting long-term secure leases to tenants, the State hinders the willingness of smallholders to invest in reforestation. Tree species selection for forest restoration depends on expected benefits and local conditions, therefore it varies between communities and agro-climatic zones, as the literature review and household survey confirmed. Results of the survey showed a preference for native species over exotics. Considering both local environment and preferred characteristics, the most desired and suitable tree species for restoration appeared to be *Cordia africana, Croton macrostachyus, Acacia* spp. and *Olea europea*. To better understand the potential of these species for restoration, more aspects should be investigated, specifically: the amount of natural regeneration of these species in farmers' plots, the availability of seed for planting, seed behaviour, farmers' capacity to handle seed and establish nurseries.

Keywords: Ethiopia, Oromia, sustainable forest restoration, tree species selection, tree species uses

Contact Address: Déglise Guillaume, Bern University of Applied Sciences (BFH), School of Agricultural, Forest and Food Sciences (HAFL), Im Altried 1b, 8051 Zurich, Switzerland, e-mail: guillaume.deglise@gmail.com

Challenges for Cocoa Farms Diversification towards a Sustainable Scheme in the Peruvian Amazon

Jeimy Katherin Feo Mahecha[1], Gerald Kapp[2], Jeisson Rodriguez Valenzuela[3]

[1] *Technische Universität Dresden, CIPSEM, Germany*
[2] *Technische Universität Dresden, Inst. of International Forestry and Forest Products, Germany*
[3] *Corporación Colombiana de Investigación Agropecuaria Agrosavia, Obonuco Research Center, Colombia*

Cocoa (*Theobroma cacao* L.) is a plant from the Latin American rainforest and the main economic resource for the welfare of many farmers in such areas. In fact, the Juanjui municipality in the San Martin department is characterised as one of the most important centres for storing, drying, and marketing cocoa beans in the Peruvian Amazon, and this place is the meeting point for farmers who sell their products. However, the high dependency on this crop makes the smallholders vulnerable, affecting their economy and food security, a situation which, under the Covid-19 pandemic became even worse. Therefore, this study aimed to analyse the agricultural practices for cocoa production and the species available to produce annual and perennial fruits, timber, meat and eggs, and non-timber forest products within the farms of small producers in the municipality of Juanjui. The methodology was based on a focus group discussion with a total of 20 farmers to analyse the effects of the production and sale of cocoa beans in the context of the pandemic, a survey to 20 cocoa producers, and 2 interviews with the association manager. The results obtained show that although farmers have some other species associated with cocoa on their farms (citrus, mango, coconut, medicinal plants, fast- and medium-growing timber species), as well as a few farm animals (guinea pigs, chickens, pigs, and ducks), these alternatives are not being used towards a diversified production model at a commercial level. Another outstanding result was that the Covid-19 pandemic has delayed cultivation tasks, and the price of cocoa has been quite affected. In consequence, the group of farmers decided to create a cocoa association with the target to apply for credits, incentives, and access to better markets. The focus group discussions revealed the eagerness of the farmers to know more about the productive, commercial, and environmental potential of the species other than cocoa present in their farms. Therefore, this study proposes models and strategies for diversification as alternatives to achieve a more profitable and sustainable production.

Keywords: Agroforestry systems, farm diversification, pandemic uncertainty, Peruvian Amazon, *Theobroma cacao*

Contact Address: Jeimy Katherin Feo Mahecha, Technische Universität Dresden, CIPSEM, 01217 Dresden, Germany, e-mail: jkatherinfm@gmail.com

Cameroon's Cocoa Farmers Intensify and Expand Production Yet Retain Shade Trees: Evidence from two Decades

Stan Blomme[1], Johannes Pirker[2], Bart Muys[1], Dominique Endamana[3], Jean-Remi Ayo Ayo[4], Lindsey Norgrove[5]

[1] *KU Leuven, Dept. of Earth and Environmental Sciences, Division Forest, Nature and Landscape, Belgium*
[2] *UNIQUE Forestry and Land Use, Freiburg, Germany*
[3] *IUCN - Central and West Africa Program (PACO), Cameroon*
[4] *International Institute of Tropical Agriculture (IITA), IPM, Cameroon*
[5] *Bern University of Applied Sciences, School of Agricultural, Forest and Food Sciences (HAFL), Switzerland*

More than 80 % of the world's cocoa is grown in West Africa, traditionally under thinned forest where timber, fruit, or nut trees were retained. Nowadays, farmers, particularly in Côte d'Ivoire, are reportedly shifting to full-sun cocoa. Cameroon is the world's 5th largest cocoa producer, where it is grown by 70 % of farmers in the humid south, yet yields are low and constrained by blackpod disease (*Phytophthora megakarya*). Our objective was to assess how management changes have impacted yields between 2001 and 2018. We hypothesised that shade trees had been removed and intensification has occurred fueled by increased inputs. In 2001, we conducted surveys in seven villages with 210 farmers, then again in 2018 with 126 farmers. In 2018, 69 % of farmers had extended their cocoa fields, compared with 28 % in 2001. 54 % of farmers had a nursery compared with 25 % in 2001. In 2018, 61 % of the farmers said that cocoa sales were their largest revenue sources, compared with 40 % in 2001. In 2001, no farmer used herbicide or fertiliser whereas by 2018 this had increased to 9 % for both products. Numbers of farmers using insecticides quadrupled from 18 % in 2001 to 69 % in 2018. Fungicides remained the most used pesticide with 65 % of farmers using them in 2001 compared with 86 % in 2018. Walking time to the field more than doubled and more farms had been established after partial clearance of secondary forest than after short fallow. The global trend towards full-sun systems was not observed as there were less farmers in 2018 indicating they used full-sun systems and more reported using higher shade levels. Average reported yields were higher in 2018 (176 kg ha^{-1}) than in 2001 (115 kg ha^{-1}). Yield was positively and significantly correlated with total costs and labour invested, yet negatively correlated with total area of cocoa holding. Bucking the regional trend, Cameroon's cocoa farmers have simultaneously pursued expansion into secondary forest and intensification by increasing chemical inputs and labour investment, and yields have increased by 50 %. Full-sun systems are rare; farmers still cultivate traditional, carbon-rich agroforestry systems albeit with more inputs.

Keywords: Agroforestry, blackpod, cocoa, intensification, West Africa

Contact Address: Lindsey Norgrove, Bern University of Applied Sciences, School of Agricultural, Forest and Food Sciences (HAFL), Laenggasse 85, 3052 Zollikofen, Switzerland, e-mail: lindsey.norgrove@bfh.ch

Eucalyptus Woodlot Adoption and its Determinants in Mecha District, Northern Ethiopia

Asabeneh Alemayehu Munuyee[1], Yoseph Melka Ako[2], Abeje Eshete Wassie[1]

[1] *Ethiopian Enviornment and Forest Research Institute, Bahir Dar Enviornment and Forest Research Center, Forest economics , Ethiopia*

[2] *Hawassa Univeristy, Wondo Genet Collage of Forestry and Natural Resource, Natural Resource Economics, Ethiopia*

The area of Eucalyptus plantations has now expanded greatly and growing Eucalyptus at a farm level in a form of woodlot primarily for income generation has become popular among Ethiopian smallholder farmers. Despite its wider use/practice, studies and systematic documentation on the adoption and economic significance are scarce to inform evidence-based policy making. The aim of this study was to investigate factors affecting adoption of Eucalyptus woodlot in Mecha District, northern Ethiopia. Multistage sampling procedure was used to select 186 sample respondent households from three purposively selected kebeles of the district. For the collection of primary data key informant and in-depth interviews, focus group discussions, market assessment and direct observations were used and complemented by secondary data. Descriptive statistics were employed to analyse the collected data. Double-hurdle econometric model was used to identify factors influencing households' adoption decision and adoption intensity of Eucalyptus woodlot. Parameter estimates from the double-hurdle econometric model revealed that educational level, income of the household head, number of parcels of land, off farm work engagement, farmers perception towards Eucalyptus woodlot production and credit availability significantly and positively influence household adoption decision. Whereas, family size and income of the household (negatively), land holding size, number of parcels of land, off farm work engagement, adjacent land, farmers perception towards Eucalyptus woodlot production and access to market (positively) were found to be significantly influencing the adoption intensity of Eucalyptus woodlot. Moreover, lack of support, lack of road access, lack of storage space for selling, lack of segregation of land, disease as well as limited technologies were the major constraints of tree growers. In general, the household's decision to plant Eucalyptus has been influenced by different demographic, socioeconomic, institutional and physiological factors. Improving farmers' level of education, productivity of land, cluster planting, providing alternative options for the farmers, easing credit access, training on silvicultural management, technologies adoption, implementing the existing policies and enforcing rules and regulations are areas that need policy attentions to improve the livelihood of the communities.

Keywords: Adoption, determinants, double hurdle, Eucalyptus, woodlot

Contact Address: Asabeneh Alemayehu Munuyee, Ethiopian Enviornment and Forest Research Institute, Bahir Dar Enviornment and Forest Research Center, Forest economics , Addis Alem, Bahir Dar, Ethiopia, e-mail: alemayehuasabeneh@yahoo.com

The Assessment of Relation between Market Forces and Deforestation in Indonesia and Brazil: A Meta-Analysis

TEREZA LYSAKOVA

Czech University of Life Sciences Prague, Fac. of Tropical AgriSciences; Dept. of Economics and Development, Czech Republic

High targets towards lowering emissions of GHGs set in developed markets, e.g. in the European Union, and increased share of renewable resources within the energy mix or high world demand for agricultural imports of commodities such as palm oil or soya beans and grains from developing countries influences the agricultural production and worsen the state of the environment resulting in further removals of tropical forests in these countries.
Deforestation in Indonesia and Brazil connected with the agricultural production represents a significant problem. Especially due to the fact, that the agricultural land is gained via direct conversion of forest area into the production areas mainly to the detriment of the environment. With regards to exploitation of natural resources, loss of biodiversity and contribution towards global GHG emissions, the research focuses on deforestation as a significant indicator of environmental degradation in case study countries.
Supported by the scientific literature, trade activities in developing countries associated with increased agricultural exports represent the main driver of deforestation since the beginning of 2000s, along with the pressures on economic development and population growth as well as insufficient institutional capabilities to decrease the forest degradation.
The main purpose of this research therefore was to compile scientific papers, policies, governmental documents, special reports and scientific journals to conduct meta-analysis of the scientific and practical publications on the deforestation and protection of rainforest in relation to high world demand for palm oil, soya beans and grains – to provide an assessment of relation between market forces and deforestation in Indonesia and Brazil.

Keywords: Deforestation, export, market forces, palm oil, renewable resources, soya

Contact Address: Tereza Lysakova, Czech University of Life Sciences Prague, Fac. of Tropical AgriSciences; Dept. of Economics and Development, Kamýcká 129, 165 00 Praha - Suchdol, Czech Republic, e-mail: lysakova@ftz.czu.cz

Briquette Production from Baobab (*Adansonia digitata*) Fruit Shells

Arshidin Khassiyarov, Rolf Rheinschmidt, Dietrich Darr, Matthias Kleinke
Rhine-Waal University of Applied Sciences, Fac. of Life Sciences, Germany

The major sources of traditional biomass in the Republic of Malawi are firewood, charcoal, and agricultural residues, which contribute approximately 87 %, 6.4 %, and 6.6 % to total energy supply. The intense utilisation of charcoal and firewood can contribute to forest degradation and deforestation, particularly around densely populated areas. With an annual deforestation rate of 2.4 percent, forest loss in Malawi is among the highest in Southern Africa. Forest conservation in the country is hampered by the interaction of high population density, poverty, and rural dependence on forests, as well as poor forest management institutions. Agricultural expansion and high demand for wood fuels are two major forest threats.
The use of organic waste material as an alternative fuel can help alleviate this problem. Considering the area-specific biomass availability, related transportation cost and lack of attractive alternative uses, baobab (*Adansonia digitata*) fruit shells are a suitable agricultural residue. These shells are abundant due to the baobabs' wide distribution in the southern region of Malawi, and along the lakeshore in the central and northern regions of the country.
The purpose of this study is to analyse the physical properties of baobab shells, to determine the technological feasibility and fuel efficiency of baobab shell briquettes and crushed shells, and to assess any health risk related to using baobab shell briquette as an alternative fuel.
A briquette screw press machine with a motor capacity of 15 kWh and a capacity of 109.5 kg/h was used in Malawi to produce the baobab shell briquettes. The samples were characterised with regard to major physical properties, i.e. energy content, bulk density, ash content, and inflammability; and an exhaust gas analysis was performed. The baobab shells briquette samples were compared with other fuels typically used in Malawi to identify their respective advantages and disadvantages. The technology will be evaluated and recommendations for practice will be presented.

Keywords: *Adansonia digitata*, baobab, biomass, briquettes, energy, fuel, Malawi

---

Contact Address: Matthias Kleinke, Rhine-Waal University of Applied Sciences, Fac. of Life Sciences; Sustainable Food Systems Research Group, Marie-Curie-Str. 1, 47533 Kleve, Germany, e-mail: matthias.kleinke@hochschule-rhein-waal.de

Timber from Organic Cacao Agroforestry Systems, an Additional Source of Income for Farmers in Bolivia

Matthias Baumann[1], Lukas Brönnimann[2], Ulf Schneidewind[3], Laura Armengot[1], Joachim Milz[3], Johanna Rüegg[1], Monika Schneider[1]

[1]*Research Institute of Organic Agriculture (FiBL), International Cooperation, Switzerland*

[2]*Bern University of Applied Sciences (BFH), School of Agricultural, Forest and Food Sciences (HAFL),*

[3]*Ecotop, Bolivia*

Unlike cacao monocultures, cacao agroforestry systems offer a wide range of additional ecosystem services and values to the farmers, such as (i.a. regulating pests and diseases, balancing the understory climate and carbon sequestration) and enable a higher independence from the main crop by generating a potential additional income. To assess the standing timber volume and value, a tree inventory was conducted in 2017 in 16 smallholder cacao agroforestry plots in Alto Beni, Bolivia. Farmers and experts were interviewed to identified the challenges for the timber production in these agroforestry systems. The timber trees on the plots had mainly an age of about 10–20 years (mean 15.5 years) but some trees were also in the range between 2–40 years. A total number of 2'941 trees were counted on all the plots and 20 % of it were *Swietenia macrophylla*, which makes it the most popular timber species. Other very common species were *Myroxylon balsamum* (12 %), *Amburana cearensis* (11 %) and *Centrolobium ochroxylum* (10 %). The average timber tree density was 230 trees ha^{-1} and the standing timber volume was 46 m^3 ha^{-1}. The standing timber per plot had an estimated average value of 12 947 USD ha^{-1} at the time. Because of lack of professional timber processing, such as timber transportation and sawmill, the loss in timber volume is estimated around 40 %. Additionally, farmers are challenged with trimming and pruning of trees as well as with the legal requirements. With the aim to increase farmer's income from timber trees we suggest the following measures at three levels: (1) improving plantation layout (density, layout, species) and tree management (criteria for selection for cut off trees, trimming and pruning); (2) to support a more professional timber logging and processing to decrease losses and (3) to create service providers such as farmer owned cooperatives for logging, sawing, registration of trees and logging permits.

More information on the project can be found here: https://systems-comparison.fibl.org

Keywords: Bolivia, income, organic cocoa-agroforestry, timber, tree inventory

Contact Address: Monika Schneider, Research Institute of Organic Agriculture (FiBL), Department of International Cooperation , Ackerstrasse 113, 5070 Frick, Switzerland, e-mail: monika.schneider@fibl.org

Environmental Effects of Predominant Practices for the Passion Fruit Production in the Huila Department, Colombia

Luis Augusto Ocampo Osorio, Jeisson Rodriguez Valenzuela

Corporación Colombiana de Investigación Agropecuaria Agrosavia, Nataima Research Center, Colombia

Due to its environmental conditions, Colombia is characterised as one of the countries with high diversity in the production of passifloras crops in the tropics. Among the fruit species from this group, passion fruit (*Passiflora edulis*) stands out for its high demand. Huila is one of three departments in Colombia with the highest production of this crop, however, the most predominant management practices for its cultivation have been traditional, without considering technological recommendations under the One Health approach (food, human, animal, and ecosystem safety). Therefore, the purpose of this study was to analyse the management practices of this crop that are employed by producers in the municipalities of La Plata, Guadalupe, and Suaza in the department of Huila, and consequently to observe the potential effects and impacts of these traditional technologies. A participatory research process (workshops) was carried out with passion fruit producers in the three municipalities described above, in addition to inspection visits to the farms producing this species. The results showed that one of the major limitations of this crop is the high incidence and severity of pests, and consequently, the phytosanitary control implemented by the farmer is through the inappropriate use of chemical pesticides. Among the impacts of these practices, farmers report contamination of water sources and soil. Furthermore, there is evidence of loss of insect pollinators, which has caused the necessity of manual pollination in this crop. Concerning the quality and safety of the fruit, this does not meet the requirements to expand its commercialisation at the international level. Another result to highlight is that because the crop requires a tutoring system of trellising for its installation, the producers use woody species of the area to carry out this work, which has caused significant deforestation in areas of protection of water sources. Therefore, despite the profitability and high demand for this fruit at the national level, some agricultural practices of the crop have a significant impact on the environment in the study area.

Keywords: Deforestation, losses of pollinating insects, One Health approach, soil and water pollution, tropical fruit, unsuitable agricultural practices

Contact Address: Jeisson Rodriguez Valenzuela, Corporación Colombiana de Investigación Agropecuaria Agrosavia, Obonuco Research Center, Km 5 Vía Pasto - Obonuco Nariño, Pasto, Colombia, e-mail: jrodriguezv@agrosavia.co

Food and Medicinal Uses of *Annona senegalensis* Pers: A Country-Wide Assessment of Traditional Theoretical Knowledge and Actual Uses in Benin, West-Africa

Janine Conforte Fifonssi Donhouede, Valère Salako, Achille Ephrem Assogbadjo

University of Abomey-Calavi, Lab. of Biomathematics and Forest Estimations, Benin

The growing interest in more natural products in food and health industries has led to increasing research on traditional knowledge related to plants. While theoretical knowledge (TK) informs on a wide spectrum of potential uses of species, actual uses (AU) highlight their potential being actually used. Distinguishing between the two is important when reporting ethnobotanical studies but has often been confused possibly misleading decision-making. This study assessed TK and AU of Annona senegalensis focusing on its food and medicinal uses in Benin. We further assessed how factors such as age, sex, sociolinguistic group, and main activity determine the distribution of TK and AU among local communities. Data were collected through semi-structured individual interviews (n = 755) and analysed using relative frequency of citation (RFC) and reported use- value (RUV). A total of 167 theoretical uses were recorded but only 92 were actually practised of which 3 were food and 88 medicinal. The average difference between TK (3.16±0.04) and AU (2.53±0.04) was 0.64±0.03 and was the highest for Bariba sociolinguistic group and did not vary with age, sex, and main activity. TK and AU were high for traditional healers and low for Non- farmers. The highest TK was found with Bariba sociolinguistic group and the highest AU with Otamari sociolinguistic group. Fruits (100 %) and flowers (10 %) were the most commonly used organs for food while leaves (40 %) and roots (7 %) were mostly used for medicinal purposes. The most common food uses were consumption of the ripe fruits (100 %) and the use of flowers to season foods (10 %). The most cited diseases were malaria (28 %), and intestinal worms (8 %). Our study shows the importance of differentiation between TK and AU. The study documents ranges of uses of *A. senegalensis* and highlights the most common uses of the species, providing ways for its valorisation.

Keywords: *Annona senegalensis*, knowledge gap, socio-demographic factors, sociolinguistic groups, use patterns

Contact Address: Janine Conforte Fifonssi Donhouede, University of Abomey-Calavi, Lab. of Biomathematics and Forest Estimations, Abomey Calavi, Cotonou, Benin, e-mail: jdonhouede@gmail.com

Morphological Diversity of *Allanblackia parviflora* in Ghana

Dennis Kyereh[1], Prasad Hendre[2], Alice Muchugi[2], Bohdan Lojka[1]
[1]*Czech University of Life Sciences Prague, Department of Crop Sciences and Agroforestry, Czech Republic*
[2]*World Agroforestry (ICRAF), Kenya*

Allanblackia parviflora A. Chev. is an indigenous tree species which is found in the rain forest zones of West Africa. It is an under-utilised fruit tree species that has been targeted for improvement as part of efforts to domesticate high-value indigenous multi-purpose trees for fruits and seeds production in Africa. *Allanblackia* has several benefits which include provision of shade, timber, and medicine, but production of edible oil from the seeds is the economically most important use. The *Allanblackia* seed oil is currently being developed as a new agri-business in Ghana, Nigeria, Cameroon and Tanzania. This rural based enterprise would not only increase livelihood opportunities for farmers but also ensures retention of trees on farms for environmental sustainability. There is limited information on natural variation in the phenotypic characteristics of the species in Ghana. The objective of the study was to evaluate morphological diversity in natural and domestic stands of the species. The study was conducted across four provenances in Ghana: The Wet Evergreen, Moist Evergreen, Moist Semi-deciduous South East and Moist Semi-deciduous North West. Data were collected from a total of 100 individuals of *A. parviflora* from natural and on-farm stands, with adequate representation of male and female species, in each ecological area. Morphological data collected includes tree height, trunk height, crown diameter, DBH, age of tree, length of leaf blade, fruit, and seeds, width of leaf blade, fruit and seeds, weight of fruits and seeds, and number of seeds per fruit.
Morphological diversity among and within natural and domestic populations will be tested using ANOVA in STATISTICA 12 software. The study is expected to provide understanding on natural variation in phenotypic diversity of *A. parviflora* in the four provenances in Ghana. The knowledge and understanding of morphological diversity will guide effective decision making towards domestication of the species in Agroforestry systems.

Keywords: Agroforestry systems, *Allanblackia parviflora*, diversity, morphology

Contact Address: Dennis Kyereh, Czech University of Life Sciences Prague, Department of Crop Sciences and Agroforestry, Kamycka 129, 165 21 Prague, Czech Republic, e-mail: kyereh@ftz.czu.cz

Population Structure of *Garcinia kola* Heckel in Central Region of Cameroon

Irikidzai Prosper Chinheya, Marie Kalousová, Anna Manourova, Bohdan Lojka

Czech University of Life Sciences Prague, Fac. of Tropical Agrisciences, Dept. of Crop Sciences and Agroforestry, Czech Republic

Garcinia kola is a multipurpose fruit tree species indigenous to West African communities, where it is of significant ethnomedicinal, cultural and economic importance. Faced with the threat of declining population numbers, the species was selected for conservation and participatory domestication programmes however, a lack of adequate information on genetic diversity is widely reported as a limiting factor in both processes. The aim of this study was to assess the genetic diversity of 96 *G. kola* individuals from eight existing population groups in the Central region of Cameroon using Amplified Fragment Polymorphism (AFLP) markers. A total of 1176 fragments were amplified using four primer combinations with 98.6 % polymorphism at the species level and a mean number of 261.9 fragments per individual. The computed values for Nei's gene diversity within populations (Hj), Total gene diversity (Ht), and the Wright's fixation index (F_{ST}) were 0.1894, 0.1922 and 0.0145 respectively. The obtained results revealed a higher genetic diversity within the assessed populations than among them. Bayesian analysis of sampling groups revealed the existence of two differentiable but admixed genetic clusters, implying a weak population structuring. Attempts to assess for correspondence between clustering and geographic distances revealed no clear patterns. The study revealed that AFLP markers are a useful tool for assessing the genetic diversity of G. kola. Results suggest possible human-mediated gene flow events, potentially attributed to the selection of kernels for trade or natural selection through the adaptation of the species to local environmental. This study may open the door for advancing participatory tree domestication programme and conservations programmes within the study area. However, it is recommended that initiatives be undertaken to safeguard the existing genetic diversity such as the use of gene banks, sustainable utilisation of genetic diversity in PTD or the protection of important individuals within their stands.

Keywords: AFLP marker, AFTPs, agroforestry, bitter kola, genetic diversity, provenance, tree domestication

Contact Address: Irikidzai Prosper Chinheya, Czech University of Life Sciences Prague, Fac. of Tropical Agrisciences, Dept. of Crop Sciences and Agroforestry, Kamycká 129, 165 21 Prague, Czech Republic, e-mail: irikidzaic@gmail.com

Medicinal Use Patterns of *Parkia biglobosa* (Jacq.) Benth. and *Vitellaria paradoxa* (Gaertn. F), two Important Traditional Agroforestry Species in Benin, West-Africa

Olouwatoyin Grâce Ricardine Odounharo, Setonde Constant Gnansounou, Valère Salako, Rodrigue Idohou, Guy Apollinaire Mensah, Romain Glèlè Kakaï, Achille Assogbadjo

University of Abomey-Calavi, Lab. of Biomathematics and Forest Estimations, Benin

In West Africa, African locust bean (*Parkia biglobosa* (Jacq.) Benth.) and Shea (*Vitellaria paradoxa* (Gaertn. F) are among the most important multipurpose plant species commonly found in traditional agroforestry systems. Most of research on these species are dominated by patterns and properties of their food uses, and additionally cosmetics for the shea. Yet, the species also have interesting medicinal properties that have been little explicitly explored. Using an ethnobotanical survey, we explored the patterns of diseases and other human disorders healed by the species, the different plants parts involved in diseases treatment, the recipes adapted for the treatment of the diseases and disorders and the other species involved in recipes composition in northern Benin where they are widely distributed and used. Plants parts used by respondents were subjected to a Principal Component Analysis together with the ethnic groups. Alpha diversity indices were used to compute disease diversity while the Intraspecific Use Values index was applied to assess the frequency of utilization of each plant part. Results showed that 11 categories comprising 51 diseases and disorders were listed by respondents for the two species, with the predominance of gastro-intestinal diseases (RFC = 31.10 % for *P. biglobosa* and RFC = 31.81 % for *V. paradoxa*) and Infectious diseases (RFC = 26.82 % for *P. biglobosa* and RFC = 27.27 % for *V. paradoxa*). Most used plants parts were nuts (IVU = 90.90 %) and roots (IVU = 90.90 %) for V. paradoxa and roots (IVU = 90.24 %) and bark (IVU = 70.73 %) for *P. biglobosa*. The PCA analysis showed a strong variation in the used plants parts across the sociocultural groups.

Keywords: Benin, medicinal uses, Natitingou, *Parkia biglobosa*, *Vitellaria paradoxa*

Contact Address: Olouwatoyin Grâce Ricardine Odounharo, University of Abomey-Calavi, Lab. of Biomathematics and Forest Estimations, Lot 22 Gbodjo, BP 5225 Abomey-Calavi, Benin, e-mail: ricardinodounharo97@gmail.com

Assessment of Local and Morphological Descriptors of Local Accessions of *Mangifera indica* L. in North Benin

Dowo Michée Adjacou[1], Thierry Dèhouégnon Houehanou[1], Gérard Nounagnon Gouwakinnou[1], Taffa Moussa[1], Abdul-Raouf Mama[1], Kathleen Prinz[2], Armand Natta[1]

[1]*University of Parakou, Lab. of Ecology, Botany and Plant Biology, Benin*

[2]*Friedrich-Schiller-University Jena, Inst. for Ecology and Evolution, Germany*

The local mango tree (*Mangifera indica* L.) occupies an important place in the family consumption in West Africa. The characterisation of this local fruit crop is neglected comparatively to improved varieties. Therefore the characterisation of local accessions of *Mangifera indica* L. is necessary to guarantee a sustainable future through its diversity conservation. Thus, this study was carried out in North Benin and aims to (i) identify local knowledge discriminating local accessions of *M. indica* (ii) assess the morphological descriptors and (iii) evaluate the convergence between the local and morphological descriptors. Sixty five (65) people were questioned using a structured guide on the knowledge of morphological characteristics discriminating the different types of accessions that occur in their production systems. In addition, 65 individuals of *M. indica* were sampled and characterised by 56 traits with 21 of which were quantitative and 35 qualitative. From each tree individual, 10 healthy mature fruits and 10 leaves were randomly collected, whose characteristics were evaluated. The data of the questionnaire and morphological characterisation were processed and analysed with univariate and multivariate statistical tools. The population recognised two different local accessions of *M. indica* distinguished generally by the shape and fibrousness of the fruit. An elliptical shape and a very fibrous character of the fruit were recognised locally for one accession while for the second the fruit has a rounded shape and is less fibrous. The morphological categorisation allowed us to distinguish three groups of accessions which differ essentially in fruit weight and leaf length. The morphological traits evaluated are almost all discriminating between the three morphological groups. The morphological characterisation of the two accessions converges globally with the local descriptors recognised by the population. The knowledge of these superior traits is fundamental to guide the sustainable management and conservation of local *Mangifera indica* accessions in Benin.

Keywords: Benin, conservation, local descriptors, *Mangifera indica*, morphological traits

---

Contact Address: Dowo Michée Adjacou, University of Parakou, Lab. of Ecology, Botany and Plant Biology, Zongo, 229 Parakou, Benin, e-mail: micheadjacou94206471@gmail.com

A Probabilistic Framework for the Cost-Benefit Evaluation of Restoration Outcomes for a Dry Afromontane Forest in Northern Ethiopia

Yvonne Tamba[1], Cory Whitney[2], Eike Luedeling[2], Keith Shepherd[1], Ermias Aynekulu[1], Aklilu Negussie[3], Negusse Yigzaw[3], Joshua Wafula[1], Yemane Gebru[3], Caroline Muchiri[1]

[1]*World Agroforestry (ICRAF), Land Health Decisions, Kenya*

[2]*University of Bonn, Inst. Crop Sci. and Res. Conserv. (INRES), Germany*

[3]*WeForest, Ethiopia*

Forest and Landscape Restoration (FLR) is carried out with the objective of regaining ecological functions and enhancing human well-being through intervention in degrading ecosystems. However, uncertainties and risks related to FLR make it difficult to predict long-term outcomes and inform management plans. We applied a Stochastic Impact Evaluation framework (SIE) to simulate returns on investment in the case of FLR interventions in a degraded dry Afromontane forest while accounting for uncertainties. We ran 10,000 iterations of a Monte Carlo simulation that projected FLR outcomes over a period of 25 years. Our simulations show that investments in assisted natural regeneration, enrichment planting, exclosure establishment and soil-water conservation structures all have a greater than 77 % chance of positive returns. Sensitivity analysis of these outcomes indicated that the greatest threat to positive cashflows is the time required to achieve the targeted ecological outcomes. Value of Information (VOI) analysis indicated that the biggest priority for further measurement in this case is the maturity age of exclosures at which maximum biomass accumulation is achieved. The SIE framework enabled us to clearly define the objectives of FLR activities and quantify the expected impacts on land use and land cover trends. With the framework, we translated outcomes into economic impacts for easy integration into decision-making processes. For FLR implementers, initial costs incurred to establish livelihood interventions, mobilise communities, strengthen social governance structures, and provide capacity building can result in net losses in the first few years. Implementers will therefore need significant financial support to see their interventions through to the medium and long term.

Keywords: Decision analysis, forest and landscape restoration, Monte Carlo risk analysis, uncertainty, value of information

Contact Address: Yvonne Tamba, World Agroforestry (ICRAF), Land Health Decisions, 30677 U.n. Avenue, 00100 Nairobi, Kenya, e-mail: Y.Tamba@cgiar.org

Economic Performance of Five Different Cacao Production Systems

Johanna Rüegg[1], Laura Armengot[1], Joachim Milz[2], Ulf Schneidewind[2], Monika Schneider[1]

[1]*Research Inst. of Organic Agriculture (FiBL), International Cooperation, Switzerland*
[2]*Ecotop, Bolivia*

Agroforestry systems for cocoa production are commonly promoted for biodiversity conservation, climate change mitigation and adaptation as well as for food security and risk mitigation. Generally, these systems include timber, legume or fruit trees, and sometimes additional crops. Knowledge gaps exist about the economic performance of cocoa based agroforestry systems, including the by-crops.

Here we present the economic performance of 5 cocoa production systems from planting to entering the mature stage (11 years). In a long-term trial in Bolivia, a gradient of complexity from monocultures, agroforestry systems to successional agroforestry systems (SAFS) is studied. Additionally, for monocultures and agroforestry, conventional and organic management are compared, while SAFS are managed organically. Income was calculated taking into account yields of cocoa, fruit trees and by-crops with farm gate prices. Only for cocoa organic premium prices were reached and taken into account. Labour time was registered for management, input preparation and post-harvest.

Cocoa yields were lower in the agroforestry systems compared with monocultures, and lowest in SAFS. For monocultures, they were higher under conventional management, while in agroforestry systems management had no influence. Total system yields in agroforestry systems (dry matter) were 3–4 times higher than in monocultures. This was mainly due to banana production in agroforestry systems and from a diversity of by-crops in SAFS.

Income over all years was comparable among all systems. In agroforestry systems, cocoa was responsible for more than 50 % of the income, while in SAFS the share of cocoa was smaller, as some crops like pineapples or peach palm had good markets in the region. The income generated per workday invested during the whole period did not differ between the systems.

Income analysis shows the importance of cocoa as a cash crop, but also the potential of by-crops depending on the development of their markets. On the other hand, agroforestry systems contribute to food security and mitigate risks of price or yield fluctuations in the cash crop. In conclusion, the data show that with different strategies and plantation design, the same level of income and income per work day invested can be reached.

More information on the trial layout can be found here: https://systems-comparison.fibl.org

Keywords: Agroforestry, cocoa, diversification, economic, income, organic, systems comparison, *Theobroma cacao*

Contact Address: Johanna Rüegg, Research Inst. of Organic Agriculture (FiBL), International Cooperation, Ackerstrasse 113, 5070 Frick, Switzerland, e-mail: johanna.rueegg@fibl.org

The Impacts of COVID-19 on Deforestation in Colombia

Mehran Khodadadi[1], Jonnathan Cespedes[2], Katharina Löhr[1], Augusto Castro-Nunez[2], Stefan Sieber[1]

[1]*Leibniz-Zentrum für Agrarlandschaftsforschung (ZALF) e. V., Sustainable Land Use in Developing Countries (SusLAND), Germany*
[2]*International Center for Tropical Agriculture (CIAT), Colombia*

Considering the high deforestation rates in recent years and the land management challenges to halt and reduce forest loss around the world, we are looking to analyse how the current coronavirus pandemic impacts on deforestation in order to determine if the pandemic adds further complexity to forest conservation. To achieve this purpose, Amazon region in Colombia is selected as the case study. It is predictable that deforestation could increase in forest sectors due to lack of control in some regions and put international forest-based industries that developed sustainable products and livelihoods of local people at risk. According to WWF (2020), about 645 thousand ha lost in March 2020, where Indonesia led by 130 thousand ha (almost three times more than March 2019). The study reports The Democratic Republic of Congo in the second place (almost 100 thousand ha), followed by Brazil with 95 thousand ha. In Colombia, contrary to other countries in the Amazon region, the trend in 2019 showed a reduction in deforestation compared to 2018. However, 2020 started with an increasing tendency, and the quarantine seems to have worsened the situation (FCDS, April 2020). According to the Amazon Institute for Scientific Research (SINCHI) around 13,000 hot points were recorded in the Amazon region (with presence of FARC) in March. This number is almost 275 percent higher than in the same month in 2019.
The limitations or lack of monitoring and controlling parties during the pandemic could encourage the rural people and regional mafias to illegal cultivation and overgrazing, which causes increasing deforestation rates over the country. Therefore, this study aims to analyse the impacts of COVID-19 on deforestation rates in Colombia using Remote Sensing (RS) techniques (Landsat 8 OLI/TIRS C1 Level-1 with a cloud cover of less than 40 %.). Key questions this study aims to answer are i) what are the impacts of COVID-19 on deforestation in Amazon region of Colombia? (Does COVID-19 enhance deforestation in Colombia?), ii) how do restriction measures on mobility impact deforestation? And iii) how Landsat-8 data perform over the selected region compare to the terra-i results.

Keywords: ArcGIS, Ccassification, i-terra, Landsat-8, remote sensing

Contact Address: Mehran Khodadadi, Leibniz-Zentrum für Agrarlandschaftsforschung (ZALF) e. V., Sustainable Land Use in Developing Countries (SusLAND), Eberswalder Straße 84, 15374 Müncheberg, Germany, e-mail: mehran.khodadadi@zalf.de

Intercropping Maize with *Gliricidia sepium* and Pigeonpea Enhances Productivity and Resilience of Cropping Systems in Semiarid Tanzania

Anthony Kimaro[1], Johannes Michael Hafner[2], Christine Lamanna[3], Peter Steward[3], Todd Rosenstock[4]

[1]*World Agroforestry (ICRAF), Tanzania Country Programme, Tanzania*
[2]*Leibniz Centre for Agric. Landscape Res. (ZALF), Sustainable Land Use in Developing Countries (SusLAND), Germany*
[3]*World Agroforestry (ICRAF), Kenya*
[4]*World Agroforestry (ICRAF), Land Health Decisions, DR Congo*

Low and sporadic precipitations adversely affect cropping system resilience in semiarid tropics. However, crop diversification through intercropping enhance agroecosystem resilience. We tested whether intercropping maize with *Gliciridia sepium* and/or pigeonpea improves productivity and drought resistance of maize under semiarid conditions. Our study adopted a split-split-plot experiment to test the effects of intercropping (maize monoculture, sole pigeonpea, maize-Gliricidia, maize-pigeonpea and maize-pigeonpea-Gliricidia), fertilisers (with and without) and rainfall (ambient and drought). Drought was induced using the above-canopy rainout shelters which intercepts 50 % of the ambient rainfall. No significant effects of rainout shelters was note on air temperature and relative humidity above the maize canopy, but photosynthetically radiation (PAR) was reduced by 15.8 %. Gravimetric soil moisture was also reduced by 12.5 %. As expected, the drought treatment reduced soil moisture, but without creating artificial growing conditions for crops. Significant effect of intercropping was noted in 2019 and 2020 seasons, which is attributed to the combined impacts of weather variability and interactions of maize with *G. sepium* and pigeonpea. Significant interactions between treatments (intercropping, fertiliser and drought) was noted in 2019 with *G. sepium*-pigeonpea intercropping under fertilisation and ambient rainfall significantly suppressing maize yield relative maize monoculture. However, maize- *G. sepium* and maize-pigeonpea intercropping did not reduce maize grain yield relative, suggest that farmers can diversify their fields with either *G. sepium* or pigeonpea without compromising yields of maize even in drought prone seasons like 2019. In 2020, fertilisation improved maize grain yields under drought by 63 % compared to ambient rainfall. Heavy and well distributed precipitations throughout the 2020 cropping season (961 mm) account for the reduction of maize yields under ambient as the rainfall amount received in the drought treatment (790 mm) was above the long-term average

Contact Address: Anthony Kimaro, World Agroforestry (ICRAF), Tanzania Country Programme, P.O. Box 6226, Dar-es-Salaam, Tanzania, e-mail: a.kimaro@cgiar.org

(425 mm) for semiarid areas of Tanzania. Overall maize yield in intercropping in 2020 was similar to the yield obtained with maize monoculture. Also, maize resistance, measured as yield loss due to drought, was higher in intercropping and land equivalent ratios were above 1 in both growing seasons. This study show intercropping maize with *G. sepium* or pigeonpea sustains productivity and improve resilience of maize cropping systems.

Keywords: Agroforestry, climate change, crop diversification, drought

Climate, soils and more

Climate change impacts, adaptation and mitigation in the agricultural sector

Extreme Weather Events and Permanent Internal Migration: Evidence from Mongolia

Julian Röckert, Kati Krähnert

Potsdam Institute for Climate Impact Research, Research Department II: Climate Resilience, Germany

Mongolia is among the most vulnerable regions in the world to climate change. Aside from temperature rises and changes in precipitation patterns, extreme weather events have increased both in their intensity and frequency in recent decades. The national livestock economy, which constitutes the main source of income for the majority of the rural population is put under increased pressure by this development. As extreme weather events are accompanied by high livestock mortality, they endanger the traditional livelihood of herding families. Through this vein, natural calamities may contribute to forced climate-induced migration decisions. The accelerating urbanisation notwithstanding, there is to-date little to no empirical evidence on the question to which extend extreme weather events matter for population mobility in the Mongolian context.

In this article we study the effect of weather shocks on internal migration figures in Mongolia. While previous studies mostly focus on migration effects of floods, droughts, and storms in tropical or dry climate areas, our focus is on extremely harsh winter events. The causal impacts of extreme winter events on internal migration dynamics are identified by exploiting exogenous variation in their intensity across time and space. We exploit an unusually long time series of migration data spanning the 1990–2018 period in a two-way fixed-effects panel estimator. Results show that extreme winter events cause significant and sizeable permanent out-migration from affected areas. These effects are also mirrored in the overall population figures on the provincial as well as the district-level of Mongolia.

In addition, the occurrence of extreme weather events is found to be a strong predictor for negative changes in the local population of pastoralist households. This suggests that the abandonment of traditional livelihoods is one channel through which climate affects permanent within-country migration in the Mongolian context.

Keywords: Extreme weather events, internal migration, Mongolia, pastoralism

Contact Address: Julian Röckert, Potsdam Institute for Climate Impact Research, Research Department II: Climate Resilience, Weichselstr. 54, 12045 Berlin, Germany, e-mail: roeckert@pik-potsdam.de

Prediction of the Key Determinants of Climate-Resilient Technology Adoption: A Case study of Odisha

Pradyot Jena, Purna Tanti
National Institute of Technology Karnataka, Surathkal, School of Management, India

Climate change, poverty, and inequality are the major issues of developing countries. Climate change threatens rural livelihoods by adversely affecting crop yields. Developing countries are the most vulnerable to climate change due to their lack of adaptive capacity. Climate-smart agricultural (CSA) practices are advanced as a possible solution. However, resource-poor farmers often face financial constraints to adopt practices that could sustainably increase their crop yields. The current paper using a structured questionnaire survey among the farming households of an Eastern Indian state, namely, Odisha, explores the key determinants of CSA adoption. Two districts with one each from the coastal and the inland regions of the state are chosen for the study. The majority of the respondents (95 %) perceive the effects of climate change in the region. The respondents have adopted practices such as rescheduling planting (79 %), crop rotation (50 %), micro-irrigation (19 %), and early maturity seeds (18 %). Farmer's perception of climate change has been analysed to assess the knowledge of farmers on climate change. To explore the key determinants of adopting these five major practices, a probit model is estimated. Understanding the role of women farmers to upscale the CSA practices has been described qualitatively. The result shows that factors such as farmer field school participation, subsidies, access to energy use, and perception of climate shocks are the major determinants. Further, the interaction between landholding and credit availability has positively affected the decision to adopt. Government extension services have a substantial impact on various adaptation practices. Farmers' access to govt extension is more likely to do rescheduling planting by 11 %, crop diversification by 14 %, crop rotation by 18 %, and early maturity variety seeds by 28 %. Lack of training, poor access to the market, lack of land rights to females, and resource constraints are barriers for women to adopt CSA. Region-specific policies such as farmers' field schools, subsidies on farm machinery, and resource endowments can upscale CSA adoption in the region.

Keywords: Agricultural extension, climate-smart agriculture, perception of climate change, Probit model

Contact Address: Pradyot Jena, National Institute of Technology Karnataka, Surathkal, School of Management, Srinivasnagar, 575025 Mangalore, India, e-mail: jpradyot@gmail.com

Analysis of Rainfall Variability and Trends for Better Climate Risk Management in the Major Agro-Ecological Zones in Tanzania

Jacob Emanuel Joseph[1], Karuturi P.c. Rao[1], Reimund P. Rötter[2], Anthony M. Whitbread[3]

[1]*International Crops Research Institute for the Semi-Arid Tropics (ICRISAT), Innovation Systems for the Drylands, Tanzania*

[2]*University of Goettingen, Dept. of Crop Sciences; Tropical Plant Production and Agricultural Systems Modelling (TROPAGS), Germany*

[3]*International Crops Research Institute for the Semi-arid Tropics (ICRISAT), India*

Managing climate risk in agriculture requires a proper understanding of climatic conditions, regional and global climatic drivers, as well as major agricultural activities at the particular location of interest. Critical analyses of variability and trends in the historical climatic conditions are crucial in designing and implementing action plans to improve resilience and reduce the risks of exposure to harsh climatic conditions. However, in Tanzania, less is known about the variability and trends in the recent climatological conditions. The current study examined variability and trends in rainfall of major agro-ecological zones in Tanzania using station data from ten locations i.e. Hombolo, Igeri, Ilonga, Lyamungu, Naliendele, Mlingano, Tumbi, and Ukiliguru which had records from 1981 to 2020 and two locations, Dodoma and Tanga, having records from 1958 to 2020. The variability in annual rainfall was high in Hombolo and Tanga locations (CV $\geq$ 28 %) and low in Igeri (CV = 16 %). The OND season showed the highest variability in rainfall (34 % to 83 %) as compared to the MAM (26 % to 36 %) and DJFMA (20 % to 31 %) seasons. We found increasing and decreasing trends in the number of rainy days in Ukiliguru and Tanga respectively, and a decreasing trend in the MAM rainfall in Mlingano. The trends in other locations were statistically insignificant. We assessed the forecast skills of seasonal rainfall forecasts issued by the Tanzania Meteorological Authority (TMA) and IGAD (Intergovernmental Authority on Development) Climate Prediction and Application Center (ICPAC). We found TMA forecasts had higher skills compared to ICPAC forecasts. However, our assessment was limited to MAM and OND seasons due to the unavailability of seasonal forecasts of the DJFMA season issued by ICPAC. We also examined the influence of teleconnection phenomena, i.e. the El Nino Southern Oscillation(ENSO) and the Indian Ocean Dipole (IOD) on the variability and predictability of seasonal rainfall in Tanzania. Our analyses showed that ENSO and IOD phases on seasonal rainfall variability are weak and decrease from north to south of Tanzania. However, the Sea Surface Temperature anomalies (SSTa) in the Tropical Indian Ocean correlated strongly with the OND season rainfall in the northern part of Tanzania.

Keywords: Climate variability, El Nino southern oscillation, Indian Ocean dipole

Contact Address: Jacob Emanuel Joseph, International Crops Research Institute for the Semi-Arid Tropics (ICRISAT), Innovation Systems for the Drylands, Mwenge Coca-Cola Road Mikocheni, Dar Es Salaam, Tanzania, e-mail: j.emanuel@cgiar.org

Framing Indigenous Farming Practices for Adaptation to Climate Change: Evidence from the Rwenzori (Uganda)

Bosco Bwambale[1], Tibasiima Kahigwa Thaddeo[2]

[1] *Vrije Universiteit Brussel (VUB), Geography, Belgium*

[2] *University of Natural Resources and Life Sciences (BOKU), Sustainable Agricultural Systems, Austria*

The limited institutional capacities in the tropics which is prone to climate hazards have led to the wider interest of related scholars and institutions into the role of indigenous knowledge. Indigenous knowledge is seen from the lens that it can (enable to) produce locally adapted interventions. While research on indigenous knowledge for farmers exposed to climate change risk is increasing, it remains focused on content. There is scarce focus on the socio-epistemic processes through which indigenous practices are crafted by farmers, thereby limiting the grasp of the best ways to engage with indigenous knowledge practices. In this study, ethnographic field surveys were conducted to interact with farmers and grasp the processes that enables them to design the indigenous practices to adapt. The study based on a case of indigenous flood risk reduction since flooding is regular and can induce perspectives on a regular basis. Findings indicate several interesting aspects. First, farmers have learnt from experience, tradition, and the local culture. Second, they follow the regularities of local ontologies to develop indigenous theorisations that enhance their lived experience and then grounded into their localised farming practices. Thirdly, indigenous theorisations are enabled by the cultural structures that favour openness to production and applying the most appropriate knowledge. Overtime, this enable farmers to learn from one event to another how to adapt and live with floods. The capacity to learn from each flood event is illustrated by developing "a basket of adapted options" from rescue to prevention and management of subsequent floods. Key examples include shifting from degrading to non-degrading farming practices, e.g., obuwathirira (slash and plant, without ploughing, practised mainly practised in the higher slopes and along close to the river), shift from crop to tree farming and/or pastoralism in the downstream; but also agronomic adjustment like intensive cultivation during flood free periods and land use planning (e.g., plant indigenous trees close to the river, then followed by crops), agroforestry where specific flood 'immunizing' trees are interplanted with crops. This study highlights the ability of indigenous people to question their knowledge foundations to evolve practices that enable living with floods.

Keywords: Climate change adaptation, disaster risk reduction, indigenous farming, indigenous knowledge

Contact Address: Bosco Bwambale, Vrije Universiteit Brussel (VUB), Geography, Pleinlaan 2, 1050 Brussel, Belgium, e-mail: bosco.bwambale@vub.be

Multiple Shades of Climate Change Impacts and Responses among Smallholder Farmers: Case Studies from Nigeria: 2012–2021

Oluwafunmiso Adeola Olajide

University of Ibadan, Dept. of Agricultural Economics, Nigeria

There is a need for information that will support Africa's climate change response at the grass-root level. This requires evidence of its impact at the grass-roots level on crop production as well as farmer's response; as different from traditional production challenges and farm management practices. In order to close this gap, this study synthesizes cross-sectional studies which examined farmers' perception and understanding of climate change, the different effects on crop production pattern, yield, and income, and food security. It also examined the adaptation strategies adopted by male and female farmers in order to ensure a sustainable livelihood and food security. The data used were both secondary and primary data. The secondary data were collected from NIMET and the GHS data collected by the World Bank and the National Bureau of Statistics. The primary data were collected through a multistage sampling technique to select 100–150 respondents across different agro-ecological zones in Nigeria. The cross-sectional studies were carried out at different locations within a 9 year period. The data were analysed using descriptive statistics, inferential statistics, and regression models at α 0.05. The results show that over 50 percent of farmers have an idea of what climate change is; frequent adaptation strategies are mainly traditional farm management practice and either of change in land size, changes in crops, and change in the planting season. The results also show that climate change presents both challenges and opportunities for expansion. But on the general level, information and training to enhance farmer's adaptation strategies as well as households' adaptive capacity are required. The potentials to take advantage of migration for agricultural development across the zones also exist. The findings from this study can inform regional policy support for climate change adaptation and mitigation in sub-Saharan Africa.

Keywords: Climate variability, farming, gender, households, migration, sustainable livelihoods, synthesis

Contact Address: Oluwafunmiso Adeola Olajide, University of Ibadan, Dept. of Agricultural Economics, Ibadan, Nigeria, e-mail: funso.olajide@ui.edu.ng

Sustainable Management of Coffee By-products and Determination of Emission Factors

San Martin Ruiz Macarena[1], Reiser Martin[1], Kranert Martin[2]
[1]*University of Stuttgart, Emissions, Germany*
[2]*University of Stuttgart, Waste Managment, Germany*

Agriculture is the fourth largest contributor to global greenhouse gas (GHG) emissions, which include non-CO_2 gases, including CH_4. In the coffee sector, one important criteria to consider in future assessment options is to understand the formation and determination of GHG emissions that contribute to climate change. Coffee by-products is also a contributor, such as coffee pulp which generates air, water and soil pollution, especially when coffee berries are ripe and processed by the wet method and are not properly treated generating CH_4 emissions.
Due this, the detection of CH_4 during the treatment of coffee by-products is one of the most important tools to meet the objective of methane reduction. An experimental methodology for the sustainable management this residue was developed in Costa Rica, for the transformation from a residue into a soil amendment that can be used in agricultural crops including coffee plantations. This may reflect its impact on soil fertility and agricultural productivity in the future. Within the framework of the project, tests were conducted to explore and optimise the utilisation of coffee by-products and other organic waste materials. Consequently, this project aims to improve the waste management from the coffee processing currently applied in the country, as well as the verification of CH_4 during composting, giving the community the opportunity to receive a positive environmental impact, emissions reduction and suggest emission factors for the coffee sector and its soil improvement. The project will generate significant environmental, economic, technological and commercial impacts, without omitting the main objective of obtaining a product from waste, which will contribute to the concept of bio-economy and circular economy. These types of initiatives enhance a sustainable development and at the same time provide a paradigm shift in high impact productive sectors and in the coffee sector and the reduction of its emissions.

Keywords: Coffee by-products, emission factors, emissions, methane

---

Contact Address: San Martin Ruiz Macarena, University of Stuttgart, Emissions, Am Bandtaele 2, 70569 Stuttgart, Germany, e-mail: macarena.sanmartin@iswa.uni-stuttgart.de

Delivering Climate Change Outcomes with Agroecology: Evidence and Actions Needed

Sieglinde Snapp[1], Yodit Kebede[2], Eva Wollenberg[3]

[1]*Michigan State University, Plant Soil and Microbial Sciences, United States*

[2]*French National Institute of Research for Sustainable Development (IRD), France*

[3]*Research Program of the CGIAR on Climate Change, Agriculture and Food Security (CCAFS) and University of Vermont, United States*

We conducted a rapid evidence-based review to assess the quality and strength of evidence regarding (i) the impact of agroecological approaches on climate change mitigation and adaptation in low and middle income countries, and (ii) the programming approaches and conditions supporting large-scale transitions to agroecology. We found the following:

Substantial evidence exists for climate change adaptation associated with practices and systems aligned with agroecology; e.g., farm diversification, agroforestry and organic agriculture. The agroecological approach with the strongest body of evidence for impacts on climate change adaptation was farm diversification (strong evidence and high agreement). This included positive impacts of diversification on crop yield, pollination, pest control, nutrient cycling, water regulation and soil fertility. Agroforestry had a positive impact on biodiversity, water regulation, soil carbon, nitrogen and fertility and for buffering temperature extremes. Organic agriculture improved regulating (pest, water, nutrient) and supporting services (soils, biodiversity).

Evidence suggests that agroecology provides more climate change adaptation and mitigation than conventional agriculture by emphasising locally relevant solutions, participatory processes and co-creation of knowledge. Specifically, co-creation and sharing of knowledge supported farmers' capacity to adapt to local conditions (strong evidence, medium agreement). Multiple lines of evidence show that engaging with local knowledge through participatory and educational approaches are effective at adapting technologies to local contexts and thereby delivering improved climate change adaptation and mitigation. No evidence was available for impacts related to response to extreme climate events and limited evidence for greenhouse gas emmissions, as well as for agroecological livestock systems. There is clear need for more evidence from developing countries.

Recommendations include a focus on sustainablity outcomes and tradeoffs, rather than contesting the meaning of labels. Although current evidence for agroecology is partial, these findings show the importance of the following to achieve climate change outcomes:

Contact Address: Sieglinde Snapp, Michigan State University, Plant Soil and Microbial Sciences, 1066 Bogue St., 48824 East Lansing, United States, e-mail: snapp@msu.edu

- Diversification of field and farm products and practices.
- Processes that support farmer innovation, co-learning and adaptation of innovations to local contexts

Keywords: Agroecology systematic literature review, climate change outcomes, diversification, recommendations

Adoption of Agro-Climate Services

Thi Thu Giang Luu[1], Eike Luedeling[1], Lisa Biber-Freudenberger[2], Cory Whitney[1]

[1] *University of Bonn, Inst. Crop Sci. and Res. Conserv. (INRES) - Horticultural Science, Germany*

[2] *University of Bonn, Center for Development Research (ZEF), Germany*

Agro-climate services (ACS) play a vital role in supporting agricultural planning and practices in a climate change context. In many developing country settings, however, the effectiveness of ACS is compromised by "last-mile" problems that prevent smallholder farmers from receiving crucial information. To address this gap, CARE in Vietnam, a non-government organisation, has implemented three continuous projects since 2015 to improve the provision of actionable climate-informed agricultural advice to smallholder farmers. In these projects, weather forecasters, agricultural workers and farmer champions co-generate climate-informed agricultural advice. Farmer champions consist of village leaders and women group leaders who then transfer ACS to farmers via two platforms. On one platform, they meet farmers weekly as part of their saving and credit activity. On another platform, they meet farmers on an ad-hoc basis in conventional village meetings.

We sought to support the scaling of ACS by better understanding the factors that make farmers adopt in different platform settings. To do this we developed impact pathways of farmers' adoption of ACS through focus group discussions. We surveyed 82 randomly selected farmers to validate the impact pathways involved in ACS as well as ACS adoption dynamics. Farmers who met regularly were very likely to access, read, discuss, understand, positively perceive, intend to adopt and adopt ACS. Seventeen percent of these farmers remain to have challenges to understand ACS. In the meanwhile, fifty-one percent of farmers who did not meet regularly in village meetings had difficulties understanding ACS. Personal exchange did not resolve this difficulty but still had positive impacts on the decision to adopt. As a result, adoption rates gradually increased throughout the project, peaking at around 97 % in both farmer platforms after the five-year project intervention. Our results suggest that interpersonal relations play a crucial role in promoting the adoption and peer-to-peer scaling of ACS. For now, the role of the farmer champions is also critical to the success of the intervention. However, if the intervention is to be scaled up, the effort and resources required to engage farmer champions may be a challenge.

Keywords: Agricultural advice, climate, farmers, scaling

Contact Address: Thi Thu Giang Luu, University of Bonn, Inst. Crop Sci. and Res. Conserv. (INRES) - Horticultural Science, Auf Dem Hügel 6, D-53121 Bonn, Germany, e-mail: s7thluuu@uni-bonn.de

Carbon and Nitrogen Flows within Traditional Cocoa Agroforests in the Eastern Region of Ghana

Deogratias Kofi Agbotui, Mariko Ingold, Andreas Buerkert

University of Kassel, Organic Plant Production and Agroecosystems Research in the Tropics and Subtropics, Germany

Cocoa (*Theobroma cacao* L.) is an important export commodity which employs millions of small-scale farmers in the humid tropics. Growing cocoa in agroforestry systems under shade helps to diversify farmers' income, offers ecosystem services, and maintains soil fertility. Although West Africa produces more than 70 % of the world's cocoa, data regarding the influence of organic and conventional management on traditional agroforestry systems on nutrient cycling and soil fertility is scarce. Hence our study aims to determine the carbon (C) and nitrogen (N) flows within traditional cocoa agroforests. It was conducted in four villages near Suhum in the Eastern Region of Ghana, whereby two were under organic and two under conventional management. Within each village three farms were selected for litterfall collection, soil sampling and soil gas flux measurements using a portable photo-acoustic multi-gas analyzer (Innova 1312). In each of the 12 farms, soil was taken at depths of 0 – 10 cm and 10 – 30 cm for analysis of soil chemical and biological properties. The results showed no significant difference between organic and conventional management C and N fluxes through litterfall, which contributed 327 ± 22 kg C ha^{-1} $month^{-1}$ and 10 ± 0.7 kg N ha^{-1} $month^{-1}$ to the soil. However, soil organic C and total N were 46 % and 33 % higher in organic farms than in conventional farms. In addition, microbial biomass C (Cmic) and N (Nmic) in the topsoils of organic farms were two-fold greater than in those of conventional farms, while no management effects were observed in the subsoils. Soil organic C was positively correlated ($p < 0.01$) with N, Cmic and Nmic in all soil depths. Management had no effect on CO_2-C emissions. We conclude that soil fertility is higher in organic farms since they had higher SOC, total N and microbial biomass C and N.

Keywords: Carbon dioxide emissions, land use systems, soil fertility, soil organic matter, sustainability

---

Contact Address: Andreas Buerkert, University of Kassel, Organic Plant Production and Agroecosystems Research in the Tropics and Subtropics, Steinstraße 19, 37213 Witzenhausen, Germany, e-mail: buerkert@uni-kassel.de

Agriculture Is the Main Tropical Deforestation Driver, Responsible for Carbon Emissions and Ecosystem Service Losses

Tobias Seydewitz, David Landholm, Prajal Pradhan
Potsdam Institute for Climate Impact Research, RDII: Climate Resilience, Germany

Globally, deforestation results in a large amount of anthropogenic greenhouse gas (GHG) emissions, contributing to climate change substantially. Forest cover changes also have large impacts on ecosystem services. Deforestation is the dominant cause of land cover change in the tropical region. This land cover change relates to distinct causes recognised as direct deforestation drivers. Understanding these drivers requires a significant effort. The current studies on these drivers are mostly limited to coarse resolution. Further, GHG emissions due to deforestation are quantified only in biomass removal, while linking emissions from soil organic carbon (SOC) loss to deforestation is lacking. Additionally, deforestation impacts on ecosystem services are only assessed regarding loss without a closer picture of ecosystem service changes. Therefore, we analyse for 2001–2010: (1) the magnitudes of deforestation drivers, (2) the related carbon loss resulting from the removal of biomass and change in SOC, and (3) the ecosystem service value (ESV) change due to tropical deforestation. On the global scale, agriculture (90.3 %) is the primary deforestation driver, where the expansion of grassland contributed the most (37.5 %). The deforestation drivers differ in magnitude and spatial distribution on the continental scale. During our study period, the total carbon loss by biomass removal and SOC loss accounted for 8,797.32 Mt C and 1,185.16 Mt C, respectively. Furthermore, tropical deforestation caused a gross ESV loss of 408.6 billion Int.$/yr, while the net loss represented 63.2 billion Int.$/yr. Our findings highlight that agriculture represents a substantial contribution to global carbon loss and ecosystem service loss due to deforestation. Additionally, we show that the drivers differ in magnitude and distribution for different continents. Further, we highlight the underlying danger of putting the monetary value of nature by presenting estimates on the ESV gain through land cover change, together with the ESV loss.

Keywords: Carbon emissions, deforestation, drivers, ecosystem services

Contact Address: Tobias Seydewitz, Potsdam Institute for Climate Impact Research, RDII: Climate Resilience, P.O. Box 60 12 03, 14412 Potsdam, Germany, e-mail: seydewitz@pik-potsdam.de

Climate Variability and Adoption of Climate-Smart Agriculture - Implications for Sustainability in Zimbabwe

Beaula Fadzayi Chipoyera, William Nkomoki
Czech University of Life Sciences, Faculty of Tropical Agrisciences, Czech Republic

The Sustainable Development Goals of the United Nations, among other goals, advocate for climate action responses, the pursuit of zero hunger status, and poverty reduction. The concurrent variations especially, on the onset and duration of precipitation during farming seasons, are widely perceived to increase the risks of agricultural productivity and thereby threaten food security, particularly for smallholder farmers who mostly rely on rainfed cropping systems. Climate-smart agriculture practices offer alternative strategies for mitigating the effects of climate variability on agriculture and hence, safeguard sustainable agricultural production and food security both in the short and long term. However, little is understood regarding the decision-making processes of smallholder farmers when considering the adoption of climate-smart agricultural practices in Zimbabwe. Therefore, this study aimed to investigate the adoption of climate-smart agriculture by smallholder farmers in Zimbabwe. The study answered the following i) the role of gender in adopting climate-smart agriculture, ii) the main reasons for adopting climate-smart agriculture, iii) the primary information sources for promoting climate-smart agricultural practices. A survey of 112 smallholder farmers was conducted using a structured questionnaire from two distinctive Agro-ecological regions. The findings from the chi-square tests revealed a significant difference in the adoption of crop rotation and reduced tillage between male and female farmers. On the reasons for adopting climate-smart agriculture practices, respondents indicated soil protection accounting for 63.4 %, increased crop yields 34.8 %, and climate variability 25.9 %. On the other hand, extensional services, radios, and phones are the prominent information sources with 67 %, 51 %, and 44 %, respectively. Therefore, we recommend that policymakers consider the more vulnerable women groups in their support and educational awareness. Further, stakeholders must tailor information and knowledge about climate-smart agriculture practices towards farmers' preferential needs to enhance adoption and sustainability.

Keywords: Adaptation, adoption, climate-smart, gender, sustainability, variability

Contact Address: William Nkomoki, Czech University of Life Sciences Prague, Fac. of Tropical AgriSciences; Dept. of Economics and Development, Kamycká 129, 16500 Prague, Czech Republic, e-mail: nkomoki@ftz.czu.cz

Impacts of Climate Changes on Food Production in Brazil: What Are We Going to Eat?

Thaisa Toscano Tanus
Salgado de Oliveira University, Law Course, Brazil

Climate changes are one of the most complex challenges of this century, and no country is immune to their consequences. Defined by IPCC, climate changes refer to any change in climate over time, whether due to natural variability or as a result of human activity. Climate changes scenarios point to an average temperature increase of over 2 °C on planet Earth, including major imbalances in ecosystems that are fundamental for the survival of humanity. It is estimated that global warming will cause environmental, social, economic and cultural changes in urban areas. In the agricultural sector, there may be impacts throughout the food chain, leading to changes in food practices and food safety. The fact that Brazil is one of the largest food producers on the planet raises its level of importance in global discussions. The effects of climate changes in Brazil will be perhaps most damaging because of the country's historical vulnerability to natural disasters such as droughts, floods and landslides. There is also the prediction of greater frequency of extreme phenomena that can be harmful to agriculture. The agricultural scenarios point to a reduction in the cultivable area of low risk and high potential in Brazil, which could lose approximately 11 million hectares of land suitable for agriculture due to climate changes by 2030. All this could put food production in Brazil at risk if no mitigation and adaptation measures are taken. In this framework, an exploratory research was carried out on articles in Google Scholar, in order to elucidate the main impacts of climate changes on food production in Brazil in general. The preliminary results show that such impacts will be concentrated mainly in the poorest regions of the country. And, due to the country's history and high engagement with conventional farming, and the slow and insufficient rhythm of the process of transition to an agroecological model, there are strong indications that in the short and medium term the environmental horizon in Brazil will remain uncertain, with real threats to the climate, biodiversity and environment, directly influencing food production in the country and in the world.

Keywords: Brazil, climate changes, food production, food security, socioenvironmental vulnerability

Contact Address: Thaisa Toscano Tanus, Salgado de Oliveira University, Law Course, Rua 250 nÃ□Âºmero 59 Setor Coimbra, 74535350 GoiÃ□Â□nia, Brazil, e-mail: toscanothaisa@gmail.com

Traditional Ecological Knowledge Transmission. New Agricultures, New Religions, New Education

Helga Gruberg[1,2], Joost Dessein[2], Jean Paul Benavides[1], Alba Eliana[1], Marijke D´Haese[2]

[1]*Universidad Católica Boliviana San Pablo, Centro de Investigación en Ciencias Exactas e Ingenierías, Bolivia*

[2]*Ghent University, Dept. of Agricultural Economics, Belgium*

There is a growing interest in the study of traditional ecological knowledge due to its great capacity to contribute to the resilience of rural communities in the face of socio-ecological changes. The international academic community urge us to study the ways this knowledge is transmitted, transformed, and eroded in order to contribute to its revitalisation in favour of resilience. This knowledge is closely linked to indigenous peoples' worldviews and it is experienced and reproduced through practices and rites. This is the case of weather forecasting through the observation of natural indicators: phyto-indicators, zoo indicators, astronomical and atmospheric indicators. This study explores the transmission of this knowledge and the main socio-ecological factors that affect it. We present a case study in a rural municipality of Bolivia (Plurinational State) where data was collected during the period 2017–2021 through a series of transdisciplinary action-research experiences. The study followed a mixed methods approach with a combination of qualitative methods (i.e., participant observation, in-depth interviews, participatory community mapping and analysis of stories, myths, and narratives) and quantitative methods (surveys). Although technical reports argue that natural indicators are no longer used due to extreme climate variability from climate change, results show that there are other main factors affecting the transmission of knowledge: migration, education, agricultural restructuring, and politicisation of agrarian unions. Although the general outlook is discouraging, there are various revaluation actions in Bolivia that strengthen the bridges for the transmission of knowledge and aim to increase the resilience of rural communities, which can be used as a reference for other scenarios.

Keywords: Andes, hybrid knowledge, revitalisation, socio-ecological systems, TEK, transmission

Contact Address: Helga Gruberg, Universidad Católica Boliviana San Pablo, Centro de Investigación en Ciencias Exactas e Ingenierías, Calle Murguia 1835, Cochabamba, Bolivia, e-mail: hgruberg@ucb.edu.bo

Adoption of Climate Smart Agriculture among Smallholder Farmers in the Techiman Municipality, Ghana

Dorcas Gyan

University of Cape Coast, Dept. of Geography and Regional Planning, Ghana

Globally, agriculture is one of the major sources of livelihood for most rural communities. However, climate change over the years has been a problem to farmers in most rural communities. Due to the vulnerability of smallholder farmers to climate change, an adaptive and mitigating initiative known as Climate Smart Agriculture was introduced to help reduce the adverse effect of climate change and variability on agriculture. Climate Smart Agriculture (CSA) has been proposed as one of the best ways forward to mitigate the impact of climate change on agriculture. The benefits of Climate Smart Agriculture are hinged on three main pillars thus, to sustainably increase the income of farmers, reduce the emission of Green House Gases (GHG), and to strengthen resilience against Climate Change and Variability. Despite the benefits of Climate Smart Agriculture to farmers, literature has it that its adoption rate is low especially in the Techiman Municipality. Employing the Innovation Diffusion Model, this study was conducted to assess the adoption of Climate Smart Agriculture among smallholder crop farmers in the Techiman Municipality. The study will employ a pragmatic research design guided by both the interpretivist and positivist philosophies. 337 farmers were selected to represent 2,695 farmers that were trained by GIZ on CSA innovations, practices and technologies within the Techiman Municipality. Quantitative data was gathered by employing Simple Random sampling method whiles purposive and convenient sampling technique were used to gather qualitative data. The quantitative data was analyzed using T-test, regression and correlation. The qualitative data on the other hand was analyzed using the thematic analysis. All ethical issues that protect the researcher and participants were duly considered. The study revealed that the concept of Climate Smart Agriculture is widely known in the Municipality and its adoption rate is increasing with time. Farmers who adopted CSA were found to earn more income and were more food secured as compared to non-adopters. The study recommends that, government should consider the integration of CSA practices into the sector's policy and employ more extension officer to carry out such duties.

Keywords: Climate change and variability, innovation diffusion model

Contact Address: Dorcas Gyan, University of Cape Coast, Dept. of Geography and Regional Planning, Cape Coast, Ghana, e-mail: dorcas.gyan@stu.ucc.edu.gh

Methane Emission from Rice Production as Affected by Rice Variety Selection

Thi Bach Thuong Vo[1], Reiner Wassmann[2], Bjoern Ole Sander[3], Nha Duong Van[4], Folkard Asch[5]

[1]*University of Hohenheim, Inst. of Agric. Sci. in the Tropics (Hans-Ruthenberg-Institute), Germany*

[2]*Karlsruhe Institute of Technology, Germany*

[3]*International Rice Research Institute (IRRI), Vietnam Office, Viet Nam*

[4]*Kien Giang University, Vietnam, Agriculture and Rural Development Faculty, Viet Nam*

[5]*University of Hohenheim, Inst. of Agric. Sci. in the Tropics (Hans-Ruthenberg-Institute), Germany*

Rice production is a primary source of greenhouse gases (GHG) which is attributed to emission of methane generated in flooded soils. In order to meet the 1.5-/ 2-degree goal of the Paris Agreement, several studies have assessed the potential reduction of GHG emissions from rice production by changing farming practices namely water, nutrient and straw management. The impact of selecting different rice varieties, however, is still poorly understood. This study aims to quantify the mitigation potential of variety selection in combination with Alternate-Wetting-and-Drying, a water management proven to reduce GHG emissions. A field experiment has been conducted in the Mekong Delta, Vietnam in the early-year season of 2020 using the closed chamber method. The emission potentials of different rice varieties have to be quantified in form of a Scaling Factor (SF) to be used in the IPCC Tier[2] guidelines for regional extrapolations. Conceptually, the SF quantifies a deviation from the baseline emissions whereas SF's <1 correspond to lower and >1 to higher emission rates. Strictly speaking, regional emission estimates should also be specified for one 'Baseline Variety' (BV), but this aspect has previously been taken into account. This lack of information can largely be attributed to the cumbersome field sampling for a high number of rice varieties, i.e. we had 120 plots (based on 20 varieties, 3 replicates and 2 water managements) requiring sampling in weekly intervals. Due to the pronounced changes in plant physiology under AWD, it can be assumed that the variety effects will change as a function of water management. The available results confirm that the mean emission rates show large differences between two water management practices. Applying AWD reduced methane emission by more than 50 % across all varieties tested whereas inter-varietal differences were lower. Thus, we can already conclude that water management determines of the magnitude of GHG emissions and that variety selection modulates these emissions within a smaller range. Nevertheless, it is fairly easy for farmers to change rice varieties – as compared to changes in water management – so that this approach should be considered in view of optimising mitigation impacts.

Keywords: Alternate-Wetting-and-Drying, IPCC Tier 2, methane

Contact Address: Thi Bach Thuong Vo, University of Hohenheim, Inst. of Agric. Sci. in the Tropics (Hans-Ruthenberg-Institute), Garbenstr. 13, 70599 Stuttgart, Germany, e-mail: thibachthuong.vo@uni-hohenheim.de

Agro-Ecological Approaches in Climate Resilient Agriculture: Local Farming Models in Coastal Bangladesh

Shilpi Kundu[1], Rajesh S. Kumar[2], Bishwajit Kundu[3]

[1]*Griffith University, School of Environment & Science, Australia*
[2]*Member, Indian Forest Service(IFS), New Delhi, India, India*
[3]*Bangladesh Jute Research Institute, Jute Farming Systems, Bangladesh*

The dynamics at the farm level particularly among the smallholder agriculturists influence the implementation of Climate Resilient Agriculture (CRA) strategies in highly fragmented agricultural landscapes in South Asia countries. Challenges for CRA are further exacerbated in the coastal areas due to geographic, agricultural, socioeconomic, institutional and governance linked reasons as well as pre-dispositions. However, the availability of evidenced knowledge on CRA practices at the holding level reflecting on the key areas such as technology adoption, institutional and governance measures, socio-economics are vital in scaling up CRA practices to adapt to climate change at the landscape level. In the current paper, we analyse and assess the CRA practices in coastal Bangladesh through the window of the Farms of the Future framework. The study specifically focuses on CRA practices, institution and governance, socio-economic and gender aspects influencing the adaptation actions at farm unit level to landscape level. The case study was carried out using questionnaire surveys, practitioner interviews, key informant interviews, and focus group discussions in the coastal agricultural landscape of Bagerhat District in Bangladesh. We present our results of the analysis with recommendations cross-cutting the integral dimension of CRA such as a) sustainable natural resource management practice, b) adaptation of agricultural technologies, c) adaptation actions for productive social safety nets and d) adaptation actions to improve the institution and governance to manage climate risks. The results indicate that agroecological aspects have been potentially influencing the prospects for CRA in the study localities. However, we also noted the need for collaborative and social networking in community learning and technology adoption, closing the gaps in the availability of locally/regionally appropriate technologies, policy arrangements, institutional strengthening, and governance arrangements coupled with enhanced space for equity. We conclude by presenting a framework of suggestions across the critical domains of CRA to scale it up at the landscape level. We also hold that these findings have elements that can find relevance in promoting CRA in the coastal areas of Bangladesh, and in similar production landscapes in coastal South Asia.

Keywords: Climate resilient agriculture, coastal Bangladesh, farms of the future framework, food security

Contact Address: Shilpi Kundu, Griffith University, School of Environment & Science, 170 Kessels Road, Nathan, 4111 Brisbane, Australia, e-mail: shilpi.agro@gmail.com

Soil quality and soil health

Oral Presentations

Posters

A Soil Health Learning Platform in Malawi

Sieglinde Snapp[1], Regis Chikowo[2,1], Mateete Bekunda[3]

[1]*Michigan State University, Plant Soil and Microbial Sciences, United States*
[2]*University of Zimbabwe, Zimbabwe*
[3]*International Institute of Tropical Agriculture (IITA), Tanzania*

Agricultural development and sustainable soil management on smallholder farms is challenged by complexity, as well as high uncertainty and value conflicts. There is widespread interest in how local adaptation can address this challenge, through co-learning with farmers, hand-held sensors, and remote sensed data analytical approaches. We are testing approaches through a learning platform in Central Malawi http://globalchangescience.org/eastafricanode. This was established in 2013 with soil scientists, agronomists, geographers and agricultural economists of national and international institutions of the Africa RISING partnership in Malawi, supported by IITA and USAID. Our soil learning platform involves annual monitoring of 1000s of fields, participatory action research on integrated nutrient management and sustainable intensification, carried out with hundreds of farmers. Malawi government has released technologies such as the doubled up legume rotation for soil rehabilitation, based on this research platform. Recent findings highlight the role of crop diversification, weeds and residues in soil carbon status, conditioned by soil texture. This talk will highlight the new decision guide that is being scaled out by Malawi extension using handheld sensors to characterise soil carbon and soil profiles. An on-going question being addressed is the 'value add' and scalability of locally specific adaptive management versus traditional fertiliser recommendations. Overall, this platform is an opportunity to test technologies and extension methods, and to develop practical ways to link science to local knowledge. We hope that you will consider joining this learning platform with farmers and extension systems in Malawi, to answer your soil and crop nutrient response questions under real world conditions for smallholders.

Keywords: Action research, agroecology, farmer participatory, soil health

Contact Address: Sieglinde Snapp, Michigan State University, Plant Soil and Microbial Sciences, 1066 Bogue St., 48824 East Lansing, United States, e-mail: snapp@msu.edu

Integrated Soil Quality Assessment within two Contrasting Catchments of the Southern Ethiopian Rift Valley

Tizita Endale[1,3], Jan Diels[1], Dereje Tsegaye Salfako[2], Ann Verdoodt[3]
[1] *KU Leuven, Earth and Environmental Sciences, Belgium*
[2] *Arbaminch University, Dept. of Plant Science, Ethiopia*
[3] *Ghent University, Dept. of Environment, Belgium*

This research focuses on two science-based soil quality assessments, i.e. visual soil assessment (VSA) and standard physico-chemical soil analyses. The VSA, especially when conducted by farmers, offers a cost-effective approach to study soil in extended study areas. However, little is known on the applicability of VSA in areas subjected to different degradation processes.

Our study takes place in the lowland, midland, and highland agro-ecological zones (AEZs) of the Shafe and Sile-Elgo catchments, in the Southern Ethiopian Rift Valley. It aims to contribute to the state-of-art knowledge on VSA by (1) a comparison of scores assigned by experts versus farmers and (2) their correlation to standard physico-chemical analyses. Our integrated soil quality assessment is furthermore applied to (3) develop a contextualized understanding of soil quality under different soil management levels in the AEZs within this region.

To that aim, 36 cropland fields, six within each AEZ of both catchments, were selected. Half of these fields were collected in cropland receiving soil and water conservation (SWC) strategies ("good management"), while no SWC ("poor management") was applied in the others. On each field, the VSA was conducted by farmers and experts, followed by laboratory analyses of standard physico-chemical soil properties.

The overall VSA scores assigned by both groups were very strongly correlated (r=0.905, $p < 0.001$), albeit with higher correspondence in Shafe (r=0.940, $p < 0.001$) than in Sile-Elgo (r=0.807, $p < 0.001$). We observed differences between expert and farmer groups with respect to particular VSA indicators. The overall score of the farmers' VSA proved very strongly negatively correlated (r=-0.835, $p < 0.001$) to bulk density, which is an important soil quality indicator in all three AEZs. Aggregate stability, organic carbon and nitrogen contents correlated strongly to the VSA overall score. Both the VSA and physico-chemical data revealed a higher soil quality under good management.

Integrating local knowledge with science-based, systematic approaches of soil quality can thus provide a platform to exchange experiences, foster an in-depth understanding of degradation processes, and promote sustainable agricultural use.

Keywords: Agro-ecology, chemical soil properties, physical soil properties, soil management, soil quality, visual soil assessment

---

Contact Address: Tizita Endale, KU Leuven, Earth and Environmental Sciences, 3000 Leuven, Belgium, e-mail: tiztaendale@gmail.com

Biomass Contribution and Nutrient Balances under Different Organic Matter Management Practices on Smallholder Farming Systems

Deous Mary Ekyaligonza[1], Thaddeo Kahigwa Tibasiima[1], Bendicto Akoraebirungi[2], Bernhard Freyer[1]

[1]*University of Natural Resources and Life Sciences, Vienna (BOKU), Division of Organic Farming, Austria*

[2]*Mountains of the Moon University, Dept. of Agriculture, Uganda*

This study aimed at modifying the cropping systems through the application of comprehensive OMM practices to improve biomass and nutrient balances in annual farming systems including a combination of leguminous forage cover crops, alley crops, crop residues (stalk and leaves) and farmyard manure (FYM) into a rotation to improve soil fertility, nutrient balances and biomass development. A 2-year field experiment was set up in the Rwenzori region of Uganda following a randomised complete block design with five treatments replicated on ten smallholder farms. The treatments include T1: maize monocrop with DAP application at a rate of 50 kg ha^{-1} (9N - 10.12P kg ha^{-1}); T2: grain legume-maize rotation; T3: grain legume-maize/*Mucuna pruriens* rotation + FYM at a rate of 2.5 t ha^{-1} (13N - 6P – 18K kg ha^{-1}); T4: *Faidherbia albida* alleys with incorporation of leaves + grain legume-maize/*M. pruriens* rotation; and T5: *F. albida* alleys with incorporation of leaves + grain legume-maize/*M. pruriens* rotation + FYM. T2, T3, T4 and T5 are referred to as the OMM practices. For each treatment, the crop residues from the previous season were incorporated within the field. The results show that soil pH, water holding capacity, nitrogen, phosphorus and potassium levels increased as seasons progressed but were not significantly different between treatments. The recyclable biomass (dry weight of crop residues, *F. albida* leaves and *M. pruriens*) was significantly higher under T4 and T5. The highest recyclable biomass was observed in the fourth season under T4 and the amounts were 89.9 %, 89.3 %, 87.2 % and 41.9 % higher than T1, T2, T3 and T5 respectively. All treatments resulted into negative nutrient balances during the first three seasons. The negative nitrogen and potassium balances were more under *F. albida* treatments than in other treatments. This was due to the young age of the trees as the seedlings can only transfer less nutrients from the sub soil than the mature trees. In the fourth season, all the OMM practices resulted into positive N, P and K balances indicating that the OMM practices can cover the nutrient demand better than the non-organic treatment.

Keywords: Allley cropping, biomass, *Faidherbia albida*, farmyard manure, nutrient balance, organic matter management, smallholder farm

---

Contact Address: Deous Mary Ekyaligonza, University of Natural Resources and Life Sciences, Vienna (BOKU), Division of Organic Farming, Gregor-Mendel-Straße 33, 1180 Vienna, Austria, e-mail: mary.ekyaligonza@students.boku.ac.at

Investigating Synergistic Interactions between Selenium and Microbial-Based Biostimulants to Improve Drought Tolerance in Spring Wheat

Fahim Nawaz[1], Abdullah Al Mamun[1], Sadia Majeed[2], Markus Weinmann[1], Günter Neumann[1], Uwe Ludewig[1]

[1]*University of Hohenheim, Inst. of Crop Science, Germany*
[2]*The Islamia University of Bahawalpur, Dept. of Agronomy, Pakistan*

Drought stress is one of the major constraints to agricultural productivity worldwide related to global change, which requires the adaptation of farming systems to reduce this threat to global food security. Selenium (Se) is an important mineral nutrient that can stimulate plant growth and confer increased tolerance against environmental extremes including drought. Similarly, the use of microbial plant inoculants is discussed as a strategy to improve nutrient acquisition and stress resilience of crops. In this study, we investigated the potential synergistic effects of Se supplementation in combination with selected microbial consortia supplied in a protective organic fertiliser formulation (Minigran®), to mitigate the drought stress symptoms in spring wheat grown under controlled conditions in greenhouse studies. Initially, we optimised the Se (sodium selenite) level for application in wheat, which was already effective at extremely low application rates of 0.05 mg kg^{-1} soil. Our results showed that the combined Se application with microbial consortia considerably enhanced the drought tolerance in wheat through maintenance of leaf water potential (relative water content), regulation of phytohormones (abscisic acid, jasmonic acid, indole acetic acid), accumulation of stress-related osmoprotectants (proline, glycine betaine, total phenolics) and high activities of antioxidative enzymes (total antioxidant activity, ascorbate peroxidase) responsible for reduced oxidative leaf damage and increase in plant growth and biomass, compared with untreated controls but had no effects on the nutrient status. In conclusion, due to the low Se demand, selenate could be easily included in the formulation of the selected microbial inoculants providing perspectives for synergistic drought-protective effects.

Keywords: Microbial consortia, organic amendments, selenate, *Triticum aestivum*, water stress

Contact Address: Fahim Nawaz, University of Hohenheim, Inst. of Crop Science, Steckfeldstrasse 04, 70599 Stuttgart, Germany, e-mail: fahim5382@gmail.com

Soil Amendments Improve Crop Yields and Drought Tolerance on ex-Tin Mining Area

Rizki Maftukhah[1,3], Rosana Kral[2], Axel Mentler[1], Ngadisih Ngadisih[3], Murtiningrum Murtiningrum[3], Katharina Keiblinger[1], Michael Gartner[4], Rebecca Hood-Nowotny[1]

[1]*University of Natural Resources and Life Sciences, Vienna (BOKU), Inst. of Soil Research, Dept. of Forest and Soil Sciences, Austria*

[2]*University of Natural Resources and Life Sciences, Vienna (BOKU), Inst. for Development Research, Dept. of Sustainable Agricultural System, Austria*

[3]*Universitas Gadjah Mada, Fac. of Agricultural Technology, Dept. of Agricultural and Biosystem Engineering, Indonesia*

[4]*LVA GmbH, Lebensmittelversuchsanstalt (LVA), Austria*

Ex-tin mining areas in Bangka Island (Indonesia) are characterised by a high content of sand particles with low nutrient availability and low water holding capacity. To improve soil fertility and soil moisture of these areas, soil amendments application is important. We hypothesise that soil amendments will improve agricultural yields and drought tolerance. The study was established in 2018 on the ex-tin mining area in Bangka Island. Different soil amendments were applied in a randomised block design with a plot size of 2×2 m in four replicates. The soil amendments included (i) Lime, (ii) Compost; (iii) Charcoal; and combinations of (iv) Charcoal and Compost, and (v) Charcoal and sawdust. The soil was amended with 10 t ha^{-1} for the single amendments (treatments i-iii), and with rate 20 t ha^{-1} for combined amendments (treatments iv and v). Cassava (*Manihot esculenta*) was planted as the main crop and *Centrosema pubescens* used as a cover crop. Crop yield was determined at harvest. Plant material was dried, homogenized by milling and analysed for carbon isotope discrimination ($\Delta^{13}C$) to determine drought tolerance. In this study, the application of soil amendments has significantly improved crop yields. The combined treatment (charcoal and compost, iv) had the highest biomass of cover crop. While the other combined treatment (charcoal and sawdust, v) resulted in the highest cassava yields. The carbon isotope discrimination ($\Delta^{13}C$) was not significantly different among treatments in *Centrosema* plants. However, the application of soil amendments has slightly increased $\Delta^{13}C$ values. The combined treatment (iv) had the highest $\Delta^{13}C$ of *Centrosema*, which indicates a better drought tolerance. Significant differences were found in $\Delta^{13}C$ of cassava amongst treatments, where charcoal treatment showed the highest $\Delta^{13}C$ values. Yields varied with crop type and soil amendments, likely due to different nutrient requirements. Charcoal and its combination (such as compost or sawdust) were most beneficial to improve yields and drought tolerance in degraded sandy soil studied here.

Keywords: Carbon isotope discrimination, cassava, soil amendments

---

Contact Address: Rizki Maftukhah, University of Natural Resources and Life Sciences, Vienna (BOKU), Inst. of Soil Research, Dept. of Forest and Soil Sciences, Peter-Jordan-Straße 82, 1190 Vienna, Austria, e-mail: maftukhah.rizki@ugm.ac.id

Biochar to Enhance Nutrient Availability in *Theobroma cacao* Systems: A Greenhouse Trial

Johannes Meyer Zu Drewer[1], Mareike Koester[2]

[1]*Ithaka Institut for Carbon Strategies, Germany*
[2]*Georg August University of Goettingen, Tropical Plant Production and Agricultural Systems Modelling, Germany*

The agricultural sector with its strong dependencies on natural resources is inherently vulnerable to -and affected by climate change and, at the same time affects the climate system as a considerable contributor to greenhouse gas emissions. The application of biochar-based fertilisers (BBF) in a tropical agronomic context posses the potential of mobilising native soil-nutrients, enhancing the nutrient uptake of mineral fertilisers in conjunction with high carbon sequestration. Dominating, perennial agricultural systems, such as *Theobroma cacao*, are characterised by high production potentials of underutilised biochar feedstock. Yet nexus research targeting biochar application in *T. cacao* systems is limited. In 2020 greenhouse experiments were caried out to investigate the potential of BBFs to alleviate nutrient stress and sustainably intensify system productivity. Within the framework of a pot-trial the impact of BBFs onto T.cacao seedlings, planted in an Oxisol with critically low phosphorus (P) levels (slightly acidic, high aluminium), was investigated. 16 g of milled biochar were applied per pot - representing a 2 t ha^{-1} application rate. Four different soil amendments including BBFs, were deployed at three-placement levels (i.a. topsoil- and rootzone-application). The topsoil application of mineral fertiliser, i.e. farmer practice, served as the reference point for comparisons. The rootzone application of biochar, charged with mineral fertiliser increased the aboveground biomass, total leaf area and chlorophyll content index by 56 %, 222 % and 140 % respectively. To explain this observations mechanistically, the foliar stoichiometry was analysed as a proxy for soil-nutrient availability. The seedlings having received a rootzone application of nutrient charged biochar had significantly higher level of foliar P levels (+ 53 %) compared to farmer practice. The N:P ratio of the foliar tissue was optimised indicating the potential of BBF to alleviate P availability constrains as the systems limiting factor. A strategy of large scale BBF application to tropical perennial systems can contribute to achieving a range of sustainable development goals (SDGs). In the first place, by providing more income and higher yields to farmers in conjunction with high carbon sequestration while reducing the fossil-fuel dependence on industrial fertiliser production.

Keywords: Biochar, climate change mitigation, cocoa-agroforestry, nutrient-availability, phosphorus

Contact Address: Mareike Koester, Georg August University of Goettingen, Tropical Plant Production and Agricultural Systems Modelling, Grisebachstraße 6, 37077 Göttingen, Germany, e-mail: mareike.koester@uni-goettingen.de

Effect of Biochar and Legume Biomass on *Brachiaria brizantha* cv. *Xaraés* Growth Parameters in Benin

Amos Baninwain Nambima Dene[1,2], Rodrigue V. Cao Diogo[2], Thierry Dèhouégnon Houehanou[1,2], Birthe Paul[3]

[1]*University of Parakou, Laboratory of Ecology, Botany and Plant Biology, Benin*
[2]*University of Parakou, Dep. of Sci. and Techn. of Animal Prod. and Fisheries, Benin*
[3]*International Center for Tropical Agriculture (CIAT), Tropical Forages Program, Kenya*

Climate change and soil degradation are negatively impacting productivity of food and forage crops which may hinder animal production in West Africa. It is crucial to improve forage production in Benin with locally available inputs to ensure high quality nutrition of livestock and maintain and improve soil quality. We tested the effects of organic amendments on *Brachiaria brizantha* cv. *Xaraés* productivity, including biochar produced from corn cob biomass and green manure from *Gliricidia sepium* and *Mucuna pruriens* and their combination. The trial was installed at the experimental site of the Faculty of Agronomy, University of Parakou in Benin under a tropical sub-humid climate. The experiment was set up as a randomised complete block design consisting of three blocks, each containing eight experimental treatments: T0: *Brachiaria* alone; T1: Biomass of *Mucuna* 2 t ha^{-1}; T2: Biomass of *Gliricidia* 2 t ha^{-1}; T3: Mixture of the two legumes at 2 t ha^{-1} (1 t ha^{-1} *Mucuna* + 1 t ha^{-1} Gliricidia); T4: 300 kg dry matter (DM) ha^{-1} biochar; T5: 60 kg DM ha^{-1} biochar + 2 t ha^{-1} *Mucuna* ; T6: 60 kg DM ha^{-1} biochar + 2 kg ha^{-1} *Gliricidia*; T7: 60 kg DM ha^{-1} biochar + 2 t ha^{-1} (1 t ha^{-1} *Mucuna* + 1 t ha^{-1} *Gliricidia*). The growth parameters measured included plant height, leaf production, total biomass, leaf length and width, internodes, secondary roots, and tiller production, and were sampled on five randomly selected plants of *B. ruziziensis* per plot three months after planting. The results showed that the combination of biochar and green manure performed best (T7) with 57 % higher plants (103.8 ± 17.12 cm) compared to the control plots (T0) ($p < 0.05$). Moreover, the average number of secondary roots (18.47 ± 5.72); leaf length (40.63 ± 6.16 cm); leaf width (2.27 ± 0.29 cm); number of internodes (7.47 ± 1.19); number of tillers (3.07 ± 1.44), and number of secondary roots (3.80 ± 1.57) and total biomass (20.3 ± 5.65 t DM ha^{-1}) were significantly higher for the T7 treatment than un-amended and sole treatments ($p < 0.05$). We conclude that locally available organic amendments such as biochar and legume biomass can increase productivity of forage *B. brizantha* cv. *Xaraés*. More research is needed to determine the potential contribution of improved forage crops including Brachiaria hybrids on smallholder milk production and soil quality.

Keywords: *Gliricidia sepium*, livestock feeds, livestock production, *Mucuna pruriens*, organic amendments, West Africa

Contact Address: Rodrigue V. Cao Diogo, University of Parakou, Dep. of Sci. and Techn. of Animal Prod. and Fisheries, 123 Parakou, Benin, e-mail: dcao_bj@yahoo.fr

Selection of Carrier Materials for Biofertiliser Application into Seedlings in Organic Rice (*Oryza sativa* L.) Nurseries

Priyanga Dissanayake[1], Milani Gunathilake[2], Chandi Rajapaksha[3], Sanjeewani Ginigaddara[2]

[1]*Sustainable Agriculture Research and Development Centre, Dept.of Agric., Sri Lanka*
[2]*Rajarata Universit of Sri Lanka, Fac. of Agriculture, Sri Lanka*
[3]*University of Peradeniya, Fac. of Agriculture, Sri Lanka*

Nitrogen (N) is one of the most yield limiting factor of rice, alternative N sources should be supplied especially in organic farming systems. N fixing bacteria in rice rhizosphere show diurnal cycles that mimic plant behaviour, and tend to supply more fixed Nitrogen (N) during growth stages when plant exhibits a high N demand as a biofertiliser. Biofertilisers can be inoculated as seed dipping, seedling root dipping, soil and foliar application and as incorporated to pellets or capsules. As staple food crop in Sri Lanka, rice is growing mainly in submerged conditions, direct contact of inoculant with rhizosphere is more effective and losses can be minimised. Identification of efficient and effective way of inoculation using suitable carrier materials in to the root zone is important and this study was conducted to identify suitable carrier material for N biofertiliser; to be used in parachute method which is an ideal and cheeper seedling broadcasting method in organic rice cultivation. *Azotobacter* sp. was isolated from traditional rice variety: Madathawulu by using Ashby Mannitol media, was inoculated to rice seeds (Bg 250 - a two and half months variety) in parachute trays. Biofertiliser was mixed into carrier materials as rice soil, compared with cattle manure, compost and biochar as carriet materials (T1, T2, T3, T4, T5) and non-inoculated as (T6, T7, T8, T9) were used. The biofertiliser added and non-added treatments had a significant difference ($p < 0.05$) in plant height, dry shoot weight, plant nitrogen percentage and number of tillers per plant over the control (T0). Among three carrier materials, biofertiliser with biochar (T4) gained the highest significant ($p < 0.05$) mean number of tillers per plant (7) and higher average dry shoot weight (2.45g) and mean tissue nitrogen percentage (2.47 %) than other treatments. It can be concluded that biochar can be identified as suitable carrier material in parachute for biofertiliser with nitrogen fixing bacteria using parachute seedling trays for paddy cultivation and further field experiments are needed to validate results with growth and yield of rice with economic analysis.

Keywords: Biofertiliser, carrier materials, nitrogen, parachute method, rice

Contact Address: Priyanga Dissanayake, Sustainable Agriculture Research and Development Centre, Dept.of Agric., Makandura, Gonawila (NWP), 60170 Kurunegala, Sri Lanka, e-mail: priyangadmps@gmail.com

Bio-Organic Microbial Consortia with Soil Conditioner Promotes Drought Tolerance to Improve Tuber Yield of Potato

Abdullah Al Mamun[1], Günter Neumann[1], Uwe Ludewig[1], Markus Weinmann[1], Mithun Kumar Samardar[1], Fahim Nawaz[2], Aneesh Ahmed[1], Narges Moradtalab[1]

[1]*University of Hohenheim, Inst. of Crop Science: Nutritional Crop Physiology, Germany*
[2]*MNS- University of Agriculture, Multan, Pakistan, Agronomy, Pakistan*

Application of silicatic soil conditioner to improve soil water relationships under drought stress may be strategy to support rhizosphere establishment of beneficial microbial inoculants with drought-protective potential. In this study, we investigated drought-protective effects of plant-beneficial microbial consortia based on selected strains of Pseudomonas (Cons1), Burkholderia (Cons2) and Burkholderia + Trichoderma (Cons3) in combination with arbuscular mycorrhizal fungi (*Rhizophagus irregularis*) applied in protective organic fertiliser formulations (Minigran®1–3) with or without application of a silicatic organic soil conditioner (Sanoplant®) on plant performance and tuber formation of potato (hybrid) under controlled conditions in greenhouse experiments . Our results showed that at the end of a two-weeks recovery period from 6 weeks severe drought stress treatments at a soil moisture level of only 25 % soil water-holding capacity, all tested Minigran-Consortia formulations reduced the proportion of irreversibly drought-damaged leaves by 35–88 %, irrespective of the soil conditioner treatments, with the strongest expression of protective effects by the Cons3. Interestingly, already the Minigran blank formulations had a certain beneficial effect on enzymatic (ascorbate peroxidase) and particularly non-enzymatic (total antioxidants) detoxification of reactive oxygen species (ROS), indicated by a significantly reduced ROS (H_2O_2) accumulation in the leaf tissue, which was further improved by introduction of the microbial Consortia. Both, the Mingran blank formulations and the Minigran-Consortia formulations, affected the hormonal status in the leaf tissue towards increased ABA/Gibberellic acid (GA) ratios and increased IAA levels, known to support tuber initiation and tuber growth and increased the jasmonic acid (JA) concentrations involved in abiotic stress signalling. The Minigran blank formulations also increased the root colonisation with native arbuscular mycorrhizal fungi, which was further increased by inoculation with the Minigran-Consortia combinations and the soil conditioner. In contrast to the effects on vegetative growth, beneficial effects of the consortia on tuber biomass were mainly recorded in combination with the soil conditioner and even reached the values of well-watered controls with NPK fertilisation in case of the Consortium1. Our findings suggest that application of microbial consortia formulations in combination with silicatic soil conditioner has potential to improve the drought tolerance and tuber yield of potato.

Keywords: Drought, microbial consortia, organic fertilisers, Sanoplant, *Solanum tuberosum*, stress metabolites

Contact Address: Abdullah Al Mamun, University of Hohenheim, Nutritional Crop Physiology, Fruwirthstraße 20, 70599 Stuttgart, Germany, e-mail: mamun31130@gmail.com

Plant Growth Promoting or Inhibiting Effect of Mycorrhizal Fungi Applied with Sugarcane Ashes to Cuban Soils

Onelio Fundora Herrera[1], Pedro de la Fé Rodriguez[1], Katja Rodríguez Rodríguez[2], Georgina Gálvez Concepción[1], Bettina Eichler-Löbermann[3]

[1]*University of Santa Clara, Fac. of Agriculture, Cuba*
[2]*Reserch Institute of Tropical Plants, Santo Domingo, V.c., Cuba,*
[3]*Rostock University, Agricultural and Environmental Faculty, Germany*

In Cuba, sugar cane ash is produced in enormous quantities as a residue of the sugar industry. A lack of research prevents the adequate use of cane ash as soil fertiliser. In order to determine the effect of the application of sugar cane ash alone or accompanied with mycorrhizal fungi a field study and a pot experiment were carried out with two different soils (Cambisol and Ferralsol) and maize as indicator plant. The soil PK concentrations (extraction with H_2SO_4 0.1 N) were suboptimal. The sugar cane ash had a nutrient concentration of 0.45 % P and 2.11 % K. The mycorrhizal fungi (*Rhizoglomus irregulare* and *Glomus cubense* strains) were applied as commercial product EcoMic®. In the Cambisol the best result was obtained with an ash application of 10 t ha^{-1} followed by the chemical NPK fertilisation. The application of the mycorrhiza product together with 10 t ha^{-1} ash did not have any positive effect on the plant growth and in tendency even reduced biomass yields were measured. This can be explained by the high nutrient supply with the ash, which might have counterbalanced possible benefits of the mycorrhiza product while the symbioses with the fungi cost the plants carbohydrates. In the pot experiment with the Ferralsol, however, a remarkable yield effect of the mycorrhizal product was obtained when combined with a moderate quantity of ash corresponding to 5 t ha^{-1}. The maize biomass in this treatment was almost similar to that in the treatment with 10 t ha^{-1}ash alone and significantly higher than the biomass after application of 5 t ash per ha alone. The results showed a clear fertiliser effect of sugar cane ashes in both soils while the effect of the mycorrhizal product was divers, depending on soil characteristics and amount of ash supply.

Keywords: Biomass ash, fertiliser, mycorrhiza, sustainable agriculture

---

Contact Address: Bettina Eichler-Löbermann, Rostock University, Agricultural and Environmental Faculty, Justus-von-Liebig-Weg 6, 18059 Rostock, Germany, e-mail: bettina.eichler@uni-rostock.de

Manure Application Negatively Affected Arbuscular Mycorrhizal Fungal Diversity on Enset (*Ensete ventricosum*) Roots in Ethiopia

Gezahegn Garo[1], Maarten Van Geel[2], Fassil Eshetu[3], Rony Swennen[4], Olivier Honnay[2], Karen Vancampenhout[1]

[1]*KU Leuven, Dept. of Earth and Environmental Sciences, Belgium*

[2]*KU Leuven, Plant Conservation and Population Biology, Belgium*

[3]*Arba Minch University, Ethiopia, Biology, Ethiopia*

[4]*KU Leuven, Dept. of Biosystems, Belgium*

In low-input agricultural systems with low soil fertility, arbuscular mycorrhizal fungi (AMF) play an important role in plant nutrition, protection and water use. Evaluating how common agricultural practices in such systems affect the composition of AMF communities is therefore an important step towards sustainable intensification. Here, we characterised the AMF communities in enset roots in smallholder enset-based farming systems in South Ethiopia and assessed the effects of manure on those communities. We used Illumina MiSeq amplicon sequencing to quantify AMF diversity and community composition in the roots of 181 individual enset plants from 23 farms. Roots were collected from both the inner garden of the farm, which regularly receive manure applications and from the outer garden at the margin, which receive manure only very occasionally. AMF communities in both inner and outer gardens were comprised primarily by species belonging to Glomeraceae, which accounted for 67 % of the total operational taxonomic units (OTUs) recorded. Inner gardens receiving regular manure application were associated with higher soil pH, available P, organic carbon, total N and C:N ratio, and with significantly lower AMF richness and diversity. Whereas available P, total N and organic carbon explained the decrease in AMF richness and diversity, shifts in AMF community composition between inner and outer gardens were explained by soil pH and organic carbon. Therefore, intensified manure application enhances soil nutrient availability and soil organic carbon but results in lower enset AMF richness and diversity, and in a shift in AMF community composition. More optimal allocation of manure across the whole farm area may benefit both AMF communities and overall enset production.

Keywords: Enset, Glomeraceae, high-throughput sequencing, manure, mycorrhizal fungi

Contact Address: Gezahegn Garo, KU Leuven, Dept. of Earth and Environmental Sciences, Brussiles Street 159, 3000 Leuven, Belgium, e-mail: gezahegn_garo@yahoo.com

Root Habitat Is the Key Factor to Differentiate the Distribution of AMF of *Pueraria phaseoloides*

Guo Yaqin

University of Hohenheim, Inst. of Agric. Sci. in the Tropics (Hans-Ruthenberg-Institute), Germany

Arbuscular mycorrhizal fungi (AMF) represent one of the key determinants of plant performance and tolerance by providing a plethora of functional capacities such as access to immobile nutrients, suppression of pathogens, and resistance biotic and abiotic stress. Given these benefits, application of AMF has great potential to promote plant growth in ecosystem restoration as well as to improve soil health and ecosystem quality. Also, AMF establishment is considered to be a critical factor in the success of ecosystem restoration and for sustainable agriculture. However, A robust understanding of the structural composition of AMF in different microenvironments and the factors which shape these communities is still poor. Therefore, we conducted a study to analyse the variability and niche differentiation of AMF communities in the rhizosphere soils and the roots of *Pueraria phaseoloides* in Ghana. As study sites, we selected sites which were disturbed by mining activities and are characterised by spontaneous growth of Pueraria. We applied AMF-specific primer amplicon sequencing via Illumina MiSeq platform. The root habitat showed significantly smaller numbers of Amplicon sequence Variants (ASVs), a lower community richness, and lower within-sample diversity (□-diversity) compared to the rhizosphere soils of the plants. Non-parametric multidimensional scaling (NMDS) of weighted and unweighted UniFrac distance were performed to investigate patterns of separation between AMF communities. We found that the structural AMF composition of the rhizosphere is significantly different from the root. This is largely explained by the dominance of different AMF genera in these compartments, which is Rhizophagus in the roots and Acaulospora in rhizosphere. These results contribute to a better understanding of the complex host-AMF interactions of Pueraria as an important requirement for developing management options for ecosystem restoration (nutrient efficiency) and agriculture production by using specific AMF strains.

Keywords: Arbuscular mycorrhizal fungi, ecosystem restoration, plant-AMF interaction, sustainable method

---

Contact Address: Guo Yaqin, University of Hohenheim, Inst. of Agric. Sci. in the Tropics (Hans-Ruthenberg-Institute), Garbenstr.13, 70593 Stuttgart, Germany, e-mail: yaqin.guo@uni-hohenheim.de

Reuse of Pineapple Residue in Philippine Agriculture: Pot-Scale Determination of the Net Ecosystem C Balance for a CAM Plant

Reena Macagga[1], Shrijana Vaidya[1], Danica Antonijevic[1], Matthias Lueck[1], Marten Schmidt[1], Juergen Augustin[1], Sanchez Pearl[2], Mathias Hoffmann[1], Mihály Jancsó[3]

[1]*Leibniz Center for Agricultural Landscape Research (ZALF), Isotope Biogeochemistry and Gas Fluxes, Germany*

[2]*University of the Philippines Los Baños, Agricultural Systems Institute, College of Agriculture and Food Science, Philippines*

[3]*National Agricultural Research and Innovation Center, Hungary*

The Philippines is one of the world's leading producers of pineapples, wherein production is comprised mostly of small family farms that are less than 2 hectares in size. As by-product, they generate a large amount of plant residues (e.g., crowns and stems) that are commonly left at the edge of the field. This practice releases substantial amount of greenhouse gas (GHG) emissions and neglects the potential value of pineapple residue. Enabling a waste treatment by returning them to the field through incorporation or mulching holds the potential to maintain soil fertility, reduce climate impact, secure yield stability, and achieving a high resource efficiency by closing material cycles locally. It may also increase soil organic carbon stock (SOC) and reduce greenhouse gas (GHG) emissions. To date, however, the knowledge about this is still very sparse.

The rePRISING project aims to demonstrate that returning pineapple residue either through mulching or incorporation to the field may help promote the closing of nutrient-cycles (C/N/P/K) locally, thus helping to increase soil fertility and soil C sequestration, while reducing GHG emissions. Within the project, the recycling of pineapple residue together with various local organic and inorganic amendments will be studied during a two-year field experiment using the manual closed chamber method. The field study will be supplemented by pot-scale greenhouse and incubation experiments, used inter alia to determine baseline GHG emissions and carbon budgets of pineapple cultivation systems and residue treatments.

Here we present first results of a pot experiment performed during winter 2020–2021 used to develop a suitable procedure for the in-situ determination of dynamic net ecosystem C balances (NECB) for pineapple cultivation systems. This will be further utilised for upcoming field study. This is challenging in so far as pineapple plants use the Crassulacean acid metabolism (CAM photosynthesis) and the manual closed chamber method has not yet been applied to determine NECB from CAM plants.

Keywords: Carbon sequestration, climate change mitigation, greenhouse gas (GHG) emissions, nutrient-cycling, pineapple residue

Contact Address: Reena Macagga, Leibniz Center for Agricultural Landscape Research (ZALF), Isotope Biogeochemistry and Gas Fluxes, Müncheberg, Germany, e-mail: reena.macagga@zalf.de

Digested Horticultural Residues as Alternative to Conventional Fertilisers

Avilés-Tamayo Yanelis[1], Guardia-Puebla Yans[1], López-Sánchez Raúl[1], Yero-Montoya Alejandro[1], Pichardo-Lesme Dariannis[1], Mijail Bullain[1], Bettina Eichler-Löbermann[2]

[1]*University of Granma, Dept. of Agronomy, Cuba*
[2]*Rostock University, Agricultural and Environmental Faculty, Germany*

The use of digestates as fertiliser in agriculture provides many advantages in comparison to chemical fertilisers, like effects against fungi and insects, better water retention, pH buffering effects and lower oxidative stress. The present study aimed to evaluate the effect of digestates based on residues of horticultural crops on the growth and development of pepper (*Capsicum annuum*) under controlled conditions. Digestate : soil mixtures of 50:50, 40:60, 30:70, 20:80 and 10:90 were tested applying statistical process optimisation thought surface response methodology (RSM), to model the morphometric variables of growth of the pepper plant. Additionally, one treatment with compost and a treatment with mineral fertilisers (NPK) were established. Application of digestates had a positive effect on growth dynamics on all morphometric variables evaluated: plant emergence percentage, length of the roots and stem, number of leaves, fresh and dry root weights, and fresh and dry stem weights. The results of the statistical - mathematical models obtained by RSM show that a linear relationship between the proportions of digested applied and the length of the root and stem; meanwhile, the number of leaves and fresh and dry weights of the roots and stems had a non-linear behaviour. The use of the 10:90 ratio (digestate: soil) had better characteristics compared to the application of compost and mineral fertiliser. The results showed that the application of digested horticultural residues can provide adequate amount of nutrients and can have a better effect on plant growth than mineral fertilisers while contributing to nutrient cycles in agriculture. The use of the 10:90 ratio (digestate: soil) had better characteristics compared to the application of compost and mineral fertiliser. The results showed that the application of digested horticultural residues can positively influence the growth and development of the plant, as a soil improver and biostimulant, at the same time contributing to nutrient cycles in agriculture.

Keywords: Digestate, horticultural residues, plant growth

Contact Address: Bettina Eichler-Löbermann, Rostock University, Agricultural and Environmental Faculty, Justus-von-Liebig-Weg 6, 18059 Rostock, Germany, e-mail: bettina.eichler@uni-rostock.de

Alkalisation as Efficient Slurry Treatment Reduces Ammonia and Methane Emissions and Enables P-Recycling

Felix Holtkamp[1], Veronika Overmeyer[2], Joachim Clemens[3], Manfred Trimborn[2]

[1]*University of Bonn, Inst. Crop Sci. and Res. Conserv. (INRES), Germany*

[2]*University of Bonn, Inst. for Agricultural Engineering, Germany*

[3]*SF-Soepenberg Gmbh, Research and Development, Germany*

Slurry has become a substance of concern, especially in regions with high animal density. The conventional way of fertilising arable land with slurry is no longer feasible, as too large nutrient loads are applied per unit of area. The excessive addition of nutrients pollutes the soil, the groundwater, and releases high quantities of methane (CH_4), carbon dioxide (CO_2), and nitrous oxide (N_2O) into the atmosphere. Slurry management is responsible for 0.7 Gt CO_2-eq. a^{-1}, which is more than 25 % of the emissions generated by the whole livestock production sector.

Therefore, this study investigates aspects of an innovative and decentralised solution strategy that is based on the alkalisation of slurry to reduce emissions and to separate nutrients via precipitation processes. This involves the characterisation of buffer capacities that are present in the slurry to optimise the application of bases or alkaline acting additives. We have found that buffers functioning in the acidic milieu are subject to a microbial decomposition process, which paradoxically can lead to an increase in the consumption of bases in the alkaline milieu. Consequently, buffers must be considered as complex and interacting systems. Furthermore, alkalisation sanitises the slurry, which can inhibit or eliminate microbial activity, resulting in lower emission rates. This study showed that a pH of 10 reduces CH_4 and CO_2 emissions by about 99 % and N_2O emissions by about 60 % after a storage period of 8 weeks.

The alkalisation via calcium additives allows Ca-phosphates to be precipitated and separated so that they can be used as mineral fertiliser. Furthermore, the alkalized slurry can be combined with a stripping technology, allowing the rate of ammonia removal to be increased. The released ammonia can be further processed to create ammonium sulfate. Plant trials will finally be conducted with ryegrass (*Lolium perenne*) and maize (*Zea mays*) to evaluate the produced fertiliser and the treated slurry in terms of their value as a fertiliser and to determine if alkalized slurry poses a potential risk to these crops and the soil.

Keywords: Ammonia emissions, buffer capacities, fertiliser, greenhouse gas emissions, nutrient recycling, phosphorus elimination, slurry alkalisation, waste management

Contact Address: Felix Holtkamp, University of Bonn, Inst. Crop Sci. and Res. Conserv. (INRES), Karlrobert-Kreiten-Straße 13, 53115 Bonn, Germany, e-mail: holtkamp@uni-bonn.de

Perception and Determinants of Utilisation of Urban Household Organic Waste for Home Garden in Kumasi

NANA FENYI FORSON[1], MAAME ARKO ABABIO[2], FRED NIMOH[3]

[1]*Czech University of Life Sciences Prague, Fac. of Tropical AgriSciences; Dept. of Economics and Development, Czech Republic*
[2]*Audencia Business School, Department of Food and Agribusiness Management, France*
[3]*Kwame Nkrumah University of Science and Technology, Dept. of Agricultural Economics, Agribusiness and Extension, Ghana*

Households in urban towns and metropolitan areas can convert urban waste into resources such as organic fertiliser or compost for agricultural production to help reduce their cost of disposal, save resources, reduce health risk and improve productivity. Unfortunately, most households do not consider the waste they generate as a resource, but as waste with no economic value. Thus, the objective of the study was to assess urban households' perceptions and factors affecting their utilisation of organic solid waste for home gardens in Kumasi. A multistage sampling technique was used to select communities and respondents for the study. The stratified random sampling technique was also used to select 6 communities from each stratum since they were clustered into low, middle, and high-income groups. The visual reconnaissance survey and systematic random sampling approach was employed to select 180 households involved in home gardens from the communities. Urban households' knowledge was sought to explore their awareness on the use of organic solid waste and to investigate their perception and determinants of the reuse of organic solid waste for their home gardens. The results showed that most of the respondents were aware of the reuse of organic solid waste but few of them had adequate knowledge on the application of organic solid waste for home garden. It was found that urban households had an agreeing perception on the use of organic solid waste for home gardens. A binary probit regression model was employed to estimate the determinants of utilisation of organic solid waste for home gardens. Empirical results from the probit model showed that socio-economic characteristics such as age, occupational status and perception on the ability of organic solid waste to help improve the soil structure and perception of organic waste as an important resource had a significant influence on utilisation of organic solid waste for home garden. Insufficient knowledge on application was identified as the main constraint to the utilisation of solid waste for home gardens. The study recommends that educational and media institutions should educate urban households to increase the scope of knowledge on the use of organic solid waste.

Keywords: Organic waste, perception index, probit model, random-utility theory

Contact Address: Nana Fenyi Forson, Czech University of Life Sciences Prague, Fac. of Tropical AgriSciences; Dept. of Economics and Development, Olympijska 1903/3 Brevnov, 16900 Prague, Czech Republic, e-mail: fnanafenyi@gmail.com

Dynamics of Soil Microarthropod Populations Affected by a Combination of Extreme Climatic Events in Tropical Home Gardens of Kerala, India

Lakshmi Gopakumar[1], Francesca Beggi[2], Cristina Menta[3], Kumar Nallur Krishna[4], Jayesh Puthumana[5]

[1]*School of Environmental Studies, CUSAT, School of Environmental Studies, India*
[2]*Bioversity International / CIM, Bengaluru Office, India*
[3]*University of Parma, Dept. of Chemistry, Life Sciences and Environmental Sustainability, Italy*
[4]*Bioversity International, India*
[5]*Cochin University of Science and Technology, National Centre for Aquatic Animal Health, India*

Climate change is predicted to increase drought and flooding events in the coming decades, especially in the tropics. Soil microarthropods represent an important component of soil living communities and play a role in maintaining soil quality and health. We need to understand the short-term responses of both plant community and soil microarthropods populations to these extreme events. Yet most of our current knowledge on this comes from manipulative studies in temperate regions. Some microarthropod groups present specific adaptations to soil conditions which make them suitable candidates for estimating soil biological quality.
The present study reports the impact of a severe summer drought followed by a catastrophic flood in 2018 on soil microarthropod density and biological soil quality across 25 tropical home gardens in Southern India. Drought reduced the density of organisms six-fold and flooding caused a further threefold reduction. Biological soil quality in the home gardens that were not flooded was restored during the monsoon season but this was just partially possible for microarthropod population density. Three months after flooding, the microarthropod groups that we encountered were generally able to re-establish their population to the condition prior to flooding. However, none of the group recovered to the mean seasonal values. The severity of previous drought seemed to have overridden any possible adaptation to annual summer water scarcity. Plant species richness in home gardens played a key role towards mitigation and possible resilience to disturbance of these ancient agroecosystems. The capacity of the soil microarthropod communities to recover over the long term after extreme climatic events and its relevance to maintaining ecosystem functions and services in home gardens through soil fertility warrant further investigation.

Keywords: Drought, flood, plant community, QBS-ar index, resilience, soil fauna

Contact Address: Francesca Beggi, Bioversity International / CIM, Bengaluru Office, Bengaluru, India, e-mail: f.beggi@cgiar.org

Nitrogen Forms Differentially Affect pH and Response of Rice on Contrasting Soil Types

Jolibe Llido, Shyam Pariyar, Mathias Becker

University of Bonn, Inst. Crop Sci. and Res. Conserv. (INRES) - Plant Nutrition, Germany

Rice is grown in two known systems in the Philippines, irrigated which is characterised by the presence of water throughout the growing period and rainfed that depends solely on rainwater. Theshift in rhizosphere systems may occur under rainfed conditions from anaerobic to aerobic and vice-versa. These changes may create an environment, which can affect physicochemical properties of soils and consequently altering nutrients availability (macro andmicro) and their uptake. In this study, we hypothesized that the pH and the performance of rice will be affected differentially by the applications of different nitrogen forms on contrasting soil types. This study was conducted in the greenhouse condition of the Institute of Crop Science and Research Conservation in the University of Bonn, Germany. A pot experiment with three soil types with contrasting inherent pH (acidic, neutral and alkaline), three nitrogen forms (NH_4^+, NO_3^-, NH_4NO_3) and two rice genotypes (IR64 and Nipponbare). Bromocresol purple staining showed that rhizosphere of ammonium-fed rice was acidified with pH value of 4. Whereas, alkalinization was observed on the nitrate-fed rice with pH value of 6 – 7. All nitrogen forms were able to increase the pH level of acidic soil (from 4.0 increased up to 4.5) and neutral soil (from 6.9 increased up to 7.3). Shoot dry biomass of both genotypes were higher on acidic soil regardless of the nitrogen forms. Moreover, biomass of rice genotypes was lowest on alkaline soil and NO_3^--fed plants showed relatively lower dry biomass (10 % less) than other nitrogen forms in acidic soil. Rice on alkaline soil has the lowest uptake of Zn with less than 5 mg plant-1 and highest in acidic soil with more than 60 mg plant^{-1} in all nitrogen forms. This study showed that the inherent soil pH was influenced by the different nitrogen forms and thus affects rice performance.

Keywords: Ammonium, nitrate, nutrient management, *Oryza sativa*, system shift

Contact Address: Jolibe Llido, University of Bonn, Inst. Crop Sci. and Res. Conserv. (INRES) - Plant Nutrition, Karlrobert-kreiten Strasse 13, 53115 Bonn, Germany, e-mail: jolibellido@gmail.com

What Kind of Incentives Matter for Farmers' Adoption of Sustainable Soil Practices?

Maria Eugenia Silva Carrazzone[1], Juan Francisco Rosas[2], Jorge Sellare[1]
[1]*University of Bonn, Center for Development Research (ZEF), Economic and Technological Change, Germany*
[2]*University Ort, Uruguay*

Agriculture is key to the livelihoods of the population in developing countries and investments in this sector are often seen as effective mechanisms for economic growth and food security. However, increased agricultural production puts additional pressure on natural resources. Agri-environmental policies can contribute to a paradigm shift towards more sustainable agricultural systems, thus minimising the trade-offs between agricultural production and environmental conservation. Previous studies analysed the effects of diverse policy mechanisms on land use and on the adoption of sustainable agricultural practices. Yet, how different incentives are perceived by farmers and how they affect the success of these policies remains largely under-researched. In this paper, we analyse whether farmers' point of reference regarding their potential losses/gains (i.e. the context from which they face the incentive) and whether their "present bias" (i.e. preference for short-term gains) affect their compliance with a command-and-control policy. Our empirical analyses build on data from Uruguay's soil conservation policy. Under this policy, farmers must submit land-use plans for each harvest season. These policies can create trade-offs between agricultural outcomes and environmental goals. In the short term, these constraints lead to profit losses, as a less intensive soil use is promoted. In the longer term, a healthier soil can increase productivity and income gains. These data constitute a unique database for the assessment of decision-making on land use at the farmer level. The objectives of this paper are (i) assess how the "point of reference" influences the willingness of farmers to accept a profit loss; (ii) assess to what extent the "present bias" can prevent the effective implementation of the policy; and (iii) identify the implications of these findings on the agri-environmental policies. Using General Algebraic Modelling System (GAMS), we build models in different scenarios that represent different points of reference and time distributions of losses and gains. We expect to obtain results on how the context of crop prices can be used as an adequate point of reference and attenuate the "present bias" to promote higher incentives of the farmers to comply with the policy.

Keywords: Environmental sustainability, soil conservation, trade-offs

Contact Address: Maria Eugenia Silva Carrazzone, University of Bonn, Center for Development Research (ZEF), Economic and Technological Change, Genscherallee 3, D-53113 Bonn, Germany, e-mail: s7masilv@uni-bonn.de

How Integrated Land Use Affects Seasonal and Spatial Soil Moisture in the Brazilian Cerrado Biome

SARAH GLATZLE[1], SABINE STÜRZ[1], MARIANA PEREIRA[1], ROBERTO G. ALMEIDA[2], DAVI JOSE BUNGENSTAB[2], MANUEL C. M. MACEDO[2], MARCUS GIESE[1], FOLKARD ASCH[1]

[1]*University of Hohenheim, Inst. of Agric. Sci. in the Tropics (Hans-Ruthenberg-Institute), Germany*

[2]*EMBRAPA Beef Cattle, Integrated Production Systems, Brazil*

The Cerrado biome, constituting the native Brazilian semi-arid savannah vegetation types, has been replaced to a large extent by crop and pastureland in the past decades with pronounced effects on key agro-ecosystem functions. With regard to sustainable intensification of agricultural production, integrated systems, such as crop-livestock systems (ICL) or crop-livestock-forestry systems (ICLF) are discussed to serve as an alternative land use to improve or maintain soil health and water related functions. As compared to conventional pasture or mono-cropping systems, especially the introduction of trees via an integrated land-use raises the question if these systems hold the potential to function more akin natural Cerrado with regard to soil moisture dynamics and water cycles.

Our study compares measured soil moisture up to 100 cm depth for natural Cerrado forest (cerradão), continuous pasture (COP), ICL, and ICLF systems for almost two years on a long term monitoring site with different land use plots and relict Cerrado vegetation. In both dry and rainy season, mean soil moisture (SM) in the upper 100 cm was highest in ICL followed by degraded continuous pasture and the land use types including trees, ICLF and Cerrado (CER). However, spatial analysis of SM showed differences between soil layers and land use systems. While in COP and in ICL water was mainly taken up from the upper 30 cm, in ICLF, strongest soil moisture depletion was observed between 40 and 100 cm depth with SM values close to permanent wilting point (PWP) being reached during the dry season. In CER, SM was lower than in the other land use types in the upper 40 cm, but water was conserved below 60 cm depth. Compared to COP in both integrated systems, soil properties such as bulk density and soil organic carbon were improved, and biomass productivity increased indicating the benefits of both ICL and ICLF over the traditional grazing system. However, in comparison to CER, all secondary land use systems studied represent a significant intervention, both in terms of changes in soil properties and in water availability in the observed soil layers.

Keywords: Agroforestry, Eucalyptus, land use change, pasture, Savannah, soil water dynamics, water cycle

---

Contact Address: Sarah Glatzle, University of Hohenheim, Inst. of Agric. Sci. in the Tropics (Hans-Ruthenberg-Institute), Stuttgart, Germany, e-mail: sarah.glatzle@uni-hohenheim.de

Abiotic stresses in plants

Posters

Application of Soil Resistivity Measurements to Evaluate Subsoil Salinity in Rice Production Systems of the Vietnam Mekong Delta

Van Hong Nguyen[1], Jörn Germer[1], Nha Duong Van[2], Folkard Asch[1]

[1]*University of Hohenheim, Inst. of Agric. Sci. in the Tropics (Hans-Ruthenberg-Institute), Germany*

[2]*Kien Giang University, Vietnam, Agriculture and Rural Development Faculty, Viet Nam*

Rice is a major crop in the Vietnam Mekong Delta (VMD), contributing more than half the rice production of the whole country. Due to climate change induced sea level rise, soil salinity is increasingly threatening rice production in the VMD. Soil salinity in the Mekong Delta can be caused by saltwater intrusion into lowland areas through the canal system, or by capillary rise of salty water from near surface saline water table, both resulting in salt accumulation in the top soil. In order to develop appropriate management strategies for the adaptation of rice production systems of the Mekong Delta to climate change both in terms of water and salinity management, it is important to disentangle the two effects and their relative importance. Here we report on the possibility to use geoelectrical methods to evaluate the subsoil salinity threat to rice production. A case study was conducted in the Tra Vinh province of the VMD measuring subsoil electrical resistivity using an ARES II to a depth of 40 m to detect potential salinity threats to different rice production system. The electrical resistivity measurements were validated on a 1 m scale by drilling down to 40 meters depth and analysing soil type and electrical conductivity of the different layers. Preliminary results show no link between the production system (single, double or triple rice cropping) with the depth of the saline water table, but a clear link with the proximity to the sea. We discuss the applicability of the ARES II method on agricultural field scale and the implications for salinity management in VMD rice production systems.

Keywords: ARES II, electrical conductivity, electrical resistivity

Contact Address: Van Hong Nguyen, University of Hohenheim, Inst. of Agric. Sci. in the Tropics (Hans-Ruthenberg-Institute), Garbenstr. 13, 70599 Stuttgart, Germany, e-mail: van.nguyen@uni-hohenheim.de

Physiological Responses and Ion Accumulation of Twelve Sweet Potato Clones Subjected to Salinity

Shimul Mondal[1], Ebna Habib Md Sofiur Rahaman[2], Folkard Asch[1]

[1]*University of Hohenheim, Inst. of Agric. Sci. in the Tropics (Hans-Ruthenberg-Institute), Management of Crop Water Stress in the Tropics and Subtropics, Germany*

[2]*International Potato Center (CIP), Sweet potato breeding in the tropics, Bangladesh*

Salinity is an increasingly severe threat to coastal agriculture, particularly in the mega deltas of the Asian large rivers. In these systems, agriculture during the dry season often includes tuber crops such as sweet potato that face the threat of salt intrusion either from inherent soil salinity or through irrigation. Sweet potato varieties with a relatively high degree of salt tolerance are urgently needed for the future resilience of such systems. Little research has been done on identifying traits for salinity tolerance in sweet potato. In this study, we subjected 12 contrasting sweet potato clones from the Bangladesh Agricultural Research Institute to varying levels of root zone salinity (0, 50, 100 and 150 mM NaCl) for 21 days in a hydroponic setup. Vines were separated into young, medium, and mature sections and the respective leaves, petioles, and stems were sampled individually. Vine length, leaf number, leaf area, SPAD values, and dry weight were determined and all samples analysed for sodium, potassium, and chloride concentrations.

We present data for the contrasting responses of the twelve clones to salinity and the severity of the stress. Based on salinity induced dry weight reductions we analysed thresholds and slopes of genotypic responses to root zone salinity and identified promising clones. Sodium, potassium and chloride uptake and distribution were analysed in view of tissue age. Salinity, in general, reduced all morphological parameters by 12 %, 30 % and 66 % at 50mM, 100mM, and 150 mM, respectively. Genotypes contrasted strongly in the severity of these reductions. Clones identified as tolerant based on the threshold and slope analyses showed smaller reductions than the sensitive ones. Chlorine uptake was strongly correlated to sodium uptake. In general, sodium was taken up to all parts of the plant, with a tendency to accumulate in older tissues more strongly. Tolerant clones showed increased potassium concentrations in the younger parts of the vine. Based on these results, potential traits for resistance to salinity in sweet potato will be discussed.

Keywords: Cl- and K+ concentrations, dry mater, Na+, salt tolerant, tissue age

Contact Address: Shimul Mondal, University of Hohenheim, Inst. of Agric. Sci. in the Tropics (Hans-Ruthenberg-Institute), Management of Crop Water Stress in the Tropics and Subtropics, Garbenstr. 13, 70599 Stuttgart, Germany, e-mail: shimul.mondal@uni-hohenheim.de

Ectomycorrhizal Fungus *Scleroderma bermudense* to Improve the Salt Tolerance in *Coccoloba uvifera* (L.)

Mijail Bullain[1], Fatoumata Fall[2], Raul Lopez[3], Ludovic Pruneau[4], Amadou Ba[5], Bettina Eichler-Löbermann[6]

[1]*Granma University, Plant Biotecnology Study Center, Cuba*
[2]*Centre de Recherche de Bel-Air, Lab. Commun de Microbiologie, Senegal*
[3]*University of Granma, Plant Biotechnology Study Center, Cuba*
[4]*Université des Antilles, Laboratoire de Biologie Marine, Guadeloupe*
[5]*Université des Antilles, Laboratoire de biologie et physiologie végétales , Guadeloupe*
[6]*Rostock University, Agricultural and Environmental Faculty, Germany*

Coccoloba uvifera (L.) L. (Polygonaceaea), also named seagrape, is a woody plant often subject to high levels of salinity along the Atlantic, Caribbean and Pacific coasts of the American tropics and subtropics. It is an important tree for edible fruits and mushrooms, ornamental plantings and coastal windbreaks along Caribbean beaches and roadsides. We studied the effect of the ectomycorrhizal (ECM) fungus *Scleroderma bermudense* on growth, photosynthesis and transpiration rates, chlorophyll fluorescence and content, K/Na and Ca/Na homeostasis, and water status of two Cuban provenances of *Coccoloba uvifera* seedlings exposed to four levels of salinity (0.0, 54.7, 164.1 and 273.5 mM NaCl equivalent to 0.02, 5, 15 and 25 dS m $^{-1}$, respectively). The results indicated that *S. bermudense* improved the salt tolerance in seagrape seedlings. There were no differences in terms of growth performance, nutritional and physiological functions between the two ECM seagrape provenances in response to salt stress. The reduction of the Na concentration and the increase of K and Ca favoured a higher K/Na and Ca/Na ratio, respectively, in the tissues of the ECM seedlings. Furthermore, the beneficial effects of ECM symbiosis on the photosynthetic and transpiration rates, chlorophyll fluorescence and content, stomatal conductance and water status resulted in the improved growth performance of the seagrape provenances exposed to salt stress. Seagrape in symbiosis with *S. bermudense* only may benefit for the individual plant but, more importantly, may result in the development of ornamental plantings and coastal windbreaks along beaches and roadsides in Cuba.

Keywords: Chlorophyll, ectomycorrhiza, photosynthesis, seagrape, water status

Contact Address: Bettina Eichler-Löbermann, Rostock University, Agricultural and Environmental Faculty, Justus-von-Liebig-Weg 6, 18059 Rostock, Germany, e-mail: bettina.eichler@uni-rostock.de

Effects of Halo- and Hydropriming on Early Rice Seedling Vigour under Salinity

Anna Tabea Mengen, Kristian Johnson, Folkard Asch
University of Hohenheim, Inst. of Agric. Sci. in the Tropics (Hans-Ruthenberg-Institute), Germany

For the food security of a growing world population, rice production needs to increase significantly. Therefore, current production will need to be maintained, intensified, and in some cases expanded to include less suitable areas. Meanwhile a rising sea level, linked to climate change, leads to impaired field emergence and salt stress from tidal flooding and saline groundwater. Priming as a seed pre-treatment, which includes soaking of the seed in a solution and subsequent redrying, has been reported to improve field establishment and plant stress tolerance in diverse species under a variety of abiotic stresses. Here, the effect of Hydro- and Halopriming on germination under saline conditions was tested in a blind test of three varieties. First, the moisture uptake pattern of the seeds during priming was assessed under three NaCl-concentrations (0, 50, 100 mM). Afterwards, seeds were primed in these solutions, followed by a germination test under the three levels of salinity. It was found that the NaCl-concentration of the priming solution has no significant effect on moisture uptake. Furthermore, priming treatments were able to accelerate germination speed, with hydropriming showing the best results. In contrast, halopriming improved endosperm use efficiency depending on the variety.. These results suggest that priming can have a beneficial effect on germination, supporting rapid field emergence and field establishment under adverse conditions. Nevertheless, the underlying mechanisms have yet to be investigated and treatment × variety interactions, as well as the effect of different treatments on different germination parameters, need to be further evaluated to develop suitable priming methods applicable on a large scale. Priming has the potential to be a cheap and easy way to increase the productivity of rice production under unfavourable conditions and to help mitigate the effects of climate change.

Keywords: Germination, priming, rice, salinity

Contact Address: Folkard Asch, University of Hohenheim, Inst. of Agric. Sci. in the Tropics (Hans-Ruthenberg-Institute), Garbenstr. 13, 70599 Stuttgart, Germany, e-mail: fa@uni-hohenheim.de

Effect of High Temperature on Boro Rice Varieties at Flowering Stage under Different Soil Moisture Levels

Md. Tariqul Islam

Bangladesh Institute of Nuclear Agriculture, Crop Physiology Division, Bangladesh

Changing climate rises air temperature due to increasing concentration of CO_2 and other atmospheric greenhouse gases. The rise in atmospheric temperature causes detrimental effects on growth, yield, and quality of the crop varieties by affecting their phenology, physiology, and yield components. Boro rice is transplanted in January-February and usually faces high temperature (36–39 °C) at its reproductive stage in April-May. Flowering stage of rice is very critical for high temperature. High temperature may cause drying of pollen and stigma and ceasing pollen tube development for fertilisation. As a result, unfilled grains are produced. The experiment was carried out at BINA, Mymensingh, Bangladesh during December 2019 to May 2020 with three boro rice varieties. The objective of the study was to estimate proper soil moisture level at flowering stage to reduce high temperature effect. So, Binadhan-5, Binadhan-10 and Binadhan-14 were grown in pots each of 8 kg soil in ambient temperature and those were kept at 38 °C at flowering stage for 24 hours under different soil moisture levels (80 % field capacity (FC), 100 % FC and 2 inches standing water) in plant growth chamber. Then all the plants were again continued to maturity under sufficient soil moisture in ambient condition. The experiment was conducted in CRD with three replications. Data on photosynthetic parameters, yield and yield attributes were recorded. The results revealed that high temperature significantly decreased photosynthetic rate and yield and increased transpiration rate and unfilled grains. Higher transpiration rate maintained T leaf of 33–34 °C during T air of 38 °C. Better yield with less sterility was found in 100 % FC and standing water of 2 inches compared to 80 % FC. So, maintain 100 % FC or standing water of 2 inches at flowering stage of rice can reduce high temperature effect.

Keywords: Boro rice, flowering stage, high temperature, photosynthesis, transpiration, yield

Contact Address: Md. Tariqul Islam, Bangladesh Institute of Nuclear Agriculture, Crop Physiology Division, Bangladesh Agricultural University Campus, 2202 Mymensingh, Bangladesh, e-mail: islamtariqul05@yahoo.com

Impact of Drought Stress on Growth Traits of Cauliflowers and Broccoli

Fatemeh Izadpanah[1], Navid Abbasi[2], Forouzande Soltani[2], Susanne Baldermann[3]

[1]*Leibniz Institute of Vegetable and Ornamental Crops (IGZ) E.v., Plant Quality and Food Security, Germany*

[2]*University of Tehran, Horticulture, Iran*

[3]*University of Bayreuth, Food Metabolome, Germany*

Global climate change affects rainfall patterns and causing plants to suffer from drought, especially in arid and semi-arid regions. Drought stress, as a limiting factor, considerably impacts plant growth and development, and thus these parameters could be used to select drought-stress tolerant plants. With the global production of 26.9 million tons cauliflower (*Brassica oleracea* L. var. *botrytis*) and broccoli (*Brassica oleracea* var. *italica*) are among the most important vegetables of the Brassicaceae family. This study aimed to elucidate the responses of three different coloured cauliflower cultivars (white; Clapton, green; Trevi and purple: Di Sicilia Violetto), and one broccoli cultivar to deficit irrigation. One week after transferring the transplants in the greenhouse (University of Tehran), four levels of irrigation including 85–100 % (control, no stress), 65–80 % (low drought stress), 45–60 % (moderate stress), 25–40 % (severe stress) were applied, till the end of the experiment. At the end of the growth period, the whole plants were harvested and different growth parameters and yield were determined. In the control treatment (no drought stress) the floret fresh weight was higher in the white cultivar (~ 500 g) rather the coloured cultivars and broccoli (~ 300 g), but the deficit irrigation affected the yield in this cultivar (Clapton). The lowest yield was obtained by severe stress (25–40 % irrigation) in the white cultivar of cauliflower; while no significant decrease was observed in the other two cauliflower cultivars. It is expected that drought stress also reduces leaf size, stem extension, and root proliferation in plants. In this study, there is no significant difference revealed between all the treatments for root fresh weight, leaf area, and floret size, and only the stem fresh weight was significantly decreased by the lowest irrigation level in the white cultivar. Furthermore, severe stress (25–40 % irrigation) just reduced the plant height in the purple cultivar, but the leaf number was not changed compared to the other cultivars. In conclusion, the research demonstrates that cauliflowers and broccoli are likely drought-resistant vegetables. Further research should implement additional quality traits, such as micronutrients.

Keywords: *Brassica oleracea* L. var. *botrytis*, *Brassica oleracea* var. *italica*, irrigation, water deficit

Contact Address: Fatemeh Izadpanah, Leibniz Institute of Vegetable and Ornamental Crops (IGZ) E.V., Plant Quality and Food Security, Theodor-Echtermeyer-Weg 1, 14979 Großbeeren, Germany, e-mail: izadpanah@igzev.de

Leaf Wettability and Leaf Angle Affect Air-Moisture Deposition in Wheat for Self-Irrigation

Sadia Hakeem[1], Zulfiqar Ali[1], Muhammad Abu Bakar Saddique[1], Muhammad Habib-ur- Rahman[2,1], Sabah Merrium[1]

[1]*MNS University of Agriculture, Multan, Inst. of Plant Breeding and Biotechnology, Pakistan*

[2]*University of Bonn, Inst. Crop Sci. and Res. Conserv. (INRES), Germany*

Climate change and the ever-increasing world population affect the water sources and reduce irrigation water supply. Means of translating scarce water sources like air moisture has gained attention to be utilised for crop irrigation. The present study was designed to explore and architect the wheat plant for self-irrigation through exploiting the leaf surface structures like leaf hydrophilicity, leaf-stem angle supported by optimum leaf rolling. For this purpose, a set of 30 wheat genotypes and four local cultivars with all possible combinations of leaf angle (droopy >90°, semi-droopy; 60–90°, semi-erect; 30–60° and erect; <30°) and leaf rolling (abaxially, adaxially, and spirally rolled) was selected from diverse germplasm (1796 genotypes). The germplasm was characterised for the leaf traits, physiological traits (stomatal conductance, net photosynthesis, water use efficiency, transpiration rate), and soil moisture content at the tillering, booting, and the anthesis stage. The yield traits like spikelets per spike, ear weight, seed weight, and plot yield were also recorded. The climate parameters including soil and air temperature, solar radiation, rainfall, wind speed, and relative humidity were recorded by the sensors installed in an automatic weather station at the study site. The core set of 10 wheat genotypes was evaluated for the interception and water channeling by the measurement of water through stem flow and leaf wettability. The biplot, heat map, and correlation analysis indicated wide diversity and traits association. The soil moisture content and collected water through stem flow showed a substantial amount of moisture in the root zone (2–3.5 ml). The genotypes coded as 1, 7, and 18 with characteristics semi-erect to erect leaf angle, spiral rolling, and hydrophilic leaf surface provided that contact angle hysteresis less than 10° had higher soil moisture content (6–8 %) and moisture harvesting efficiency (3.5 ml). These findings can provide the basis for in-depth anatomical and molecular studies to develop self-irrigating and drought-tolerant wheat cultivars as an adaptation to climate change.

Keywords: Climate change, contact angle, hydrophilicity, leaf rolling

Contact Address: Zulfiqar Ali, MNS University of Agriculture, Multan, Inst. of Plant Breeding and Biotechnology, Chungi # 21 Old Shujabad Road, 66000 Multan, Pakistan, e-mail: zulfiqar.ali@mnsuam.edu.pk

Physiological Responses of Yacon *Smallanthus sonchifolius* under Chilling Stress

Uche Okafor, Iva Viehmannová, František Hnilička

Czech University of Life Sciences Prague, Fac. of Tropical AgriSciences; Crop Sciences and Agroforestry, Czech Republic

Yacon [*Smallanthus sonchifolius* (Poeppig Endlicher) H. Robinson, *Asteraceae*] is a root crop that originated in the Andes. This plant has myriads of potentials due to its antidiabetic and nutritious properties. It also represents a rich source of inulin-type fructooligosaccharides. This ongoing study aims at assessing the cold stress response of this frost-sensitive plant. To carry out the experiment rhizomes with vegetative eyes from one octoploid (2n = 8x = 58), and one dodecaploid (2n = 12x = 87) yacon genotype was selected and pre-cultivated under semi-controlled greenhouse conditions (natural light conditions, air temperature 20±2/15±2°C day/night, relative air humidity 65%-85 %). After pre-cultivation, plants in the four-leaf stage were transferred to a climabox with controlled conditions simulating cold stresses (5 ºC) for 14 days. Leaf samples were taken in regular periods to determine the leaf water potential and leave relative water content from control and treated (climabox) plants. A non-destructive physiological evaluation of the chlorophyll fluorescence and photosynthesis was also carried. Preliminary results show significant differences between the control and treated plants in all parameters tested. The dodecaploid genotype proved to have superior cold stress tolerance in terms of stomatal conductance and fluorescence quantum yield when compared to the other genotype. The genotype with a lower ploidy level was also not able to recover after the tress treatment. This study is imperative to decipher the strategies adopted by yacon plants in response to cold stress. It will also enable the identification of specific a yacon genotype as either stress escaper, stress avoider or stress tolerance. Understanding the cold stress response of yacon is imperative to further improve the breeding strategies of new resistance cultivars.

Keywords: Chlorophyll fluorescence, cold stress, photosynthesis, *Smallanthus sonchifolius*

Contact Address: Uche Okafor, Czech University of Life Sciences Prague, Fac. of Tropical AgriSciences; Crop Sciences and Agroforestry, 16500 Prague, Czech Republic, e-mail: okafor@ftz.czu.cz

Assessing Nitrogen Level in Maize (*Zea mays*) with Infrared Thermography

Rodrigo Beltrame, Shamaila Zia-Khan, Klaus Spohrer, Zichong Wang, Yang Zhang

University Hohenheim, Inst. of Agricultural Engineering, Tropics and Subtropics Group, Germany

Intensification of agricultural production due to the higher demand of a growing population will continue to increase pressure on natural resources. Because of the important role of N in various physiological processes, such as chlorophyll synthesis, numerous techniques aim to make nitrogen utilisation more efficient, ranging from breeding programs over biological nitrogen fixiation. In addition, deficiency of N will influence grain and plant weight, harvest index, leaf area index and crop photosynthetic rate thus, reducing the supply to the ear and ultimately lower the yields

Infrared Imaging has been widely used to detect crop water status. It is a promising method that aims to substitute direct measurements such as taking soil samples for laboratory analysis. These are time-consuming, mostly not applicable to large populations simultaneously, destructive, sometimes difficult to apply on fields, and simplify/accelerate possible responses to crop stress. In Infrared imaging increase of leaf temperature is directly related to to stomatal closure (cooling effect due to evapotranspiration). And this increase of temperature is captured through infrared camera and images were later analyzed.

Previous studies identified an inverse correlation of N fertilisation levels and canopy temperature in wheat (*Triticum aestivum* L.). The objective of this study is to detect nitrogen status as early as possible via infrared imaging on maize.. a greenhouse experiment is planned where thermal images of maize (*Zea mays* variety "Amadeo") plants will be taken throughout the growing stages. Four levels of N (50, 75, 100, 150 %) will be applied and time to time soil and leaves samples will be taken for the laboratory nitrogen analysis. Simultaneously, leaf stomatal conductance (gL) will be measured to provide information on the physiological status of the plants.

Keywords: Infrared thermography, nitrogen fertilisation

Contact Address: Rodrigo Beltrame, University Hohenheim, Inst. of Agricultural Engineering, Tropics and Subtropics Group, 70593 Stuttgart, Germany, e-mail: rodrigobeltrame@yahoo.com

Bacillus-Mediated Cross-Protection against Iron Toxicity and Brown Spot Disease (*Bipolaris oryzae*) in Lowland Rice

Tanja Weinand[1], Tom Schierling[2], Julia Asch[1], Barbara Kaufmann[2], Abbas El-Hasan[2], Ralf Thomas Voegele[2], Folkard Asch[1]

[1]*University of Hohenheim, Inst. of Agricultural Sciences in the Tropics (Hans-Ruthenberg-Institute), Germany*

[2]*University of Hohenheim, Dept. of Phytopathology, Germany*

The overall fitness of plants under stress conditions has been shown to be influenced by microorganisms. In addition to their own elaborate tolerance mechanisms against biotic and abiotic stresses, plants rely on interactions with their microbiome. Systematic identification of microbial strains providing cross-protection against multiple stressors and the dissection of host-microbe interactions under diverse stress conditions is essential for integrating such microorganisms in sustainable agriculture. Iron toxicity and brown spot disease incited by *B. oryzae* (telemorph: Cochliobolus miyabeanus) are important abiotic and biotic stressors which constitute great constraints to lowland rice (*Oryza sativa*) production worldwide. Excess uptake of iron caused by high concentrations of reduced iron (Fe II) in the soil can result in complete yield losses. Yield reduction of up to 90 % - as reported in the Bengal famine - can occur in areas affected by brown spot disease.

In the present study, three different Bacillus isolates (*B. pumilus* and *B. megaterium*) were tested for their ability to provide cross-protection against iron toxicity and brown spot disease in six different rice cultivars (IR75866–17-B-12-WAB1, Suakoko 8, IR31785–58-1–2-3–3, TOX 4004–8-1–2-3, CG 14, Sahel 108). To this end, plants were inoculated with Bacillus cell suspensions and, after one week, exposed to excess iron (1,000 ppm) for 8 days or inoculated with *B. oryzae*. The effects of bacteria treatments on the plant's response to both stresses were evaluated by leaf scoring and determination of dry weight.

Leaf scoring showed that the effects of Bacillus strains on brown spot development and iron toxicity tolerance significantly differ between rice cultivars and Bacillus isolates. In most cultivars, the effects of bacteria inoculation on brown spot expression differed from those on tolerance to iron toxicity. While application of *B. pumilus* strains suppressed brown spot disease in most cultivars, it increased iron toxicity leaf symptoms in naturally tolerant cultivars. In IR31875 however, *B. pumilus* provided cross-protection against both stressors. Varietal differences in changes in symptom expression following bacteria treatment will be related to the effect the different Bacillus isolates can have on general stress responses (e.g. ROS scavenger enzyme activity, lipid peroxidation) and iron homeostasis in selected cultivars.

Keywords: Bacillus, *Bipolaris oryzae*, Brown spot disease, *Cochliobolus miyabeanus*, iron toxicity, *Oryza sativa*

Contact Address: Folkard Asch, University of Hohenheim, Inst. of Agric. Sci. in the Tropics (Hans-Ruthenberg-Institute), Garbenstr. 13, 70599 Stuttgart, Germany, e-mail: fa@uni-hohenheim.de

Processing, quality and detection

Oral Presentations

Posters

Effect of Formulation on Physicochemical and Sensory Properties of African Nightshade Leafy Sauces

Amina Ahmed, Gudrun B. Keding, Elke Pawelzik

University of Goettingen, Dept. of Crop Science, Division of Quality of Plant Products, Germany

Postharvest loss of African leafy vegetables accounts for up to 50 % of the produced vegetables in Tanzania. The reasons for high postharvest loss include field heat, poor transportation and handling facilities, and lack of knowledge on how to preserve surplus leafy vegetables. As a result, there is low vegetable consumption during the offseason. Therefore, four green sauces were formulated from African nightshade with varied ratios of tomato (6 % and 12 %), carrot (0 %, 6 %, and 12 %), and baobab powder (0 % and 6 %). Moreover, peanut paste, and spices were added in the same proportions and the sauces were cooked at 87±3°C for 20 and 30 minutes. The sauces were assessed in Morogoro region, Tanzania for total soluble solids (TSS), pH, viscosity, and sensory properties using quantitative descriptive analysis (QDA). The panel for the QDA were also used to rate the total liking of each formulation in order to select the best two formulations for the consumer acceptance test. The consumer acceptance test was conducted using a hedonic scale. The results shown that TSS and pH of the fresh and cooked sauces ranged from 6 to 17, and from 3.8 to 5.0, respectively. Generally, formulations had significant effect on TSS, pH, viscosity and sensory descriptors of the sauces, while cooking time had less effects. The QDA scores show a positive correlation between sourness and pH, mouthfeel and viscosity of the sauces. Moreover, QDA scores shown that an increased level of carrots resulted in increased roughness (mouthfeel) and colour intensity, whereas sourness and total liking scores were decreased., The consumer acceptance test showed high acceptability of all sauces in terms of colour, smell, consistency, mouthfeel, sourness, and total liking. The addition of the baobab powder and increased tomato content had a positive effect on the acceptance of the African nightshade sauces. Therefore, the production of sauces from African nightshade leaves with particular ingredients has the potential to reduce postharvest loss, increase food security, as well as creating market opportunity to increase household incomes for both rural and urban communities in Tanzania.

Keywords: African nightshade, sauces, sensory properties, viscosity

Contact Address: Amina Ahmed, University of Goettingen, Dept. of Crop Science, Division of Quality of Plant Products, Carl-Sprengel Weg 1, 37075 Goettingen, Germany, e-mail: aahmed1@gwdg.de

Design and Performance Evaluation of a Solar Hybrid Dryer for Cassava Breadflour in Benin

Schadrac Don de Dieu Agossevi[1], Thierry Godjo[2], Joseph Dossou[1]

[1]*University of Abomey-Calavi, Faculty of Agricultural Sciences, Benin*

[2]*University of Science, Technology, Engineering and Mathematics, Doctoral School of Science, Technology, Engineering and Mathematics, Benin*

Cassava is a well-known crop and a staple food in Benin. The cassava cossette derivative is used in addition to wheat flour in breadmaking. This research aims to design a hybrid solar dryer. A diagnostic survey was carried out by cassava processing cooperatives to identify the difficulties of producing good quality cossette and by the Ouidah Pain bakery (which incorporates 30 % cassava flour in the production of bread) to collect its needs in bread cossettes. Following the need analysis, a dryer meeting the expected functions was designed and tested. The dryer operating temperatures and relative humidity of the air in the drying chamber were respectively 65°C and 10 % during the day and 32°C and 88 % at night. A solar collector and a heat exchanger were used for air heating. The physico-chemical and food microbiological analyses of dried cassava chips samples were carried out. The cassava cossettes obtained from 1 mm, 3 mm and 5 mm thick flakes had a final moisture content of 9.65 $\pm$ 1.23 %, 12.05 $\pm$ 0.66 % and 14.52 $\pm$ 0.71 % respectively, compared to 14.10$\pm$0.31 % for those from the cooperatives in the wet period. In the dry period, the same cossettes have a final moisture content of 5.47$\pm$0.91 % and 9.10$\pm$4.96 % compared to 12.36 $\pm$ 0.65 % for the cooperatives cossettes. All the cossettes respect the normative values of the Beninese standard 03.06.006 of CEBENOR with water content (< 12 %), luminance (L > 93) and clarity (ΔE < 12) except for 5mm which has water content higher than 12 %. On the other hand, the acidity level of the cossettes, between 0.06 and 0.09, is slightly higher than the normative value (0.05). The sanitary quality of the 1mm cossettes in terms of yeasts and moulds < 10 g CFU is in line with the standard. The high presence of total mesophilic aerobic germs (2.5.106 $CFUg^{-1}$) had a negative impact on the cossettes during the rainy season due to airborne contamination of the air entering the dryer. It is therefore imperative to focus on air filtration techniques and thermal regulation of the hybrid solar dryer.

Keywords: Benin, cassava, cossettes, flakes, solar hybrid dryer

Contact Address: Thierry Godjo, University of Science, Technology, Engineering and Mathematics, Doctoral School of Science, Technology, Engineering and Mathematics, B.P. 2282 Goho, 01 Abomey, Benin, e-mail: thierrygodjo@hotmail.com

Foam Mat Drying of Yellow Cassava Using Egg Albumin and Properties of the Associated Products

Oluwatoyin Ayetigbo[1], Sajid Latif[1], Christina Boura[2], Joachim Müller[1]

[1]*University of Hohenheim, Inst. of Agricultural Engineering, Tropics and Subtropics Group, Germany*

[2]*University of Hohenheim, Environmental Protection & Agricultural Food Production, Germany*

The main challenges of the use of fresh cassava as food are the rapid post-harvest deterioration and the toxicity due to the presence of cyanogenic glucosides. Foam-mat drying (FMD) was considered to solve these problems. Response surface method was used to optimise the foaming variables. Five independent variables were considered including pulp concentration (20, 25, 30 %), holding time (1, 3, 5 days), concentration of foaming agent (egg albumin) (2, 16, 30 %), concentration of stabiliser (NaCMC) (0.1, 0.8, 1.5 %), and whipping time (1, 3, 5 min). Two responses were measured: foam overrun (FO) and air volume fraction (AVF). The optimum foaming variables were achieved when pulp concentration was 20 %, holding time was 1 day, concentration of egg albumin was 30 %, concentration of NaCMC was 0.1 %, and whipping time was 5 min. The predicted and validated FO were 33.9 % and 30.2 %, respectively, and the predicted and validated AVF were 0.25 and 0.23, respectively. Drying kinetics at 70 °CC showed that Page model was the best model to describe moisture removal ratio, and cassava foam dried faster than the non-foamed cassava pulp. The total carotenoids content (TCC) reduced significantly with FMD. The cyanogenic glucosides content (CGC) of the cassava foam powder was 10.6 $\mu g\, g^{-1}$. The foam air bubbles size and the pulp starch granules size had slight positive skew probability distribution. Luminosity (L*) increased with FMD. The water absorption capacity (WAC) of the foam powder was 0.89 gg^{-1} and the oil absorption capacity (OAC) was 0.73gg^{-1}. The swelling power (SP) and solubility of the foam powder was 966.9 % and 38.2 % respectively. The foam powder required a least gelation concentration of 6 % w/v and the pulp powder required 20 % w/v. Freeze thawing led to syneresis of 38.7 % water from the foam powder and 17.2 % from the pulp powder. The viscosity of the foam powder and the pulp powder at different temperatures (25, 55, 85 °C) and shear rates (100, 300, 500, 700, 900 s^{-1}) showed that the powders had pseudoplastic, non-Newtonian rheology.

Keywords: Carotenoids, cassava foam, egg albumen, foam mat drying, viscosity

Contact Address: Oluwatoyin Ayetigbo, University of Hohenheim, Inst. of Agricultural Engineering, Tropics and Subtropics Group, Garbenstr. 9, 70599 Stuttgart, Germany, e-mail: ayetigbo_oluwatoyin@uni-hohenheim.de

Hyperspectral Imaging and Chemometrics for Colour and Nutrients Determination During Hot Air Drying of Cocoyam

John Ndisya[1], Duncan Mbuge[2], Arman Arefi[1], Ayub Gitau[2], Liliana Bădulescu[3], Elke Pawelzik[4], Oliver Hensel[1], Barbara Sturm[1]

[1]*University of Kassel, Agricultural & Biosystems Engineering, Germany*
[2]*University of Nairobi, Environmental & Biosystems Engineering, Kenya*
[3]*University of Agronomic Sciences and Veterinary Medicine of Bucharest, Dept. of Bio-engineering of Horti-Viticultural Systems, Romania*
[4]*University of Goettingen, Dept. of Crop Science, Division of Quality of Plant Products, Germany*

Cocoyam is a root crop grown in the tropical and subtropical regions of the world for food. The crop is valued for its starchy corm and micronutrient composition of minerals, vitamins and bioactive compounds. Hot air drying of the starchy corm is commonly practised for preservation and value addition. However, hot air drying substantially changes the colour and nutrient composition of food materials. Therefore, adequate process monitoring and dynamic adjustment of the drying process are critical to avoid degradation of material properties and the loss in final product quality. In this study, hyperspectral imaging (HSI) and chemometrics were successfully applied to develop models for online monitoring of colour and chemical attributes during hot air drying of purple-speckled Cocoyam. Drying experiments were undertaken with purple-speckled Cocoyam slices of 4 mm thickness. The slices were dried in a cabinet dryer at a temperature of 60 °C until a final moisture content of 0.11 $^{kgW}/_{kgDM}$. Hyperspectral images were acquired during drying using a camera system operating in the spectral range 400 – 1700 nm. Samples were drawn during the drying process every 30 minutes for measurement of colour and chemical content using standard laboratory techniques. The colour attributes measured included CIELAB L*, a* and b*, browning index (BI), whiteness index (WI), chroma (C*) and hue angle (h°) while chemical attributes included total phenolic content (TPC), total flavonoid content (TFC) and the total antioxidant activity (TAA). The acquired HSI images were processed to derive spectral absorbance and reflectance. The spectral data were then split into calibration and validation datasets and pre-processed by taking the first and second derivatives. Prediction models were built using PLS regression and significant wavelengths were selected based on the regression coefficients. The performance of the PLSR models was assessed using the Root Mean Square Error (RMSE), coefficient of determination (r^2) and Residual Prediction Deviation (RPD). The agreement between predictions and observations was assessed using Huber regression and Bland-Altman analysis. The measures of performance and agreement applied revealed good comparability of prediction results to the observations on BI, b*, C*, h°, TFC and TAA collected using standard laboratory measurements.

Keywords: Antioxidant activity, colour, drying, flavonoids, food quality, phenolics

Contact Address: John Ndisya, University of Kassel, Agricultural & Biosystems Engineering, 37213 Witzenhausen, Germany, e-mail: jndisyas@gmail.com

Development and Implementation of Biobased Packaging Solutions in West African Food Supply Chains to Increase Food Security

Barbara Götz[1], Mathias Hounsou[2], Sylvain Dabadé[2], Antonia Albrecht[1], Fernande Honfo[2], Harold Hounhouigan[2], Djidjoho Joseph Hounhouigan[2], Judith Kreyenschmidt[1]

[1]*University of Bonn, Inst. of Animal Sciences, Germany*
[2]*University of Abomey-Calavi, School of Nutrition, Food Science and Technology, Benin*

Food security and food losses are of major concern in sub-Saharan Africa. Active packaging solutions can prolong shelf life and increase food safety by inhibiting the growth of spoilage microorganisms and pathogens. The reduction of food losses can help to prevent malnutrition. In addition, the environmental pollution, e.g. by plastic bags etc., is a severe issue in many African countries nowadays. Therefore, the implementation of locally produced, biobased and biodegradable packaging solutions is of great interest, especially in developing countries like Benin.

The aim of the WALF-Pack project was the development of simple and locally producible packaging solutions for food products from Benin, namely the green leafy vegetable Gboman (*Solanum macrocarpon*), smoked chicken and the soft cheese Waragashi. Therefore, a holistic interdisciplinary approach was chosen. First, the supply chains of all products were investigated and characterised in order to identify the hot spots of food loss. A coordination framework of local farmers, suppliers, wholesalers etc. was established for regular exchanges upon actual developments. One main focus of the research addressed the identification of the antimicrobial potential of local plants for the incorporation into food packaging in order to prolong the shelf life. Afterwards, packaging solutions were designed and local materials were chosen to create prototypes. The suitability of the packaging was investigated in product and pilot studies. The consumer acceptance was taken into account and finally all results were made accessible in an online toolkit.

The packaging solutions include papers from different grasses, bioplastic based on PLA, local oils and fibers with an antimicrobial active additive and a biogenic active coating that can be applied on different matrixes like cloth, paper or banana leafs. The active coating and packaging can reduce microbial growth and therefore prolong the shelf life of the product.

Therefore, these packaging solutions can reduce food loss and increase food security and safety when implemented at the food loss hot spots. In addition, the sustainable biobased packages can be produced locally and are biodegradable or easy to dispose. Therefore, they will also have an environmental impact in Benin and contribute to a sustainable packaging and waste management.

Keywords: Active packaging, antimicrobial activity, Benin, biobased, biodegradable, food loss, food security, packaging, sustainability, Western Africa

Contact Address: Barbara Götz, University of Bonn, Inst. of Animal Sciences, Katzenburgweg 7-9, 53115 Bonn, Germany, e-mail: bgoetz@uni-bonn.de

Operational Improvement of a Convective Coffee Dryer by Numerical Methods and Computational Fluid Dynamics

Eduardo Duque-Dussan, Jan Banout

Czech University of Life Sciences Prague, Fac. of Tropical AgriSciences; Dept. of Sustainable Technologies, Czech Republic

Considering that the coffee harvest peaks in Colombia coincide with the rainy seasons, the coffee growers face many complications when attempting to sun dry their coffee due to the high cloudiness and low direct sun radiation; deriving in post-harvest processes delays or incomplete grain dryness, risking the quality, innocuousness and safety of the product. That is why local workshops started to fabricate low capacity mechanical dryers simulating the industrial coffee drying equipment's working principles. One of the most commercialised units is a triple tray rectangular-shaped convective dryer with a net drying capacity of 31.25 kg of dry parchment coffee per 21-hour batch, providing an acceptable solution to the drying concerns. However, the process seemed improvable, focusing mainly on the dryer's geometry, air inlet and coffee bed thickness, seeking to enhance the airflow inside the unit, and reducing the product's drying time. Hence, a new dryer design was proposed displaying a circular shape, a lower grain bed thickness and a vertical air inlet accompanied by a diffusive tray. Based on the Thompson and Michigan State University mathematical models for grain drying, both units were simulated to obtain their theoretical drying time. A computational fluid dynamics simulation was also conducted to observe the drying air's behaviour inside the units; the circular drying unit presented a notable theoretical reduction in the drying time. It also displayed a more homogeneous and uniform air distribution, optimising the dryer's performance, deriving into improving the grower's profitability, dynamising the post-harvest processes to avoid product spoilage and reducing the electrical and gas consumption.

Keywords: CFD Simulation, coffee drying, parchment coffee, porous bed

Contact Address: Eduardo Duque-Dussan, Czech University of Life Sciences Prague, Fac. of Tropical AgriSciences; Dept. of Sustainable Technologies, Kamýcká 129, 16500 Prague, Czech Republic, e-mail: duque_dussan@ftz.czu.cz

Investigation of Air Flow Resistance for Maize Cobs Bulk Using an Automatic Test Ring

Janvier Ntwali, Ziba Barati, Joachim Müller

University of Hohenheim, Inst. of Agricultural Engineering, Tropics and Subtropics Group, Germany

Maize production in Rwanda has increased in recent years, increasing the need for proper postharvest technologies. Drying is the main step in maize postharvest with maize planting seeds being preferably dried on cobs at low temperatures. The need to minimize the dryer energy consumption and increase the maize quality necessitates optimization of the drying efficiency through aeration to improve the airflow distribution. A pressure drop automatic test ring for a batch dryer was developed at the Institute of Agriculture Engineering in the Tropics and Subtropics with the aim to investigate the airflow resistance of products during drying. In this research, air flow resistance of maize cobs dried in a batch dryer was assessed and the moisture content was monitored. Three batches of 210 kg maize cobs were dried at a constant temperature of 35 °C. The dryer is equipped with pressure sensors at 0.1 to 1.15 m of the bulk height and the ventilation was regulated to vary from 0.1 to 1 m s^{-1} air velocity in the bulk. Moisture content of maize in different positions was monitored throughout the process. The results show that it took approximately 35 hours to dry maize cobs from 25 % to less than 14 % moisture content wet basis. Differences in pressure were recorded at different heights of the maize bulk for the velocity of 1 m s^{-1} with a maximum static pressure of 329 Pa recorded at the bottom of the bulk and a minimum of 55 Pa at the height of 1.115 m in the maize bulk at the beginning of the drying experiment. This pressure dropped significantly as drying continued and was 275 Pa and 32 Pa at 0.1 and 0.9 m height respectively. This difference could be explained by the significant reduction in bulk size due to the shrinkage of maize cobs as the bulk height reduced from 1.115 m at the beginning of the drying process to 0.9 m at the end of the drying experiments resulting in reduction of bulk density from 409.43 to 354.76 kg m^{-3}. Maize cobs drying takes longer than grains drying. Mechanical energy to run the fan can be supplied by photovoltaic system in order to reduce the energy cost of drying.

Funding: This research was funded by The Schaufler Foundation.

Keywords: Air flow resistance, batch drying, bulk density, grain shrinkage, maize cobs, maize drying, pressure drop

---

Contact Address: Janvier Ntwali, University of Hohenheim, Inst. of Agricultural Engineering, Tropics and Subtropics Group, Garbenstr. 9, 70599 Stuttgart, Germany, e-mail: janvier.ntwali@uni-hohenheim.de

Experimental Analysis and CFD-Based Modelling of Grain Bulk Drying Dynamics

IRIS RAMAJ, STEFFEN SCHOCK, JOACHIM MÜLLER

University of Hohenheim, Inst. of Agricultural Engineering, Tropics and Subtropics Group , Germany

Drying is of great importance in the postharvest processing of agricultural commodities. It refers to the removal of the surplus moisture responsible for biochemical, microbiological, and other moisture-related deteriorative reactions for quality preservation. However, drying is an intricate process comprised of simultaneous heat and moisture transfers, which depend on product and drying air conditions. Therefore, drying practices are oftentimes misused, resulting in serious degradation of product quality. For this reason, modeling can be used to provide a deeper understanding of the air-product interaction and to gain insights into drying processes. Thus, this study focused on developing a CFD systematic approach to model the drying dynamics of wheat bulk (Pioneer A, DSV AG) under controlled conditions. Within the model framework, a porous medium approach with tailored user-defined functions was utilized to represent the grain bulk characteristics. The drying experiments were performed using a high-precision and automated through-flow laboratory dryer. A coherent set of drying air temperatures T = 10–50°C, relative humidity RH = 20–60 % and airflow velocity v = 0.15–1 m s^{-1} were employed for model validation. Afterwards, the validated computational model was used to predict the drying performance at T = 40°C, RH = 40 % and v = 0.15 m s^{-1}, where the simulated temperature and moisture content agreed very well with the experimental results ($R^2 \geq 0.98$ and MAPE $\leq$ 14.93 %). The proposed model proved to be an efficient tool capable of simulating the temperature and moisture dynamics inside the grain bulk with high spatial and temporal resolution. It yielded rapid and in-depth information as compared to laborious physical experiments. In conclusion, the CFD-based approach has demonstrated a great potential to simulate drying processes, thus its capability should be further assessed for various drying technologies, operating conditions as well as agricultural commodities.

Keywords: CFD, drying, grain, high-precision, modelling, three-dimensional

---

Contact Address: Iris Ramaj, University of Hohenheim, Inst. of Agricultural Engineering, Tropics and Subtropics Group , Garbenstr. 9, 70599 Stuttgart, Germany, e-mail: Ramaj@uni-hohenheim.de

CO_2 Dynamics in the Pore Volume of Stored-Grain Bulk: Data Acquisition and Analysis via Machine Learning

Iris Ramaj, Steffen Schock, Joachim Müller

University of Hohenheim, Inst. of Agricultural Engineering, Tropics and Subtropics Group , Germany

Quality preservation is one of the major considerations in the conception and operation of grain storage facilities. Warm and humid conditions have a direct impact on the temperature and moisture changes of the grain mass beyond the safe-storage thresholds. Consequently, favourable conditions for the development of microorganisms, fungi and toxic contaminants can be created, leading to elevated CO_2 concentrations in localized pockets of the grain mass. In this regard, real-time monitoring can be an important tool to control incipient grain deterioration. Thus, this study aimed to monitor the grain storage conditions in real-time and investigate their influence on the intergranular CO_2 dynamics. Variables considered in the analysis were temperature, relative humidity, moisture content, atmospheric pressure, and height of grain bulk. For the data acquisition, a small-scale silo (2011 mm diameter and 5040 mm height) filled with freshly harvested wheat grains (Pioneer A, DSV AG) at a moisture content of 0.16 kg kg^{-1} dry basis was used. A sensor system network with linear topology comprised of 12 multifunctional sensor nodes designed in-house were installed equidistantly within the bulk to monitor the temporal and spatial variations of grain conditions. As an analytical approach, supervised machine learning using Gaussian process regression (GPR) was employed. A dataset of 360 consecutive days of grain storage with internment cooling and measuring resolution of 10 min was selected for the training and validation procedures. The dynamics of CO_2 within the grain bulk were found to be strongly influenced by the diurnal and seasonal changes of the external environmental conditions. Results exhibited significant interactions among the variables analysed. An increase of CO_2 was observed with the increase of temperature, relative humidity and moisture content of in-store grain. In contrast, the levels of intergranular CO_2 decreased as the atmospheric pressure and bulk height increased. The applied machine learning approach yielded a highly accurate prediction (R^2 = 0.99 and RMSE = 20.65 ppm), demonstrating its great potential to accurately predict the dynamics of CO_2 depending on the temporal and spatial changes of in-store grain conditions.

Keywords: CO_2, grain storage, machine learning, monitoring, sensors

Contact Address: Iris Ramaj, University of Hohenheim, Inst. of Agricultural Engineering, Tropics and Subtropics Group , Garbenstr. 9, 70599 Stuttgart, Germany, e-mail: Ramaj@uni-hohenheim.de

Drying Behaviour of Moldavian Dragonhead (*Dracocephalum moldavica*) and Blue Fenugreek (*Trigonella caerulea*) with Regards to Processing Temperature and their Quality

Ziba Barati, Supaporn Klaykruayat, Joachim Müller

University of Hohenheim, Inst. of Agric. Sci. in the Tropics (Hans-Ruthenberg-Institute), Germany

Moldavian dragonhead (*Dracocephalum moldavica*) and blue fenugreek (*Trigonella caerulea*), which are growing well in a subtropical climate, contain health-promoting ingredients such as phenolic compounds, flavonoids, essential oils, pigments and vitamins. In this study, the effect of different drying temperatures on the drying behaviour and some quality parameters of Moldavian dragonhead and blue fenugreek was examined. The plants were harvested at the experimental station of the University of Hohenheim (Stuttgart, Germany) in July 2020. The drying experiment was conducted at different processing temperatures of 40, 55, and 70 °C using a high precision laboratory dryer designed at the Institute of Agricultural Engineering, University of Hohenheim. Air velocity and absolute humidity of the air were held constant at 0.2 m s^{-1} and 10 g kg^{-1}, respectively. To establish drying curves, weight loss from the samples was recorded at constant intervals of 5 min. Furthermore, the quality of dried samples was investigated in terms of colour, total phenolic content and flavonoids. The results show that the moisture content decreased gradually until the desired moisture content of 10 % was reached. By increasing the temperature, the total drying time decreased remarkably. It was determined that the colour values of both plants were substantially influenced during the drying process. Moldavian dragonhead dried at 40 °C had the highest total phenolic (55.4 ± 1.7 mg g^{-1} of dried sample) and flavonoids content (42.9 $\pm$ 0.8 mg g^{-1} of dried sample) compared to those dried at 55 and 70 °C. There was a significant difference in the amount of total phenolic and flavonoids in Moldavian dragonhead dried at different temperatures ($p < 0.05$). However, there was no considerable change in total phenolic and flavonoids of blue-fenugreek dried at different temperatures. It was concluded that the drying temperature can affect quality parameters of some medicinal and aromatic plants and therefore must be chosen according to the plant and the intended processing.

Funding: This research was funded by the German Federal Ministry of Food and Agriculture (BMEL) and the Agency of Renewable Resources (FNR) under project number FKZ:22021317.

Keywords: Colour, drying, flavonoids , medicinal and aromatic plants, total phenolic content

Contact Address: Ziba Barati, University of Hohenheim, Inst. of Agric. Sci. in the Tropics (Hans-Ruthenberg-Institute), Garbenstr.9, 70599 Stuttgart, Germany, e-mail: barati@uni-hohenheim.de

Agro-Ecological Approaches in Climate Resilient Agriculture: Local Farming Models in Coastal Bangladesh

Catalina Acuna-Gutierrez[1], Steffen Schock[1], Vïctor M. Jimenez[2], Joachim Müller[1]

[1]*University of Hohenheim, Inst. of Agricultural Engineering, Tropics and Subtropics Group, Germany*

[2]*University of Costa Rica, Grain and Seed Research Center, Costa Rica*

Maize and maize based-products are the main commodity affected by fumonisins, a group of mycotoxins produced by the Fusarium genus. Although its presence can be found worldwide, it is commonly found in warm or warm tropical areas. It has been proven that fumonisin B1 (FB1) causes, among other adverse effects, nephrotoxicity and cytotoxicity in mammals and has been classified as a group 2B carcinogen. Because of the danger it represents to humans and livestock, there is an increasing demand to develop techniques that allow the fast detection of mycotoxins. Therefore, the feasibility of a rapid method using a semi-portable near-infrared spectrometer to detect FB1 in maize kernels was evaluated. For this, a submersion method was applied to perform control contaminations with standard solutions at different concentrations of FB1 (0, 1, 5, and 10 mg kg^{-1}) on the kernels. The samples were analysed milled and whole with a near-infrared spectrometer (NIRS) in the wavelength range of 900–2,500 nm. PCA results showed good separation between the different concentrations used in both milled and whole kernels. A calibration model was developed by a partial least square regression (PLSR) for each matrix type. Milled samples allowed the fitting of a better prediction model (R^2 = 0.74 and RMSECV = 1.30) than whole kernels (R^2 = 0.63 and RMSECV 1.56). Although improvements are needed to obtain a more robust model, initial results are promising. This methodology does not require extraction of mycotoxin, like traditional wet chemical analysis, allowing to perform *in situ* analysis, an advantage for the farmer. Therefore, the application of the developed method can help to carry out rapid analysis to ensure mycotoxin monitoring along the production chain, avoiding contaminated materials from entering the food chain.

Keywords: Cereals, food safety, mycotoxins, optical method

Contact Address: Catalina Acuna-Gutierrez, University of Hohenheim, Inst. of Agricultural Engineering, Tropics and Subtropics Group, Garbenstrasse 9, 70599 Stuttgart, Germany, e-mail: catalina.acunagutierrez@uni-hohenheim.de

Utilisation of Jackfruit Waste for Briquettes and Biogas Production as Alternative Cooking Fuels

Denis Nsubuga[1], Tadeo Mibuulo[1], Isaac Rubagumya[1], Filda Ayaa[1], Isa Kabenge[1], Noble Banadda[1], Kerstin Wydra[2]

[1]*Makerere University, Dept. of Agricultural and Biosystems Engineering, Uganda*

[2]*University of Applied Science Erfurt, Fac. of Landscape Architecture, Horticulture and Forestry, Germany*

Naturally, more than 60 % of a ripe jackfruit is not eaten and is usually discarded as agricultural waste. This research explores the possibility of utilising this waste for production of briquettes and biogas as alternative renewable energy options for cooking. Proximate analysis, calorific value and compositional analyses were conducted. Carbonized briquettes were produced with jackfruit waste mixed with other selected agricultural wastes. Biogas was produced through anaerobic digestion using both laboratory experimental set up and underground fixed-doom digester. The results showed that jackfruit waste had a calorific values ranging from 15.77 to 17.54 MJ kg^{-1} which is comparable to other agricultural waste like rice husks and cotton. The briquette developed from jackfruit waste mixed with banana leaves in a ratio of 70:30 had a calorific value of 21.98 MJ kg^{-1}. The water boiling test showed that briquettes made from jackfruit waste mixed with banana leaves in the same ratio took 29 minutes to boil 2.5 liters of water. The starch content of jackfruit waste ranged between 29.05 to 59.54 % while the sugar content ranged from 2.04 to 68.8 %. Jackfruit waste when co-digested with 25 % cow dung produced the highest volume of biogas compared to other co-digestion mixtures. The methane content from jackfruit waste increased from 25.9 % to 69.6 % when the 25 % cow dung was added. Jackfruit waste can be utilised in the production of briquettes whose heating values are comparable to values of briquettes from other substrates. Jackfruit waste can also be utilised for production of biogas with more gas volumes expected when it is co-digested with cow dung in an appropriate mixing ratio.

† *Dedicated to the memory of author Noble Banadda, who passed away during the preparation of this Abstract.*

Keywords: Biogas, briquettes, calorific value, co-digestion., jackfruit waste

---

Contact Address: Denis Nsubuga, Makerere University, Dept. of Agricultural and Biosystems Engineering, University Road Kampala, Kampala, Uganda, e-mail: dnsubuga5@gmail.com

Converting Coffee By-Products to a Promising Renewable Fuel for the Coffee Processing Sector

Huyen Chau Dang[1], Judy Libra[1], Marcus Fischer[1], Christina Dornack[2]

[1]*Leibniz Institute for Agricultural Engineering and Bioeconomy (ATB), Post Harvest Technology, Germany*

[2]*Dresden University of Technology, Inst. of Waste Management and Circular Economy, Germany*

Coffee is an important beverage for most cultures in the world and the interest in increasing sustainability in the coffee value chain is high throughout the world. So besides having to meet the demand for high coffee quality from customers, coffee farmers and producers currently have to face the challenge of how to sustainably deal with coffee by-products from coffee processing: (1) from coffee berries to beans; and (2) from beans to final products such as brewed coffee and instant coffee. Disposal amounts and methods for coffee by-products depend on the region and the processing step in the value chain. About ten million tons of green coffee beans are worldwide produced annually in processing facilities from coffee berries, leaving behind approximately three times that amount as by-products. Coffee consumption or processing to instant coffee produces spent coffee grounds as a waste. Recovery of the by-products as a renewable fuel in the coffee value chain is an important strategy for sustainability. This study investigates the potential of substituting the fossil fuels used in coffee processing with a carbon-rich material (hydrochar) produced by thermos-chemically converting coffee by-products. In lab experiments with hydrothermal carbonisation (HTC), coffee by-products were treated at various process conditions (temperature from 160 °C to 240 °C, holding time from 1 – 5 h, solid content from 15 % to 20 %) and post-treatment steps. The energetic and chemical properties of coffee by-products and their hydrochars were determined in terms of calorific value, ash content, major elements (e.g. C, N, S), minor elements (e.g. Cu, Ni, Zn, Cd, Cr, Pb, As, Hg) and compared with quality requirements for biomass and thermally treated biomass fuels in private stoves and industrial boilers. The presentation will discuss how the HTC-process can be optimised to increase the energetic properties of hydrochar from coffee by-products and the substitution potential of these hydrochars for use as solid fuel in coffee processing. The outcome of this project will support farmers and producers to improve the sustainability of the coffee value chain in Vietnam, and also provide a basis for its adaptation to other coffee production regions.

Keywords: Coffee by-products, coffee value chain, fuel, hydrochar, hydrothermal carbonisation (HTC), sustainability

Contact Address: Huyen Chau Dang, Leibniz Institute for Agricultural Engineering and Bioeconomy (ATB), Post Harvest Technology, Max-Eyth-Allee 100, 14469 Potsdam, Germany, e-mail: hdang@atb-potsdam.de

Design and Evaluation of Mechanised Seedball Production for Sahelian Smallholder Farmers

Sebastian Romuli[1], Achim Jesser[1], Charles Ikenna Nwankwo[2], Ludger Herrmann[2], Joachim Müller[1]

[1]*University of Hohenheim, Inst. of Agricultural Engineering, Tropics and Subtropics Group, Germany*
[2]*University of Hohenheim, Soil Chemistry and Pedology, Germany*

Seedball is a low-cost seed pelleting technique that combines sand, loam, seeds, and optionally wood ash or mineral fertiliser as an additive, in order to enhance the growth of pearl millet or sorghum seedlings in nutrient-poor soil. However, the massive production (10,000 seedballs per hectare) demands time, which poses challenges to Sahelian farmers. Therefore, mechanisation of seedball production is highly required. In this study, a seedball machine prototype was constructed and tested at the University of Hohenheim. The prototype essentially consists of a metal frame and a drum, in which seedballs are formed by an electric motor-powered rotation. It was designed to be easy to construct and operate. Using response surface methodology with a Box-Behnken experimental design, the effect of the prototype settings (substrate composition, rotational speed and residence time) on seedball formation and process efficiency was investigated. As the result, the amount of loam in the raw material composition and the rotational speed of the drum were found to be the most significant factors at $p < 0.05$. The stepwise optimisation led to a production capacity of 44 ± 7 seedballs per minute. Compared to manual production (4 ± 1 seedballs per minute), the prototype performed in a semi-continuous way and was able to reduce the production time by about 92 %. The substrate usage rate was 94.8 ± 3.5 %. The power requirement was about 140 W. The machine-made seedballs were also of high quality. Under greenhouse conditions, the seedballs reached a germination rate higher than 98 %. This study presents the proof of concept for mechanised seedball production that will facilitate the adoption of seedballs by local farmers. As a further step, field tests and demonstrations of the prototype in the Sahel are recommended.

Keywords: Cropping system, cultivation, frugal innovation, machine performance, operation strategy

Contact Address: Sebastian Romuli, University of Hohenheim, Inst. of Agricultural Engineering, Tropics and Subtropics Group, Garbenstr. 9, 70599 Stuttgart, Germany, e-mail: sebastian_romuli@uni-hohenheim.de

Optimal Drying Conditions for Production of Nutritious and Safe Cassava Leaves

Sheilla Natukunda, John Muyonga, Archileo Kaaya
Makerere University, Food Technology and Nutrition, Uganda

The current study was conducted to optimise oven drying conditions of cassava leaves to maximise total antioxidant activity, beta carotene and minimise total cyanide content using Response Surface Methodology (RSM). I-optimal design with two independent variables: temperature (50–80) °C and time (3–8) hours was employed. Processed cassava leaves were analysed for total antioxidant activity (2,2-diphenyl-1 picrylhydrazyl (DPPH) radical scavenging assay), beta carotene and total hydrogen cyanide. Shelf life of the optimised cassava leaves was determined by evaluating the microbial count for 12 weeks. Analysis of variance showed that a linear model was significant for total hydrogen cyanide ($R^2 = 0.90$, $p < 0.05$) and beta carotene ($R^2 = 0.89$, $p < 0.05$ while. quadratic model was significant for total antioxidant activity ($R^2 = 0.89$, $p < 0.05$). Response surface plots showed that an increase in temperature significantly increased total antioxidant activity and total cyanide content while total beta carotene reduced. The optimal drying conditions based on combination of all responses using desirability approach were: temperature (60°C) and time (3 hours). Under the above-mentioned optimal conditions the dried cassava leaves yielded total cyanide content 55.01 mg kg^{-1}, total antioxidant activity 54.06 mg VCE g^{-1} and beta carotene 5,967.11 μg/100 g. There was a close agreement between experimental and predicted values. Total plate count and yeasts and molds of the optimised dried cassava leaves increased during storage for 12 weeks but did not exceed maximum acceptable safety values. This processing method can be applied in the production of a safe, shelf stable, nutritious, high antioxidant product in the nutraceutical and food industries.

Keywords: Cassava leaves, drying, nutritious, optimisation

Contact Address: Sheilla Natukunda, Makerere University, Food Technology and Nutrition, Makerere, Kampala, Uganda, e-mail: n.sheilla@yahoo.com

Optical Detection of Two-Spotted Spider Mite Infestation in Cucumber Using Vegetation Indices

Boris Mandrapa[1], Friederike Espinoza[2], Ute Ruttensperger[2], Christine Dieckhoff[3], Klaus Spohrer[1], Joachim Müller[1]

[1]*University of Hohenheim, Inst. of Agricultural Engineering, Tropics and Subtropics Group, Germany*

[2]*State Horticultural College and Research Institute Heidelberg, Dept. of Horticultural Experiments, Germany*

[3]*Center for Agricultural Technology Augustenberg, Germany*

The two-spotted spider mite (*Tetranychus urticae* Koch) feeds on the underside of plant leaves by penetrating the chloroplast-containing cells. This primarily leads to yellowing in limited areas, which is visible as spots on the upper side of leaves. Spider mite infestation can lead to yield loss. In the case of cucumbers, heavy infestation even leads to complete loss of the assimilation surface on the leaves and, as a result, to the death of the plants. Therefore, it is necessary to detect a spider mite infestation as early as possible. Objective of the present work was to investigate the possibility of identifying spider mite infestation of cucumber by using a hyperspectral camera. An image capturing unit was built, which consisted of a Cubert UHD185 Firefly camera, a halogen light source and a camera mounting platform. Cucumber plants were infected with 50 spider mites each and kept under controlled conditions for two weeks. Subsequently, the leaves with spider mite damage were cut at their base and placed on the platform under the camera´s objective. During image acquisition, the distance between objective and leave was adjusted at 43 cm. As control, the same procedure was performed with healthy leaves. After image acquisition, infected and healthy leaves were compared by applying different vegetation indices (VI) such as: GNDVI, NRENDVI, REIP and TGI. The performance of the VIs was analysed in R program. The analysis indicated that it is possible to differentiate between healthy and infested leaves, as three of four investigated VIs showed significant difference within the compared data. Hence, VIs obtained with hyperspectral imaging can be used to detect damage on cucumber leaves caused by spider mites.

Keywords: Hyperspectral imaging, two-spotted spider mite, vegetation indices

Contact Address: Boris Mandrapa, University of Hohenheim, Inst. of Agricultural Engineering, Tropics and Subtropics Group, Garbenstraße 9, 70599 Stuttgart, Germany, e-mail: boris.mandrapa@uni-hohenheim.de

Ozonation to Enhance Plant Based Mining of Metals from Polluted Water

Jörn Germer, Folkard Asch

University of Hohenheim, Inst. of Agric. Sci. in the Tropics (Hans-Ruthenberg-Institute), Germany

Metals are introduced into aquatic systems through the weathering of soils and rocks, volcanic eruptions and a variety of human activities associated with agriculture, mining, processing or use of metals and/or metal-containing substances. Some metals such as manganese, iron, copper and zinc are essential or, like Na, non-essential micronutrients, but in high concentrations these elements can be toxic. Certain heavy metals such as arsenic, cadmium, chromium, copper, nickel, lead and mercury are known environmental pollutants due to their toxicity, persistence in the environment and bioaccumulative nature. Recent studies show that metal content in wastewater is not only a health and environmental problem, but also represents a significant economic loss.
Various methods have been used to extract metals from contaminated water. These include chemical precipitation, ion exchange, adsorption, membrane filtration, reverse osmosis, solvent extraction and electrochemical treatment. As with ore mining, the profitability of extracting metals from water increases with metal concentrations. Therefore, new approaches to concentrate water born metals can contribute to economic and environmental sustainability. The results of our research show that ozonation of treated wastewater significantly increases deposition of metals on the roots and their uptake by hydroponically grown plants. We postulate that ozone increases the oxidation of metals and thereby their precipitation. Roots exudes protons into the rhizosphere to reduce precipitated Fe3+ for uptake. However, residual ozone seems to neutralise these protons or it immediately re-oxidises the reduced Fe. To compensate for the constant loss of protons, the roots take up other oxidised divalent metals. As a result, the contents of Fe and Zn in the roots and of Ca, Mg, Mn, Zn and Cu in the shoots doubled or tripled. The contents of the heavy metals As, Cd, Co, Cr and Pb increased accordingly. For all metals, the root content was many times higher than the shoot content. Our poster explains the reduced Fe availability induced by ozonation and the thereby promoted metal accumulation on and in the roots and shoots of plants. The potential of this approach to unburden the environment and to mine metals from water is discussed for three tropical scenarios.

Keywords: Divalent metals, iron deficiency, ozonation, wastewater mining

---

Contact Address: Jörn Germer, University of Hohenheim, Inst. of Agric. Sci. in the Tropics (Hans-Ruthenberg-Institute), Stuttgart, Germany, e-mail: jgermer@uni-hohenheim.de

Non-Invasive, Real Time *in-situ* Techniques to Determine the Ripening Stage of Banana - Development of a Banana Ripening Index

Tillmann Ringer, Michael Blanke

Universität Bonn Landwirtschaftliche Fakultät, INRES- Horticultural Science, Germany

The objective of the work was to elaborate non-destructive, real-time, *in situ* opto-electronic techniques to determine the ripening stage of Cavendish banana. Four methods were used 1) chlorophyll degradation (DA-meter), 2) colour change (Minolta), 3) change in light reflectance (spectroscopy), and 4) peel gloss (Keyence CH72Z) and maybe develop a ripening index.
Cavendish bananas were examined starting from ripening stage R2 (green) to stage R7 (overripe), to identify suitable non-invasive, real time in-situ technologies to separate the ripening stages:

1. Chlorophyll degradation, measured by the DA meter, decreased from ca. 2.1 (R2) to 0.2 IAD units (R7), i.e. 10-fold decline.
2. Colour CIE-Lab a values dramatically increased as indication of chlorophyll breakdown and enable differentiation between all ripening stages R2 to R7. Colour angles declined from 98.7° hue (R2), 97.3° hue (R3), 92.7° hue (R4), 89.4° hue (R5); 87.5° hue (R6) until 82.0° hue (R7).
3. Spectroscopy showed two light reflectance troughs at 494 nm and 679 nm. A novel banana ripening index (BRI) was developed and is proposed to identify and distinguish the ripening stages of banana with values starting at 4 at R1 and peaking at 8.1 at ripening stage R7.
4. Peel gloss increased from stage R2 (150 a.u.) to stage R7 (220 a.u.) in the order of ca. 50 % followed by a subsequent decrease thereafter.

All the above results identified the fruit centre (rather than the tip) as a suitable candidate due to the most advanced ripening and least curved surface region of the fruit with easy access, when a carton is opened and the hands become accessible. This novel approach based on a comparison has shown the DA-meter, colorimeter and spectrometer as suitable candidates for the identification of each ripening stage. The combination of these three devices may be suitable for monitoring of banana ripening rooms in terms of temperature and humidity in addition to the present, colour-based ripening scale.

Keywords: Banana (*Musa sapientum*), chlorophyll degradation, colour, glossiness, non-invasive measurement, precision horticulture, ripening, spectral index, storage life

Contact Address: Michael Blanke, University of Bonn, INRES- Horticultural Science, Auf Dem Hügel 6, 53121 Bonn, Germany, e-mail: mmblanke@uni-bonn.de

Optimisation of Oven-Drying of Baobab Leaves Using a Central Composite Design

Ahotondji Mechak Gbaguidi[1], Flora Josiane Chadare[2], Valère Salako[3], Vanessa Idohou[1], Achille Assogbadjo[4]

[1]*National University of Agriculture, Laboratory of Food Science, Benin*

[2]*National University of Agriculture, Laboratory of Food and Bio-Resources Science and Technology and Human Nutrition, Benin*

[3]*University of Abomey-Calavi, Lab. of Biomathematics and Forest Estimations, Benin*

[4]*University of Abomey-Calavi, Lab. of Applied Ecology, Benin*

Baobab (*Adansonia digitata*) leaves represent a key nutritional resource; although their consumption is apparently restricted to local communities, mainly as a sticky sauce, cosmetics and a variety of purposes. Ready-to-use powder of oven-dried baobab leaves can improve the availability of the product on markets, and hence its utilisation and shelf life. This study was carried out to optimise the oven-drying temperature and duration for the best conservation of baobab leaf powder and its sustainable availability for multipurpose uses. Different combinations of values of temperature and drying duration were generated, using a central composite design, in response surface methodology framework. These combinations were applied to freshly collected young baobab leaves; and the derived dried leaves were analysed by assessing the dry matter, the colour and the gelling property. Findings showed that dry matter (93.4 ± 1.8 g $100\,g^{-1}$), hue ($109.6 \pm 6.1°$), chroma (14.0 ± 1.9) and lightness (51.8 ± 5.8) were significantly influenced by oven-drying temperature and duration; while the least gelation concentration (7.2 ± 1.3 g 100 mL^{-1} dw) was significantly influenced by the temperature. Significant correlations were revealed out between the least gelation concentration and the hue value, and between the lightness and the chroma. Based on leaf dry matter, hue and chroma models, the optimal oven drying conditions for baobab leaves for good preservation leading to human consumption and other purposes are set at 45 °C for 23.5 hours. Drying in these optimal conditions lead to a leaf powder with 92.0 g $100\,g^{-1}$ of dry matter and a high rate of leaves colour preservation. These drying conditions should be adapted to local populations' capacities, based on the tropical countries conditions, and the solar drying systems.

Keywords: *Adansonia digitata*, colour, dry matter

Contact Address: Flora Josiane Chadare, National University of Agriculture, Laboratory of Food and Bio-Resources Science and Technology and Human Nutrition, Sakété, Benin, e-mail: fchadare@gmail.com

Animals

Animal breeding and genetic improvement

Oral Presentations

Posters

Genetic Analysis of Hybridisation Pattern in Goat Genotypes from East Africa

Theodora Chikoko, Kiplangat Ngeno, Thomas Muasya
Egerton University, Animal Science, Kenya

East Africa is home to an estimated 150,667,482 goats in different geographical regions which are exposed to diverse climatic, production and management conditions. This has shaped the goat genome due to adaptation or selection. This study aimed to investigate hybridisation pattern between distantly isolated goat populations in the East African community using single nucleotide polymorphism (SNP) data genotyped using a 50K goat SNP chip. In Uganda, SNP data for 144 goats from six genotypes sampled from five agro-ecological zones was retrieved from the Zenodo genome archive. In Kenya, 94 goats from four genotypes sampled from three regions were used. Quality control procedures were performed in PLINK v 1.9. Data from Kenya (48,303 SNPs) and Uganda (46,105 SNPs) was then merged in Tassel software resulting in 94,408 SNPs available for joint downward analysis. Principle component and phylogenetic analysis was used to visualise relationships between the studied populations. Four well-defined clusters, two from each country were observed with some visible outliers. From the two clusters in each country, genetically mixed genotypes were identified in only one cluster. This might suggest inbreeding within the genotypes due to close proximity between them. The degree of genetic differentiation measured using Fst ranges from 0.191 to 0.324 indicating that all the genotypes within a country are to some extent isolated from each other. Only Boer genotype from Uganda was highly isolated from all genotypes in this study. The diversity shown between Kenya and Uganda goat genotypes might be due to uncommon ancestry, isolation distance or lack of capacity to use reproductive technologies. These results will be useful in the implementation of future genetic conservation, utilisation and improvement programs in the East African community.

Keywords: Genotypes, inbreeding, single nucleotide polymorphism (SNP)

Contact Address: Theodora Chikoko, Egerton University, Animal Science, Egerton Universiy -Main Campus Po Box 536-20115 Egerton Kenya, Nakuru, Kenya, e-mail: theochikoko@gmail.com

Bayesian Analyses of Growth Traits in Mecheri Sheep Reared under Tropical Climatic Conditions of India

Thiruvenkadan Aranganoor Kannan[1], Muralidharan Jaganathan[2], Rajendran Ramanujam[1], Bandeswaran Chinnaondi[1], Satish Kumar Illa[3], Kadir Kizilkaya[4], Sunday O Peters[5]

[1] *Tamil Nadu Veterinary and Animal Sciences University, Dept. of Animal Genetics and Breeding, India*

[2] *Tamil Nadu Veterinary and Animal Sciences University, Bmecheri Sheep Research Station, India*

[3] *Sri Venkateshwara Veterinary University, Animal Genetics and Breeding, India*

[4] *Aydın Adnan Menderes University, Animal Genetics and Breeding, Turkey*

[5] *Berry University, Animal Genetics and Breeding, United States*

The objective of this study was to estimate (co) variance components and genetic parameters for body weights at different ages in Mecheri sheep born during the period from 2010 to 2020 (11 years). A total of 2825 lambs with pedigree descended from 758 dams and 119 sires were included in this study. Bayesian analysis using Gibbs sampler multi-trait animal model with direct and maternal effects were studied. Traits included in this study were birth weight (BW), weaning weight (WW), six months body weight (BW6), nine months body weight (BW9), Yearling weight (BW12). The least-squares means with standard deviation for body weight at birth, weaning and 12-months of age were 2.57 ± 0.44, 11.09 ± 2.54 and 20.67 ± 4.04 kg respectively. Analysis of variance indicated that fixed effects of year of birth, season of birth, type of birth, sex of the lamb and parity of the dam significantly affected the growth traits at different stages. The direct heritability estimates for BW, WW, BW6, BW9 and BW12 based on best model were 0.21, 0.21, 0.12, 0.14 and 0.13, respectively and the corresponding maternal heritability estimates were 0.18, 0.08, 0.11, 0.13 and 0.13, respectively. Strong and positive estimates of direct additive genetic correlations were observed in growth traits and they ranged from 0.165 (BW-BW12) to 0.904 (BW6-BW9).The findings of this study have underlined the importance of maternal effects in Mecheri sheep and their influence on growth traits. Significant genetic variability suggests further scope of selection for growth traits and a moderate rate of genetic progress seems possible in the Mecheri sheep flock for live weight traits by mass selection. The strong and positive estimates of genetic correlations between WW and BW6 indicated that improvement in one trait will bring improvements in other traits. Hence, better response to selection would be seen, if selection in Mecheri sheep for improvement of body weights followed at weaning age, than the current practice of selection at six months of age.

Keywords: Body weight, direct heritability, Gibbs sampling, maternal effects, multi-trait animal model, Mecheri sheep

Contact Address: Thiruvenkadan Aranganoor Kannan, Tamil Nadu Veterinary and Animal Sciences University, Dept. of Animal Genetics and Breeding, Veterinary College and Research Institute, 636112 Salem, India, e-mail: drthirusiva@gmail.com

Genomic Copy Number Variation of the CHKB Gene Alters Gene Expression and Affects Growth Traits of Chinese Domestic Yak (*Bos grunniens*) Breeds

Habtamu Goshu[1], Chu Min[2], Yan Ping[2]

[1]*Ethiopian Biotechnology Institute, Habtamu Abera Goshu, Ethiopia*
[2]*Chinese Academy of Agricultural Science, China*

Copy number variation (CNV) influences the mRNA transcription levels and phenotypic traits through gene dosage, position effects, alteration of downstream pathways, and modulation of the structure and position of chromosomes. A previous study using the read depth approach to genome resequencing analysis revealed CNVs of the choline kinase beta (CHKB) gene in the copy number variable regions (CNVRs) of yak breeds may influence muscle development and therefore the phenotypic traits of yak breeds. Further work is required to attain a more complete understanding and validate the importance of the detected CNVR of the CHKB gene found in yak breeds, because there is no association studies of the CHKB gene with yak growth traits have been reported. The goal of this study was to determine the distribution of CHKB copy numbers in five Chinese domestic yak breeds and evaluate their impact on gene expression and growth traits. The data were analysed using real-time quantitative PCR (qPCR). In this study, the normal CNV of the CHKB gene was found to be significantly ($p < 0.05$) associated with greater chest girth and body weight for three age groups of Datong yaks. Our results indicated that the copy number of the CHKB gene is negatively correlated with the mRNA expression level. From this result, we conclude that CNVs of the CHKB gene could be novel markers for growth traits of Chinese domestic yak breeds and might therefore provide a novel opportunity to utilise data on CNVs in designing molecular markers for the selection of animal breeding programs for larger populations of various yaks.

Keywords: CHKB gene, copy number variation, expression analysis, growth traits, yaks

Contact Address: Habtamu Goshu, Ethiopian Biotechnology Institute, Habtamu Abera Goshu, Ethiopian Biotechnology Institute, 5954 Addis Ababa, Ethiopia, e-mail: habtamu2016@yahoo.com

Comparative Analysis of Variation in Humoral Immunity, Production, Fitness and Feed Efficiency Traits in Chicken Performing in the Tropical Environment

Sophie Miyumo[1], Chrilukovian Wasike[2], Evans Ilatsia[3], Joen Bennewitz[4], Mizeck Chagunda[1]

[1]*University of Hohenheim, Inst. of Agric. Sci. in the Tropics (Hans-Ruthenberg-Institute), Germany*

[2]*Maseno University, Dept. of Animal Sciences and Fisheries, Kenya*

[3]*Kenya Agricultural and Livestock Research Organization, Dairy Research Institute, Kenya*

[4]*University of Hohenheim, Dept. of Animal Genetics and Breeding, Germany*

This meta-analysis aimed to assess genetic and environmental effects on humoral immunity in comparison to production, fitness and efficiency trait classes in chicken. Three hypotheses were formulated; associations between heritability with genetic and environmental variances across the trait classes are equal; genetic and environmental effects on the trait classes are equal; study characteristics have no effect on genetic parameters in humoral immunity. Genetic parameters and production system characteristics were extracted from a total of 85 peer-reviewed articles published between 1976 and 2020. Mean standardised genetic variance (CVA) and environmental variance (CVR) were used to compare variances of traits measured in different units. Associations between the effect sizes were explored using spearman-rank correlation (non-normal distribution). For heterogeneity tests, hierarchical model was considered while study characteristics influencing genetic parameters in humoral immunity were tested by fitting hierarchical meta-regression model; log-transformed effect sizes used as response variables but back-transformed when reporting results. Associations between heritability with CVA and CVR was trait dependent. In humoral immunity and fitness traits, heritability was positively ($p < 0.05$) associated with CVA (r = 0.34 vs 0.30) For production traits, however, heritability was highly and negatively associated with CVR than with CVA (r = -0.41 vs 0.11; $p < 0.05$). Heterogeneity tests showed significant ($p < 0.05$) variation in effect sizes across trait classes. Weighted estimates of effect sizes indicate that humoral immunity had highest genetic (12 %) and residual (23 %) effects but with low heritability (0.23). In contrast to immunity traits, production traits had the highest heritability (h2 = 0.27) but the least genetic (7 %) and residual (9 %) effects. Adjusting for moderating factors on humoral immunity, the two arms of humoral immunity,

Contact Address: Sophie Miyumo, University of Hohenheim, Inst. of Agric. Sci. in the Tropics (Hans-Ruthenberg-Institute), Garben Str. 17, 70599 Stuttgart, Germany, e-mail: sophie.miyumo@uni-hohenheim.de

chicken population and production phase were able to explain the 93.1 % of the between study variation in heterogeneity in CVA. Significant differences on genetic parameters across trait classes suggest presence of sub-groups among chicken performing in different production environments. Based on humoral immunity, accounting for these sub-grouping effects gives a better indication of true genetic effects in the different traits.

Keywords: Feed efficiency, fitness, genetic parameters, humoral immunity, production

Identification of Close Relatives of Crossbred Dairy Cattle in Ethiopia

Selam Meseret[1], Asrat Tera[2], Gebregziabher Gebreyohanes[1], Chinyere Ekine-Dzivenu[3], Julie Ojango[3], Ali Mwai Okeyo[3], Raphael Mrode[3]

[1]*International Livestock Research Institute (ILRI), Ethiopia*

[2]*National Animal Genetic Improvement Institute, Ethiopia*

[3]*International Livestock Research Institute (ILRI), Kenya*

Correct pedigree is essential to produce accurate genetic evaluations and positively affect genetic gains of dairy cattle. However, close relatives were not fully documented in dairy herds registered under Ethiopian dairy cattle database. Therefore, genetic data was used to identify close relatives and parentage reassignments. African Dairy Genetic Gains (ADGG) project led by International Livestock Research Institute has generated a medium density SNP dataset with Illumina Bovine SNP50 BeadChip, which consists of 6821 animals collected from six regions and one city administration in Ethiopia. The final dataset used for analysis passed through quality control pipeline, included SNPs above 80 % genotyping rate and missing rate per animal with less than 20 %. Thus, the final dataset consisted of 46,882 SNP and 6802 animals. Kinship coefficients and inferring identity by decent segments estimated using KING software to assign degree of relationship between pairs of dairy animals across the country. The inbreeding coefficient calculated based on the observed versus expected number of homozygous genotypes in PLINK software. A total of 75 pairs were identified as potential duplicate, 2133 pairs of parent-offspring, and 689 pairs of full-sibs relationship. Whilst 125,929 pairs are sorted under second degree relationship, contributed about 0.5 % of the total pairs of 23,130,202. In addition, nearly one tenth of the population had above 10 percent inbreeding coefficient and the population average inbreeding coefficient was 0.02. In conclusion, drawing relationship inference for close relatives in the crossbred cattle population provides complementary information for genetic evaluation to estimate highly reliable breeding value. Additionally, the current inbreeding coefficient status requires action in the national breeding plan to ensure sustainable genetic improvement program.

Keywords: Dairy cattle, inbreeding coefficient, relationship

Contact Address: Selam Meseret, International Livestock Research Institute (ILRI), Addis Ababa, Ethiopia, e-mail: s.meseret@cgiar.org

Assessing the Morphological Diversity of Ethiopian Indigenous Chicken Populations Using Multivariate Analysis of Morphometric Traits

ABERRA MELESSE[1], TADELLE ABIYU[2], ASSEFA HAILU[3], KEFALA TAYE[4], TESFAHUN KEBEDE[5]

[1]*Hawassa University, Animal and Range Sciences, Ethiopia*
[2]*Bonga University, Animal Science, Ethiopia*
[3]*Mizan Agricultural Technical Vocational Education and Training College, Animal Science, Ethiopia*
[4]*Arsi University, Animal Science , Ethiopia*
[5]*Alage Agricultural Technical Vocational Education and Training College, Animal Science , Ethiopia*

In Ethiopia, there have been many morphological characterisation studies conducted to describe the existence of phenotypic variations among the indigenous chicken populations. However, most of these studies were focused on specific districts with inadequate sample size and, except a few; none of them have also attempted to morphologically differentiate them by applying a multivariate analysis. Therefore, the aim of the present study was to assess the existence of phenotypic diversity among indigenous chicken populations of four administrative zones (Kaffa, Sheka, Metekel and Bale) based on their morphometric traits by applying multivariate analysis. Data on quantitative traits were collected from 3069 adult indigenous chickens of both sexes. Traits scored were live weight (LW), body length (BL), breast circumference (BC), wingspan (WS), shank length (SL), shank circumference (SC), keel length (KL), back length (BkL) and neck length (NL). A cluster and discriminant analysis was applied to identify the combination of variables that best differentiate among chicken populations. Results indicated that Metekel chickens were characterised by higher LW, BL, KL and BkL and differed from other groups ($p < 0.05$). Sheka chickens demonstrated the highest BC, WS, SL, SC and NL being different from others ($p < 0.05$). Cluster analysis generated two distinct groups in which chickens of Bale and Sheka were clustered in one group while those of Metekel and Kaffa in another group each separated with sub-clusters. Three statistically significant ($p < 0.001$) canonical variables (CAN) were extracted of which CAN1 and CAN_2 accounted for 73.2 and 14.6 % of the total variations, respectively. The scatter plot generated by canonical discriminant analysis showed that CAN1 effectively discriminated between chickens of Metekel and Kaffa while the CAN_2 best discriminated

Contact Address: Aberra Melesse, Hawassa University, Animal and Range Sciences, Hawassa University (CLIFFOOD-DAAD Funded Project), 05 Hawassa, Ethiopia, e-mail: a_melesse@uni-hohenheim.de

against those of Bale and Sheka. The discriminant analysis correctly classified 95.3, 94.9, 92.3, and 82.2 % of Metekel, Bale, Kaffa, and Sheka chickens into their origin population, respectively. The current study revealed that multivariate analysis of morphometric traits provided a practical basis for differentiating the indigenous chicken populations into different groups. Genetic characterisation studies have been recommended to validate the observed morphometric-based differentiations.

Keywords: Administrative zone, genetic variation, indigenous chicken, morphometric trait, multivariate analysis

Supply Chain Planning of Breeding Inputs on the Quality of Animal Breeding Services of Uganda

Charles Lagu[1], Sylvester Kugonza[2], Pross Nagitta Oluka[2], Morgan Andama[3]

[1]*National Animal Genetic Resources Centre and Data Bank (NAGRC&DB), Breeding, Uganda*

[2]*Uganda Management Inst. (UMI), Civil Service and Public Administration, Uganda*

[3]*Muni University, Uganda*

The study on supply chain planning of breeding inputs (liquid nitrogen and frozen semen) on the quality of animal breeding services of Uganda aimed to explain the effect of supply chain planning of liquid nitrogen and frozen semen on the quality of animal breeding public services in the selected cattle corridor districts (Mbarara, Mubende, Luwero and Soroti) of Uganda. The research filled the gap of supply chain planning in ensuring proficient and viable supply of animal breeding inputs from supply chain points of view compared to the traditional core science orientation which tend to focus on the biological processes of reproductive technologies in animal production and management. The study adopted cross-sectional survey design embracing both qualitative and quantitative approaches. Data was captured using structured questionnaires, review of records, focus group discussions (FGDs) and key informant interviews (KIIs) for farmers, staff of the National Animal Genetic Resources and Data Bank (NAGRC&DB), artificial insemination (AI) technicians, and Field Extension workers. The study points to gender disparity as a key concern at household levels when it comes to land and animal ownership in the selected cattle corridor districts of Uganda. The study established that there was positive relationship between planning ($\chi 2=4.270$; $p = 0.039$; $\chi 2$ critical=3.841) and animal breeding services in selected cattle corridor districts of Uganda. The study recommends that genetic centres to put in place systems for coordinated and integrated planning to facilitate outreach for AI services in Uganda. Furthermore, studies on effect of supply chain sourcing on the quality of animal breeding services in the selected cattle corridor districts needs to be undertaken to guide extension workers on efficient and effective breeding services delivery.

Keywords: Animal resources, artificial insemination services, breeding inputs, gender disparity, genetic centre, quality

Contact Address: Charles Lagu, National Animal Genetic Resources Centre and Data Bank (NAGRC&DB), Breeding, P.O.Box 183, +256 Entebbe, Uganda, e-mail: chlaguu@gmail.com

Some Factors Influencing the Physiological Level of Milk Somatic Cell Count in Lactating Camels

Shehadeh Kaskous

Siliconform, Department of Research and Development, Germany

Camels are very important animals as they can produce milk in arid climates. In order to maintain healthy camel milk for consumers, we need to regularly check the camel and ensure hygiene measures in all camel farm systems. The somatic cell count (SCC) is a very good parameter for indicating milk hygiene, milk quality and udder health in dairy camels. In particular, the SCC is the most accurate, reliable diagnostic method for detecting subclinical mastitis in milk camels under field conditions. The present study sheds light on Some factors influencing the physiological level of somatic cell count of milk in camels.The somatic cells in camel milk contain the following cells: macrophages, polymorphonuclear neutrophils (PMN), lymphocytes, and a large number of cell fragments. Lymphocytes are the predominant cell type in camel milk in the healthy udder. So far there is no established physiological level for SCC in healthy camel milk. We suggest that 150×10^3 SCC cells/ml in milk is a limit value for healthy camel milk. If the SCC exceeds this limit, subclinical or clinical mastitis of the udder may occur and the milk may be contaminated with microbes. In order to maintain camel milk hygiene, proper machine milking such as StimuLactor for camels must mainly be used in the intensive housing systems. An increase in the SCC above the physiological level not only indicates a problem with the health of the udder, but also reduces milk production, changes the milk composition, affects milk processing and changes the bioactive ingredients of camel milk.

Keywords: Camel, mastitis, milk quality, physiology, somatic cell count, stimuLactor

Contact Address: Shehadeh Kaskous, Siliconform, Department of Research and Development, Blumenstrasse 8b, 86842 Türkheim, Germany, e-mail: shehadeh.kaskous@yahoo.de

Evaluation of the Effect of Coat Colour and Coat Type on Physiological Response of Zebu Cattle to Heat Stress in Kenya

Evance Ogingo, Kiplangat Ngeno, Thomas Muasya
Egerton University, Department of Animal Sciences,

Impacts of global warming such as heat stress have led to reduced beef cattle productivity in tropical, sub-tropical and arid areas. Thermoregulatory mechanisms vary depending on the exposure time to HS, with a cumulative effect on the adaptive responses and thermal strain of the cow. The aim of the study was to evaluate effect of Zebu cattle coat color and type on physiological response during heat stress conditions. This experiment was conducted in Isiolo county, (Ngaramare ward) located in northeastern part of Kenya. A total of 35 animals were evaluated at 0900 hours and 1300 hours. The following parameters were evaluated: Environmental temperature in degrees Celsius, length of hair, pigmentation level in the coat, rectal temperature, and respiration breaths per minute and heart rate beats per minute. Environmental temperature had a significant effect on rectal temperature and respiration rate ($p < 0.05$), with no significant effect on heart rate ($p > 0.05$). It was found that when Environmental temperature in the morning was 32.2 °C and in the afternoon to be 36.9 °C, respiration rate increased from 24.89 ± 3.54 to 37.81 ± 2.28 breaths per minute, Heart Rate increased from 46.86 ± 5.37 to 55.62 ± 3.66 beats per minute while Rectal Temperature increased 37.82 ± 0.19 °C to 38.54 ± 0.12°C. Coat color and type did not show any significant effect on rectal temperature and respiration rate ($p > 0.05$). Coat color provided significant effect on heart rate ($p < 0.05$), while coat type did not also show significant effect on heart rate ($p > 0.05$). Dark brown animals had got the least mean heart rate 23.85 ± 12.11 beats per minute, while fawn had got the highest mean Heart Rate 75.52 ± 7.79 beats per minute. Fawn colored animals also had lower respiration rate compared to dark brown animals with average of 29.87 ± 5.01 and 32.72 ± 7.51 breaths per minute respectively. Within the range of red color group fawn colored animals having the highest heart rate also had the lowest rectal temperature 37.85 ± 0.27 °C and the least coat length of 0.60 ± 0.24 mm. To summarize the study, Heat stress negatively impacted the Zebu cattle performance in tropical countries. Zebu coat color and length coat hair confer physiological adaptation within range of red color group, with fawn colored being more resilient. There is need to provide shade and drinking water for darker animals.

Keywords: Climate change, heat stress, mitigation, tropics

Contact Address: Thomas Muasya, Egerton University, Dept. of Animal Science, Njoro, Kenya, e-mail: thomas.muasya@egerton.ac.ke

Heritability and Genetic Evaluation of Black Bengal Goats for Growth Traits

Ishrat Jahan[1], A.K.F.H. Bhuiyan[1], M.S.A. Bhuiyan[1], N.H. Desha[2], M.A. Jalil[2]

[1]*Bangladesh Agricultural University (BAU), Animal Breeding and Genetics, Bangladesh*
[2]*Bangladesh Livestock Research Institute (BLRI), Bangladesh*

Estimation of heritability (h2) and its use in predicting animals' breeding are keys to selection and breeding programs aiming to livestock improvement. In this study, heritability of birth weight (BWT), weaning weight at 3-month (WWT), 6-months body weight (SMWT), and growth rates from birth to weaning (GR 1), birth to 6-month (GR 2) and weaning to 6-month (GR 3) were estimated using data collected from 607 pedigree recorded growing Black Bengal goats during 2017–2020 of Bangladesh Livestock Research Institute (BLRI), Savar, Dhaka, Bangladesh. After basic statistical analyses, variance components and heritability were estimated using an animal model fitted in Variance Component Estimation (VCE) 4.2.5 software and then obtained heritability values were used to predict breeding values of animals for each trait with Prediction and Estimation (PEST) software. The least squares means of BWT, WWT, SMWT were 1.16 ± 0.10, 5.55 ± 0.05, 8.76 ± 0.10 kg and of GR 1, of GR 2 and of GR 3 were 48.73 ± 0.63, 41.89 ± 0.59 and 34.57 ± 0.91 g day^{-1}, respectively. Parity of dam, type of birth, season of birth and year of birth and some interactions among them had significant effect on growth at different stages. Parity of dam significantly affected ($p < 0.01$) birth weight only and birth weight increased with the progress of parity. Season of birth significantly influenced ($p < 0.01$) all the growth traits except BWT and similarly, type of birth had significant effect ($p < 0.01$) on all growth traits except GR 3. Year of birth had significant influence on BWT ($p < 0.001$), SMWT ($p < 0.01$) and post-weaning growth rate whereas sex had no significant effect on any of the growth traits. The heritability $\pm$ standard errors were 0.46 ± 0.02, 0.42 ± 0.05, 0.47 ± 0.03, 0.43 ± 0.04, 0.48 ± 0.02 and 0.49 ± 0.02 for BWT, WWT, SMWT, GR 1, GR 2 and GR 3, respectively. Estimated higher heritability's indicates that selection would be effective if based on them. The mean of predicted breeding values were 0 .0010 for BWT, 0.0013 for WWT, 0.0027 for SMWT, -0.0033 for GR 1, 0.0127 for GR 2 and -0.0011 for GR 3, respectively. The predicted breeding values of the traits on each animal could well be used in selecting candidates in the on-going goat improvement program.

Keywords: Black Bengal goat, breeding value, growth traits, heritability

Contact Address: Ishrat Jahan, Bangladesh Agricultural University (BAU), Animal Breeding and Genetics, Bau Campus, 2202 Mymensingh, Bangladesh, e-mail: ishratshapla@gmail.com

A Marker- and Runs of Homozygosity-Based Assessment of the Inbreeding Status of South African Beef Cattle Populations

Simon Lashmar, Este Van Marle-Köster

University of Pretoria, Department of Animal Science, South Africa

Various factors including directional, within-breed selection practices may lead to a reduction in the genetic diversity (and subsequently, an increase in the autozygosity) of indigenous beef cattle populations. These local breeds form an integral part of the South African (SA) beef-producing industry, and contribute to commercial, communal and smallholder production systems. The loss in genetic diversity may have a detrimental impact on their ability to adapt to the stressors that will accompany expected changes in their environment. It is therefore crucial to monitor and manage the level of inbreeding, and consanguineous mating. The aim of this study was to perform a genomic characterisation and quantification of the inbreeding and runs of homozygosity (ROH) in four SA breeds, namely the SA Boran (n=270), Drakensberger (n=299), Nguni (n=291), and Tuli (n=155). Single nucleotide polymorphism (SNP) genotypes, generated using the Illumina® Bovine 150K panel, were subjected to standard call rate (0.05), MAF (0.02) and Hardy-Weinberg Equilibrium ($p < 0.001$) quality control procedures and a subset of 130,475 autosomal SNPs were retained for marker- and ROH-wise inbreeding estimation as well as the profiling of genome-wide ROH segments. The analysis identified 20,015 ROH segments in total, with mean population-wide numbers ranging from 3,235 for the Nguni to 4,723 for the SA Boran. The mean across-population ROH length was 6.24 Mb, and the highest frequency of ROH were observed in the shortest length category (< 4Mb), corresponding to more a more distant introduction of inbreeding, for all populations. Across all breeds the F(IS) and F(ROH) estimates were small and positive, which is indicative of low levels of inbreeding; the F(IS)values ranged from -0.05 for the SA Boran to 0.026 for the Nguni, whilst the F(ROH) values were 0.002, 0.003, 0.002 and 0.003 for the SA Boran, Drakensberger, Nguni and Tuli breeds, respectively. Overall, the degree of inbreeding was found to be at an acceptable and manageable level, posing no imminent threat to the genetic diversity, which is important to sustain the breeds' adaptive capabilities. The frequency and lengths of the ROH identified furthermore corresponded to the individual breed histories and the selection pressures to which each breed is exposed.

Keywords: Autozygosity, cattle, indigenous, single nucleotide polymorphism

Contact Address: Simon Lashmar, University of Pretoria, Department of Animal Science, Cnr Lynnwood Road and Roper Street Hatfield, 0028 Pretoria, South Africa, e-mail: simon.lashmar@up.ac.za

Effect of Exoticness and Agroecological Zones on Selected Production and Fertility Traits in Multibreed Dairy Cattle in Kenya

Richard Dooso Oloo[1], Chinyere Ekine-Dziyenu[2], Julie Ojango[2], Raphael Mrode[2], Ali Mwai Okeyo[2], Mizeck Chagunda[1]

[1]*University of Hohenheim, Inst. of Agric. Sci. in the Tropics (Hans-Ruthenberg-Institute), Germany*

[2]*International Livestock Research Institute (ILRI), Kenya*

The aim of this study was to evaluate the effect of degree of exoticness of multi-breed dairy cattle on production and fertility traits in 3 different agroecological zones of Kenya. Test-day milk yield (MY) records (n = 62321) together with fertility-trait records on age at first calving (AFC) (n = 1490) and calving interval (CI) (n = 2640) from a total of 1490 dairy cattle performing in semi-arid arable, semi-arid pasture-based, and semi-humid agroecological zones were analyzed. Animals were grouped into two breed classes based on the proportion of exoticness in their breed composition. These groups were, Exotic Class 1 (EC1) (≤50 % exotic, n = 481) and EC2 (>50 % exotic, n = 1009). A multiple linear regression model was fitted for AFC and a mixed-repeatability model for test-day MY and CI to determine the effect of exoticness on these traits and to assess whether this effect changed in the different agroecological zones. Overall, EC2 cows had lower AFC than EC1 cows (32.4, se = 0.2 vs. 34.0, se = 0.2 months). However, EC1 cows had a shorter CI than EC2 cows (452, se = 6 vs 466, se = 7 days). Within breed group comparison showed that EC1 had a larger AFC of 36.7 months (se = 0.4) in the semi-arid pasture-based agroecological zone compared to 31.0 (se = 0.6) in the semi-humid environment. This denotes delayed puberty in the semi-arid pasture-based agroecological zone. For the EC2, however, it was in the semi-arid arable agroecological zone where cows had the higher AFC compared to the semi-humid environment (34.7, se = 0.2 vs. 28.9, se = 0.3). In both breed groups, MY was highest in semi-humid and lowest in semi-arid pasture-based environment. Although the semi-humid agroecological zone seemed to favor the onset of puberty and high milk yield, this environment had the longest CI for both breed groups (478, se = 9 days for EC1 and 484, se = 7 for EC2). Genotype by environment interaction was significant for AFC and MY ($p < 0.01$). These findings demonstrated that biophysical variation in different agroecological zones affects production and fertility traits in multibreed dairy cattle differently and hence, it is an important factor to consider when designing genetic improvement programs.

Keywords: Agroecological zones, dairy cattle, degree of exoticness, genotype by environment interaction

Contact Address: Richard Dooso Oloo, University of Hohenheim, Inst. of Agric. Sci. in the Tropics (Hans-Ruthenberg-Institute), Stuttgart, Germany, e-mail: richard.oloo@uni-hohenheim.de

Evaluating the Genetic Diversity and Autozygosity of Indigenous South African Sheep Breeds Characterized by Small Population Sizes

Anel Retief, Carina Visser

University of Pretoria, Department of Animal Science, South Africa

Indigenous South African (SA) sheep populations are characterised by relatively small population sizes, with some at risk of becoming endangered. These unique genetic resources are utilised in the communal and smallholder systems, and generally disregarded in the commercial sector. The small population sizes could possibly increase the likelihood of bottleneck events and thus the loss of genetic variation. This could lead to an increased risk of adaptive traits as well as increased inbreeding levels. The aim of this study was to investigate genetic diversity parameters and marker- and runs of homozygosity-based inbreeding levels of SA indigenous sheep breeds. Single nucleotide polymorphism (SNP) genotypes were available for 158 animals representing five indigenous SA sheep populations namely the Black Headed Persian (BHP), Damara (DAM), Namakwa Afrikaner (NAM; endangered), Pedi (PED; endangered) and non-descript Fat-tailed sheep (FTT). The available SNP genotypes, generated using the Illumina Ovine 50K SNP BeadChip, were subjected to standard call rate (0.05), MAF (0.02) and Hardy-Weinberg Equilibrium ($p < 0.001$) quality control procedures and a subset of 48,753 autosomal SNPs were retained for diversity analysis. The study showed moderate levels of heterozygosity, with HO (HE) values ranging from 0.326 (0.335) for FTT sheep to 0.356 (0.374) for PED. The number of ROH ranged from 30 for DAM and 93 for PED and the majority of ROH segments were short in length (1 to 3.99 Mb). Inbreeding coefficients were all positive but low, with FROH ranging from 0.002 (DAM and NAM) to 0.007 (FTT). Lastly, the predicted effective population sizes (Ne) were extremely small for all populations; ranging from Ne=35 (BHP) to Ne=90 (PED). The moderate heterozygosity levels, relatively large within- and between breed variation, together with the low levels of overall inbreeding, indicate that despite the small population sizes, sufficient genetic variation exists to allow effective conservation and sustainable management of these populations.

Keywords: Conservation, diversity, inbreeding, indigenous

Contact Address: Anel Retief, University of Pretoria, Department of Animal Science, 50 Sanlam Street, 0181 Pretoria, South Africa, e-mail: u15032800@tuks.co.za

Genetic Diversity at Whole Genome and Chromosome 16 of Indigenous Chicken in Rwanda

Esther Mbakaya, Ngeno Kiplangat, Thomas Muasya
Egerton University, Dept. of Animal Science, Kenya

Indigenous chicken (IC) farmers in developing countries desire enhanced disease resistance alongside improvement of body weight and egg production. This study aimed at providing insight into the population structure and immunogenetic variability of the indigenous chicken using various methods. Population structure of IC was analysed through genotypic clustering, admixture analyses and phylogenetic relationship for the whole genome and at chromosome 16. A total of 150 IC sampled from five agro-ecological zones in Rwanda were phenotyped for Newcastle disease alongside body weight and genotyped with the genotyping-by-sequencing (GBS) method. After quality control procedures for SNP data, 65,945 SNPs retained for analysis. Following PCA, the IC were grouped into two genetic clusters, which were confirmed by lowest CV error (0.51) rates at K = 2. Population structure assessments based on SNPs in the MHC region indicated that the population as one with lowest CV error (0.50) rates which was confirmed at K =1. Clusters one a mean body weight and antibody titre of 1673.61 $\pm$ 237.14g and 4912.5 $\pm$ 55.35, respectively. Corresponding values for cluster 2 were 1311.34 $\pm$ 121.9g and 8832.5 $\pm$ 55.36. The clusters differed significantly ($p < 0.001$) for body weight and antibody titre. The cluster with low mean in bodyweight (Cluster 1) and high mean in titre and vice versa. The IC in Rwanda have been selected naturally for disease resistance against Newcastle. The observed genetic diversity of IC for BW and their negative association should be considered when designing a selection programme to ensure sustainability, flexibility and simultaneous improvement of the two traits.

Keywords: Disease resistance, MHC, sustainability

Contact Address: Esther Mbakaya, Egerton University, Dept. of Animal Science, 536-20115, 20100 Nakuru, Kenya, e-mail: esthermbakaya@gmail.com

Genomic and Immunogenic Variations of Indigenous Chicken in the Tropics

Esther Mbakaya, Kiplangat Ngeno, Thomas Muasya
Egerton University, Dept. of Animal Science, Kenya

Indigenous chicken (IC) farmers in developing countries desire enhanced disease resistance alongside improvement of body weight and egg production. This study aimed at providing insight into the population structure and immunogenetic variability of the indigenous chicken using various methods. Population structure of IC was analysed through genotypic clustering, admixture analyses and phylogenetic relationship for the whole genome and at chromosome 16. A total of 150 IC sampled from five agro-ecological zones in Rwanda were phenotyped for Newcastle disease titer alongside body weight and genotyped with the genotyping-by-sequencing (GBS) method. After quality control procedures for SNP data, 65,945 SNPs were retained for analysis. Following PCA, the IC were grouped into two genetic clusters, which were confirmed by lowest CV error (0.51) at K = 2. Population structure assessments based on SNPs in the MHC region indicated that the population as one with lowest CV error (0.50) which was confirmed at K =1. Clusters one mean body weight and antibody titre of 1673.61 $\pm$ 237.14g and 4912.5 $\pm$ 55.35, respectively. Corresponding values for cluster 2 were 1311.34 $\pm$ 121.9 g and 8832.5 $\pm$ 55.36. The clusters differed significantly ($p < 0.001$) for body weight and antibody titer. The cluster with high mean in bodyweight (Cluster 1) and low mean in titer and vice versa. The IC genetic clusters in Rwanda have variation disease resistance, which can be attributed to varied selection pressure. The observed genetic diversity of IC for BW and their negative association should be considered when designing a selection programme to ensure sustainability, flexibility and simultaneous improvement of the two traits.

Keywords: Disease resistance, MHC, sustainability

Contact Address: Esther Mbakaya, Egerton University, Dept. of Animal Science, 536-20115, 20100 Nakuru, Kenya, e-mail: esthermbakaya@gmail.com

DGAT1 Affects Milk Yield in Sudanese Butana × Holstein Crossbred Cattle

Salma Elzaki[1,2], Paula Korkuc[2], Danny Arends[2], Gudrun A. Brockmann[2]
[1]*University of Khartoum, Dept. of Animal Genetics and Breeding, Sudan*
[2]*Humboldt-Universität zu Berlin, Albrecht Daniel Thaer-Institute for Agricultural and Horticultural Sciences, Germany*

In Sudan, the *Bos indicus* zebu cattle Butana is known for its good milk yield and adaption to extreme environmental conditions. For higher milk yield, Butana cattle have been crossed with Holstein Friesian cattle, resulting in a crossbreed as robust as Butana. The genetic selection for higher milk yield using the well-described K232A polymorphism in *DGAT1* in Sudanese breeds would be helpful to meet the increasing demand for milk and dairy products in Sudan. As previously reported, the protein variant with lysine (K) of the K232A polymorphism was associated with an increase in fat and protein content and a decrease in protein and milk yields. In this study, we investigated the K232A polymorphism of *DGAT1* using the marker rs109234250 (14:611,019 G/A) in 93 purebred Butana and in 203 Butana × Holstein crossbred cattle from Sudan. The allele G corresponds to the protein variant with lysine (K). Allele-specific genotyping was performed using KASP assays. Association analysis with milk production traits and somatic cell score was performed using linear mixed models in R. In purebred Butana cattle, the allele G of the investigated *DGAT1* marker corresponding to the protein variant with K was much higher with a frequency of 0.929. In Butana × Holstein crossbred cattle, allele G had a frequency of 0.394, and significant associations were found with milk yield ($P=7.6 \times 10^{-20}$), fat yield ($P=2.2 \times 10^{-17}$), protein yield ($P=2.0 \times 10^{-19}$), and lactose yield ($P=4.0 \times 10^{-18}$). However, the correlation between these associated traits was high ($r \geq 0.892$). The minor allele G was disadvantageous for the associated traits. No significant associations were found for fat, protein, and lactose content, and somatic cell score. The outcome of this study could explain the higher milk fat content and lower milk yield of Butana cattle, as they had a high frequency for the allele G of the investigated *DGAT1* marker corresponding to these properties. In Butana × Holstein crossbred cattle, we confirmed previously reported effects of this polymorphism on milk and protein yield. Our results could be used for effective selection and thus genetic improvement of milk traits in Butana × Holstein crossbred cattle, which might be helpful in future breeding plans.

Keywords: Association analysis, *Bos indicus*, *DGAT1* gene, genotyping

Contact Address: Salma Elzaki, University of Khartoum, Dept. of Animal Genetics and Breeding, Khartoum North, Sudan, e-mail: elzakisa@hu-berlin.de

Prevalence/worm Load of Cestodes & Nematodes in Rural Chicken in the Northern Region, Ghana

Francis Addy, Gideon Adu-Bonsu, Osman Dufailu

University for Development Studies, Department of Biotechnology, Ghana

Rural chicken production is an important agricultural enterprise in rural communities in Ghana. Chickens are kept for purposes of providing nutrition to the family and for their socio-cultural and economic benefits. Generally, rural chickens are kept under extensive or at best semi-intensive production systems that is characterised by low productivity of birds due to poor housing and nutrition, and infections. Parasitic infection is a major problem in rural chicken production where they cause economic losses as a result of low feed intake, reduced growth rate, low eggs production, weight loss and treatment cost. We investigated the prevalence and worm load in rural chicken in the Northern Region of Ghana. A total of 22 healthy-looking rural chicks aged 5 – 8 weeks were sampled in April 2021. The birds were killed and the GIT dissected longitudinally: oesophagus-cloaca. dissected GIT was visually inspected and worms were isolated, washed in normal saline and fixed in ethanol. Worms collected were broadly grouped into cestodes and nematodes. Prevalence values of the two groups were: cestodes 77.3 %, nematodes 59.1 % and mixed cestode/nematode infections was 45.5 %. A total of 329 cestode and 72 nematode worms were recovered which translated into mean worm load of $\bar{x} = 19$ worms (range = 2–69 worms) and $\bar{x} = 6$ worms (range = 1–46 worms), respectively. In the 10 birds that suffered cestodes/nematodes mixed infections, worm load varied between 4–73. This data revealed the importance of parasitic cestode and nematodes in rural chicken production in northern Ghana. These birds would have suffered suboptimal growth rate under the high parasitic worm load seen here. We attributed the high prevalence values to the extensive production system under which the chicks were raised whereby birds feed/scavenge on contaminated feed resources or on arthropods, worms and insects that serve as intermediate hosts for various poultry worms. The cestode and nematode species involved are yet to be identified and/or characterised for good appraisal of their diversity.

Keywords: Cestodes, nematodes, northern Ghana, rural chicken

Contact Address: Francis Addy, University for Development Studies, Department of Biotechnology, Nyankpala campus, Tamale, Ghana, e-mail: francisaddy02@gmail.com

Animal feed and nutrition

Oral Presentations

Posters

Effect of Ground Perilla Seed Cake Supplementation in Pig Diets on Intramuscular Fat of Crossbred Pigs

Chaiwat Arjin[1], Prapas Mahinchai[2], Supamit Mekchay[1], Wolfram Spreer[3], Korawan Sringarm[1]

[1]*Chiang Mai University, Dept. of Animal and Aquatic Science, Thailand*

[2]*Chiang Mai Livestock Research and Breeding Center, Dept. of Livestock Development, Thailand*

[3]*Maejo University, International College, Thailand*

Ground perilla seed cake (GPC) is a by-product of *Perilla frutescens* oil extraction. It has a high content of crude protein and is a source of enriched omega-3 polyunsaturated fatty acids. Thus, GPC has a potential as an additive in pig feed to improve the intramuscular fat (IMF) content. However, the effect has not been sufficiently researched. The measurement of IMF requires to kill the animals for meat analysis. The use of real-time ultrasound to measure backfat thickness and loin muscle area represents a low-cost alternative which is accurate, safe and delivers data instantaneously. This study aimed to determine IMF in live pigs that received GPC feed supplements using real-time ultrasound.

Twenty-one crossbred pigs (Chiang Mai Black Pig) with an initial average weight of 73.91 kg, were divided into three groups. Three dietary treatments were fed during an experimental period of 12 weeks: (a) control (CON): basal diet, (b) treatment 1 (GPC 2.5): basal diet with 2.5 % supplemented by GPC, and (c) treatment 2 (GPC 5): basal diet with 5 % supplemented by GPC. Every four weeks the pigs were scanned with an ultrasound device to determine IMF percentage, backfat depth, loin area, and loin dept at the 10th to the 13th rib.

The ultrasound measurements did not exhibit significant differences between the dietary groups during week 0, 4, and 8. At week 12, both groups which had received a diet supplemented with GPC had a significantly higher IMF percentage as compared to the control group ($p < 0.05$). Also, the loin area results similar to the IMF result: The GPC 2.5 and GPC 5 were significantly higher than in the control group ($p < 0.05$). Simultaneously, no significant differences were found between GPC 2.5 and GPC 5 on the IMF and loin area.

In conclusion, real-time ultrasound is a suitable technology for monitoring the effects of feed supplements in live pigs, providing reliable data throughout the feeding experiment without the need for slaughtering the animals. It was shown that the supplementation of GPC in pig diets improves the intramuscular fat. However, more research is required to determine the suitable level of supplementation.

Keywords: Ground perilla seed cake, Longissimus dorsi, pork quality, real-time ultrasound

Contact Address: Korawan Sringarm, Chiang Mai University, Fac. of Agriculture, Dept. of Animal and Aquatic Science, 239 Huay kaew Rd., T.suthep Muang, 50200 Chiangmai, Thailand, e-mail: kanok70@hotmail.com

Potential of Integrating Organic Livestock Production in already Certified Organic Crop Production Systems in the Tropics

AIDA TESFAYE, CHRISTOPH REIBER, MIZECK CHAGUNDA

University of Hohenheim, Inst. of Agric. Sci. in the Tropics (Hans-Ruthenberg-Institute), Germany

Currently, there is an increasing shift towards safe and high quality food worldwide. This includes demand for organic animal products. In tropical countries, most who have certified organic crop production have their animal production excluded from certification. The objective of this study was to assess the potential of integrating and marketing organic livestock products in certified organic crop production systems in the tropics. Opportunities and challenges on this prospect were analysed using a systematic literature review together with seven expert interviews.

Results showed that the potential for integrating organic livestock production in already certified organic crop production systems and the marketing of organic animal products differs strongly depending on the countries' circumstances. One opportunity for the development is the rising demand for organic products. This is especially so in emerging countries like India and Brazil, where middle classes and incomes are growing, the potential for marketing organic animal products increases. In those countries organic markets, mainly for crop products, already exist which helps introducing organic animal products. In other less developed countries, markets for organic products are rare. Agricultural focus in such countries often lays on increasing the productivity, usually with conventional farming methods, therefore it is unlikely that the certification of organic animal products will play an important role there in the near future. Missing infrastructures, transportation options, and marketing options are major challenge in most of the countries. To push the production in the tropics, these obstacles need to be removed which is time and investments intensive. To promote the development of certified organic animal production in the tropics, further research focusing on organic animal production under tropical conditions is necessary. Equally important are initiatives of knowledge exchange with and among farmers. Further, training and educational work are needed to raise awareness, about the benefits of organic animal husbandry on the environment, human health and product quality, among producers and consumers.

Keywords: Certified organic crop production, infrastructure, organic livestock, tropics

Contact Address: Mizeck Chagunda, University of Hohenheim, Inst. of Agric. Sci. in the Tropics (Hans-Ruthenberg-Institute), Stuttgart, Germany, e-mail: mizeck.chagunda@uni-hohenheim.de

Farmers' Acceptance of Insects and Cassava as Alternative Feed Sources in Livestock Feeds in Kenya

Diana Wanda Odinya[1], Josiah Ateka[1], Robert Mbeche[1], Mathew Gicheha[2]

[1]*Jomo Kenyatta University of Agriculture and Technology, Dept. of Agricultural and Resource Economics, Kenya*

[2]*Jomo Kenyatta University of Agriculture and Technology, Dept. of Animal Sciences, Kenya*

Limited access to quality and adequate feeds represents a major challenge to smallholder livestock productivity in sub-Saharan Africa. To improve access to quality and adequate livestock feeds, policymakers and researchers are encouraging the utilization of non-conventional feed sources which are highly rich in proteins and energy. Cassava and insects have thus been promoted as alternative sources of energy and protein respectively. However, their uptake is low across smallholder livestock systems in sub-Saharan Africa. This study examined farmers' acceptance of insects and cassava as alternative feed sources and factors influencing their acceptance in Murang'a County in Kenya. The study applied the theory of planned behaviour and collected data from 378 dairy farming households in Kenya, identified using a simple random sampling. A multinomial logit model was used to assess factors influencing farmers' acceptance of insects and cassava as alternative feed sources in livestock feeds. The results indicate that at the time of the survey, only 11 % and 30 % of the dairy farmers were aware of insects and cassava as alternative feed sources, respectively. However, the farmers' willingness to use insects and cassava as alternative feed sources was high at 76 % and 86 %, respectively. The results of the multinomial logit regression model reveal that generally, farmers' attitudes towards the perceived benefits of using insects and cassava in livestock feed, subjective norms, perceived behavioural control, access to insurance and extension services, membership to a farmer group and farm size were significant determinants of farmers' intention to use insects and cassava as alternative feed sources. The study discusses the implications of these results in scaling up and commercialization of sustainable non-conventional feed sources.

Keywords: Energy source, multinomial logit model, non-conventional, protein source

Contact Address: Diana Wanda Odinya, Jomo Kenyatta University of Agriculture and Technology, Dept. of Agricultural and Resource Economics, P.O. Box 62000-00200, Nairobi, Kenya, e-mail: dianawanda119@yahoo.com

Assessing Knowledge, Attitudes, and Practices of Small-Scale Commercial Feed Producers in Uganda

Ben Lukuyu[1], Stella Namazzi[2], Pius Lutakome[1], Emily Ouma[1]

[1] *International Livestock Research Institute (ILRI), Feeds and Forages Program, Uganda*

[2] *National Agricultural Research Organisation (NARO), National Crops Resources Research Institute, Uganda*

In Uganda, pig production is a major source of livelihood for more than 1.1 million households. Pigs have increased from < 200,000 three decades ago to roughly 3.2 million today. Availability of well formulated and balanced rations for dairy cattle feeding is a major constraint limiting pig production. About 33 % of the dairy farmers use compounded dairy concentrate feed while nearly 56 % use feed ingredients. The commercial feed industry is dominated by small-scale feed compounders. Despite an abundance of energy based and protein-based ingredients, farmers continue to lament about the high prices, poor quality and limited access to compounded feeds. To improve quality of compounded feeds on the market, it is important to understand knowledge, attitudes, and practices (KAPs) of small-scale feed producers. A study was conducted to investigate the KAPs of small-scale commercial feed producers in four districts in Uganda. Eighty commercial feed producers were randomly selected and interviewed. Data were collected using a structured questionnaire. A Chi-square test was utilised to define the statistical difference between groups. Principal component analysis (PCA) was used to create attitudinal typologies. The results showed that over 50 % of the respondents were small-scale feed producers producing 0.2–2 tons per day. However, feed producers varied by gender, whereby 83 % were males and 24 % were females. Some 50 % of the male and female feed producers knew about protein requirement of piglets. Only 42 % demonstrated awareness about quality and safety standards. Of these, 47 % were knowledgeable about feed standards. Four attitudinal feed producer typologies i.e. quality Laissez-faire, knowledge, profit-oriented, and customer-oriented were identified. The common feed ingredients are maize bran (100 % among males and 92 % among female feed producers). Fishmeal, cottonseed cake, and sunflower cake are the dominant sources of protein. All respondents did not test for feed quality. Small-scale feed producers are aware of feed ingredients used in feed formulation. However, there are gaps in knowledge, attitudes, and practices about feeding standards, policies, regulation, production, use and marketing of feed. These findings will inform farmers, practitioners, and policy makers about potential interventions to improve feed production, use and marketing in Uganda.

Keywords: Compounded feeds, feed quality, feeding standards, pig feeding, pig production

Contact Address: Ben Lukuyu, International Livestock Research Institute (ILRI), Feeds and Forages Program, Naguru Drive, Plot 21, Kampala, Uganda, e-mail: b.lukuyu@cgiar.org

Effect of Supplementation of Air-Dried *Prosopis juliflora* Pods with Low Methane Production on Carcass Characteristics of Yearling Sheep

Assefa Tadesse[1], Aberra Melesse[1], Markus Rodehutscord[2]

[1]*Hawassa University (CLIFOOD-DAAD funded project), School of Animal and Range Sciences, Ethiopia*

[2]*University of Hohenheim, Dept. of Animal Nutrition, Germany*

Food from animal sources contributes 18 % of global calories (kcal) and 25 % of global protein consumption. However, livestock sector contributes to greenhouse gas emissions from which methane is the crucial one. In this regard, pods of unconventional feed resources were investigated through *in vitro* of which Prosopis juliflora pods (PJP) was identified as potential candidate in mitigating enteric methane (CH_4). This study was thus conducted to assess the suitability of supplementing PJP, which is known for its low CH_4 production potential, on body weight gain and carcass components in local sheep. A total of 25 yearling local sheep with initial body weight of 16.0±1.68 kg were randomly assigned to five treatment diets. All sheep received a basal diet of natural grass hay ad libitum and 270 g/head/day concentrate and the experiment lasted for 82 days exclusive of the adaptation period. The treatment diets contain a control diet without supplementation (T1) and diets supplemented with air-dried PJP at a rate of 50 (T2), 100 (T3) and 150 (T4) and 200 g/head/day (T5). The results indicated that the average slaughter weight of sheep reared on T1, T2, T3, T4 and T5 was 20.7, 21.5, 19.8, 19.7 and 19.2 kg, respectively and did not differ from each other ($p > 0.05$). Similarly, the average hot carcass weight was 8.01, 8.10, 7.73, 7.38 and 7.20 kg for yearling sheep fed with T1, T2, T3, T4 and T5 diets and did not differ significantly. Although not significant, the highest dressing percentage and lumbar weight was observed in sheep fed with T3 diet. Sheep reared in T2 diet had numerically higher rib-eye area than those fed with other diets. Moreover, no significant difference was found in average daily body weight gain between treatments. In conclusion, the supplementation of PJP to local yearling sheep improved carcass components and can be recommend to supplement poor quality forages while reducing the enteric CH_4 production from small ruminants.

Keywords: Carcass characteristics, enteric methane, local sheep, natural grass hay, *Prosopis juliflora* pod, supplementation

Contact Address: Assefa Tadesse, Hawassa University (CLIFOOD-DAAD funded project), School of Animal and Range Sciences, Hawassa, Ethiopia, e-mail: assefa.tadesse.b@gmail.com

In vitro Ruminal Fermentation, Methane Production, and Nutritional Value of Different Tropical Feedstuffs

Muhammad Khairul Bashar, Eva Haese, Markus Rodehutscord
University of Hohenheim, Dept. of Animal Nutrition, Germany

In vitro gas production (GP), methane (CH_4) production, metabolisable energy (ME), and digestibility of organic matter (dOM) from 18 tropical feedstuffs were determined using the Hohenheim Gas Test (HGT) described by Menke and Steingass (Anim. Res. Develop. 1988, 7–55). In short, approximately 200 mg of feed sample (crop residues: Rice straw, urea molasses treated straw, Maize stover; common grasses: Napier grass, German grass, Para grass; silages: Maize silage, Napier silage; leguminous fodder: Ipil Ipil, *Gliricidia*, Alfalfa hay, *Moringa* tops; concentrates: maize, wheat, wheat bran, kashari bran, rice bran, mustard oil cake) were incubated with a rumen fluid buffer solution for 72 h to measure GP. The ME and dOM were calculated using equations 12f and 43f (Menke and Steingass, 1988). Additionally, 120 mg of feed samples were incubated for 24 h to determine the CH_4 concentration in the produced gas. Among the roughages, CP concentration of leguminous fodder (166–314 g kg^{-1} DM) was the highest, followed by the common grasses (52–147 g kg^{-1} DM) and the silages (94–106 g kg^{-1} DM), but their ADF, NDF, and ADL concentration were the lowest. The crop residues showed the lowest CP (44–70 g kg^{-1}DM) and the highest cell wall concentrations. The dOM and ME of wheat (87.8 % and 14.4 MJ kg^{-1} DM), maize (90.3 % and 13.8 MJ kg^{-1} DM), wheat bran (77.3 % and 11.4 MJ/kg DM), and kashari bran (71.5 % and 9.84 MJ/kg DM) were significantly ($p < 0.05$) higher than those of other feedstuffs. The same trend was perceived ($p < 0.05$) for CH_4 concentration (% of GP) and CH_4 production (L kg^{-1} dOM). Within roughages, dOM and ME of German grass (61.6 % and 7.4 MJ kg^{-1} DM) and Ipil Ipil (58.8 % and 8.2 MJ kg^{-1} DM) were higher ($p < 0.05$), whereas the CH_4 concentration (15.7 % and 14.7 %) and CH_4 production (42.9 L kg^{-1} dOM and 34.8 L kg^{-1} dOM) were lower compared to crop residues and other common grasses. In conclusion, these results support the formulation of balanced rations for ruminants with higher digestibility and less CH_4 production when using commonly available feed resources. This may enhance animal productive performance while reducing the impact of animal production on the environment.

Keywords: Chemical composition, feedstuffs, *In vitro* digestibility, Methane emission

---

Contact Address: Markus Rodehutscord, University of Hohenheim, Dept. of Animal Nutrition, 70593 Stuttgart, Germany, e-mail: markus.rodehutscord@uni-hohenheim.de

Response of Sheep to Supplementations with *Leucanea leucocephala* Leaves on Weight Gain and Carcass Components

Aberra Melesse[1], Assefa Tadesse[1], Markus Rodehutscord[2]
[1]*Hawassa University, Animal and Range Sciences, Ethiopia*
[2]*University of Hohenheim, Dept. of Animal Nutrition, Germany*

Through *in vitro* studies, *Leucanea leucocephala* leaf (LLL) was identified as potential candidate in mitigating enteric methane (CH_4). Nevertheless, its efficiency as suitable animal feed source has not yet been determined in farm animals. This study was thus conducted to assess the suitability of supplementing LLL on body weight gain and carcass components in local sheep. Thirty yearling local sheep with initial body weight of 20.0±1.31 kg were first blocked by live weight into five groups and then individuals from each group were randomly allocated into five treatment diets with six sheep each. All sheep received a basal diet of natural grass hay ad libitum and 357 g per head and day concentrate mix. The treatment diets contain a control diet without supplementation (T1) and diets supplemented with LLL (g/head/day) at a rate of 50 (T2), 100 (T3) and 150 (T4) and 200 (T5). The experiment lasted 91 days exclusive of the adaptation period. The results revealed that the average daily weight gain did not differ ($p > 0.05$) among sheep fed with various levels of LLL supplementations. However, sheep fed with T3 diet had higher ($p < 0.05$) slaughter weight (25.8 kg) than those of T1 (22.9 kg) diet. Although insignificant, the average hot carcass weight was highest in sheep fed with T2 (10.6 kg) and T3 (10.2 kg) diets. The dressing percentage values ranged from 39.4 % in T3 to 42.0 % in T1 and were not affected by LLL supplementation. The highest foreleg carcass weight (1.97 kg) was obtained from sheep fed with T2 diet being significantly higher than those of T1 (1.80). Sheep fed with T3 diet had the highest hind leg (2.36 kg) and thorax (2.75 kg) values. The lumbar weight was highest in sheep reared in T2 and T5 diets while the lowest in T1. The highest rib-eye area (cm^2) was observed in sheep fed with T2 (14.7) followed by T3 (14.5) diets while the lowest in those of T1 (13.0). In conclusion, LLL supplementation improved carcass components of sheep compared with the control diet indicating its potential as alternative protein supplement to poor quality forages with possible reduction in enteric CH_4 production.

Keywords: Carcass components, enteric methane, local sheep

Contact Address: Aberra Melesse, Hawassa University, Animal and Range Sciences, Hawassa University (CLIFFOOD-DAAD Funded Project), 05 Hawassa, Ethiopia, e-mail: a_melesse@uni-hohenheim.de

Complete Replacement of Fish Meal with Potential Aquafeed Ingredients for Rainbow Trout in Iran

Hamed Salehi, Stefan Reiser, Ulfert Focken
Thuenen Institute, Fisheries Ecology Branch Bremerhaven, Germany

Iran as the largest producer of rainbow trout in the world needs sustainable aquafeed resources to fulfil the needs of this growing industry. Therefore, the potential feed ingredients were evaluated in fish feeding trials for their suitability and use in aquafeeds before introducing them into the market. The fish growth performance and apparent digestibility coefficients (ADCs) of nutrients were investigated for three fish meal free diets in comparison to a commercial diet and a casein-based semi-synthetic laboratory standard diet. Formulated diets were designed to meet the nutrient requirements of rainbow trout and have protein and lipid contents equal to commercial diet. TiO_2 was included as an indigestible marker to measure the digestibility of experimental diets. Two-hundred-fifty-two juvenile rainbow trouts with an initial average weight of 30 $\pm$ 3 g were stocked randomly in 18 experimental 57-l rectangular glass aquaria for 72 days. The diets were allocated to the aquaria in three replicates in a random-block-design. Feces were collected via settling method for four weeks at the end of the experiment. The results showed daily instantaneous growth rate, weight gain, feed intake and feed conversion ratio did not differ significantly among the diets. The plant-and-animal-based diet had similar protein efficiency ratio (PER) and nitrogen productive value (NPV) to commercial and laboratory standard diets. Apparent digestibility coefficients for crude protein were considerably different only between pure casein-based diet and plant-and-animal-based diet with the values of 98.7 $\pm$ 0.04 % and 90.7 $\pm$ 0.29 %, respectively. Our observations indicated that an acceptable performance could be achieved by applying poultry slaughterhouse by-products, canola meal and crystalline amino acids in carnivorous fish species without any fish meal and this may help to protect our planetary resources such as demersal fish stocks in seas.

Keywords: Aquaculture feeds, Iran, plant proteins, rainbow trout, rendered animal products

Contact Address: Hamed Salehi, Thuenen Institute, Fisheries Ecology Branch Bremerhaven, Herwigstr. 31, 27572 Bremerhaven, Germany, e-mail: hamed.salehi@thuenen.de

Effect of Vitamin E Alpha-Tocopherol Supplementation on Haemato-Biochemical Profiles of Race Stallions Horses in Khartoum State

Nahla Elhassan, Khadiga Khadig Atti

University of Khartoum, Animal Nutrition, Sudan

This study was conducted with the objective of determining the effect of vitamin E alpha-tocopherol supplementation on haemato-biochemical parameters of stallion race horses in Khartoum State, Sudan. The study was carried out using 21 thoroughbred animals with an average body weight of 300 ± 60 kg and age range of 7 – 8 years. The horses were dosed with three different levels of vitamin E (0, 2000, 3000 IU kg^{-1} feed dry matter) with 7animals assigned per treatment. The experiment comprised two periods, namely winter (November to December 2016) and summer (March to April 2017). Blood sample were collected from the jugular vein at the end of each period. Whole blood and serum samples were used to determine haemato-biochemical parameters, and statistical analysis was performed using SPSS v20.

The results indicated that serum vitamin E level and clotting time increased significantly in both seasons. Packed-cell volume (PCV), Hb (hemoglobin), TRBCS (Total Red Blood Cells), mean corpuscular volume (MCV), mean corpuscular haemoglobin (MCH) and mean corpuscular haemoglobin concentration (MCHC) increased significantly with increasing level of vitamin E in both winter and summer season, whereas white blood cells (WBC) decreased in summer season, and neutrophil- and basophil-ratios decreased in winter season. With increasing vitamin E dosage, serum concentrations of triglycerides, cholesterol, albumin, globulins and uric acid decreased significantly, while concentrations of glucose and high-density lipoprotein increased in both seasons. Significant decreases in both seasons were also recorded for the concentration of liver enzymes (aspartate aminotransferase - AST; alanine aminotransferase - ALT; alkaline phosphatase - ALP) and kidney function indicators (creatinine and urea concentrations).

Since vitamin E supplementation had a positive influence on some haematological and serum biochemical parameters in stallion race horses, it is recommended to use vitamin E to enhance the animals' capacity for race performance.

Keywords: Haemato-biochemical parameters, race horses, vitamin E alfatocopherol

Contact Address: Nahla Elhassan, University of Khartoum, Animal Nutrition, Omdurman Althora Alhara 17, 11111 Omdurman, Sudan, e-mail: nahlavetxx@hotmail.com

The Latin American Forage Seed Market: Recent Developments and Future Opportunities

Irieleth Gallo Caro, Karen Enciso Valencia, Valheria Castiblanco, Stefan Burkart

International Center for Tropical Agriculture (CIAT), Trop. Forages Program, Colombia

Cattle farming in the Latin American (LA) lower tropics is characterised by extensive production systems with low productivity levels and negative environmental impacts. Given the sector's economic and social importance, we need to understand the dynamics involved in the adoption of sustainable production technologies (e.g. improved forages). Among the limiting factors in this regard is forage seed availability. Our study has two objectives: i) analysing behaviour and potential of the LA forage seed market; and ii) identifying limitations/opportunities in forage seed commercialisation. Information was obtained in 2020–2021 through literature review, database analysis on import and export of forage seeds, and in-depth interviews with 13 principal LA forage seed suppliers. Through a descriptive market behaviour analysis, the characteristics and functioning of the forage seed market were detailed for several LA countries (i.e. involved actors, market evolution and expansion limitations, informal market and future prospects). The market potential was defined from a sustainable intensification perspective, that includes the areas currently used for cattle production requiring improvement strategies (productivity increases and environmental impact decreases). Our results show that the forage seed market, in the recent decade, followed particular dynamics occurring in the analysed countries. Major fluctuations have occurred in most countries, due to e.g. regulatory policies and taxes for seed commercialisation. Some countries (e.g. Bolivia) show a continuous increase resulting from formalisation processes strengthening forage technology adoption. Changes in seed preferences towards new varieties occurred in some countries, while in others, traditionally marketed varieties maintained their market shares. Among the limiting factors for improved forage seed adoption, cultural aspects, scarce collaboration among actors (research/development institutions, seed producers/distributors), slow bureaucratic processes and constant seed price variations (associated with a dependence on the Brazilian market dynamics) were identified as crucial. All studied countries present high market potential for improved forage seeds. Whether it can be exhausted or not is subject to each country's specific dynamics that could encourage (e.g. conservation policies) or limit (e.g. land tenure insecurity) sustainable intensification processes. Our results help policy-/decision-makers in public policy formulation aimed at forage seed market development and serve seed producers/distributors for increasing market penetration.

Keywords: Cattle, improved forages, market potential, seed systems, sustainable intensification, technology adoption

---

Contact Address: Stefan Burkart, International Center for Tropical Agriculture (CIAT), Trop. Forages Program, Km 17 Recta Cali-Palmira, Cali, Colombia, e-mail: s.burkart@cgiar.org

Feed Resource Use Efficiency in Taurine and Zebu Cattle Raised on Natural Pastures in Benin

PÉNÉLOPPE G. T. GNAVO, RODRIGUE V. CAO DIOGO, LUC HIPPOLYTE DOSSA

University of Parakou, Dep. of Sci. and Techn. of Animal Prod. and Fisheries, Benin

In sub-Saharan Africa, optimising cattle production is key for the ecological and economic profitability of farms. However, there is a controversy about whether cattle breeds consume more energy and plant protein than they produce for human consumption. To test this claim, we determined feed conversion efficiencies for energy and protein (Digestible nitrogen: DN) of four farms comprising 270 cattle in the Sudanian zone of Benin. Two taurine and two zebu herds were compared to determine which of the herds utilise available fodder resources more efficiently.

In each herd, six animals (two cows, two heifers and two bulls) were monitored during three consecutive days for pasture dry matter (DM) intake. Milk production from all milking cows in the herd was monitored at bi-weekly intervals for three months and the milk composition determined using a Milkotester device (Milkotester Ltd., Bulgaria). Calves were weighed using an electronic scale and the weights of adult animals (heavier than 100 kg) were estimated from body measurements at monthly intervals.

Feed and nutrient intakes were similar in both herds. Per day, Taurine herds consumed on average 65.0 $\pm$ 1.13 g DM kg$^{-0.75}$; 94.9 $\pm$ 1.66 kcal kg$^{-0.75}$ day^{-1} and 144.2 $\pm$ 12.79 g DN day^{-1} while for zebu it amounted to 66.4 $\pm$ 8.40 g DM kg$^{-0.75}$; 97.1 $\pm$ 12.27 kcal kg$^{-0.75}$ day^{-1} and 190.2 $\pm$ 42.98 g DN day^{-1}. Milk production was significantly higher for zebu (2.4 $\pm$ 0.60L day^{-1}) than for taurine cows (1.4 $\pm$ 0.11 L day^{-1}; $p < 0.001$). However, its composition per kg was similar ($p > 0.05$) irrespective of breed and amounted to 36.1 $\pm$ 1.55 g DN and 796.9 $\pm$ 21.49 kcal for taurine and 37 $\pm$ 0.00 g DN and 809.6 $\pm$ 0.69 kcal for zebu cows. Protein (0.1 $\pm$ 0.05) and energy (0.1 $\pm$ 0.04) feed conversion efficiencies were similar across breeds.

The results indicate that the extensive system studied is more inefficient, regardless of the herd type considered. However, a ration exclusively based on natural grazing at the end of rainy season seems insufficient to cover the daily requirements of cattle in the Sudanian zone of Benin. Our study confirms that taurine and zebu breeds consume more energy and plant protein than they produce for human consumption and therefore supplementation of the animals is necessary to improve production efficiency.

Keywords: *Bos indicus*, *Bos taurus*, energy and protein efficiencies, rangelands

Contact Address: Rodrigue V. Cao Diogo, University of Parakou, Dep. of Sci. and Techn. of Animal Prod. and Fisheries, 123 Parakou, Benin, e-mail: dcao_bj@yahoo.fr

Feed Restriction and Compensatory Growth of Giant African Snails of the Species *Archachatina marginata* (Swainson, 1821)

Youssouf Toukourou, Chadrac Tsrigbidzi
University of Parakou, Fac. of Agronomy, Dept. of Sciences and Techniques of Animal and Fisheries Production, Benin

Giant African snails of the species *Archachatina marginata* were subjected to a feed restriction for 70 days which preceded a re-feeding phase also of 70 days. The objective was to study in this species the ability to compensate for growth retardation induced by temporary dietary restriction. The study is carried out at the application farm of the Faculty of Agronomy of the University of Parakou between August 15, 2019 and January 22, 2020. In total, 90 snails, weighing on average 52.48 ± 9.03 g with an average shell length of 6.83 ± 3.38 cm were randomly distributed into three lots of 30 subjects in semi-buried enclosures, made of cinderblock and fine mesh wire mesh. Three rations containing 20.26 %, 17.18 % and 14.43 % crude protein and 2976 kcal; 2540 kcal and 2089 kcal of metabolisable energy per kg of dry matter were distributed *ad libitum* to lots 1 (control), 2 and 3 respectively. The snails showed at the end of the feed restriction period an average shell length of 8.1 ± 0.54 cm, 8.11 ± 0.43 cm and 8.13 ± 0.5 cm (p > 0.05) for an average live weight of 79.6 ± 7.3 g, 68.86 ± 11.22 g and 66 ± 10.66 g ($p < 0.05$) respectively for lots 1, 2 and 3. At the end of the re-feeding phase, the shell length was 8.81 ± 0.51 cm, 8.80 ± 0.25 cm and 8.79 ± 0.46 cm ($p > 0.05$) for an average live weight of 92.59 ± 3.32 g, 88.5 ± 5.44 and 86.63 ± 7.3 g ($p < 0.05$), respectively for lots 1, 2 and 3. It emerges from this study that the weight loss observed during a feed restriction could not be fully compensated after a certain period of re-feeding in *A. marginata*, despite a remarkable increase in weight

Keywords: Animal nutrition, Benin, feed efficiency, growth perfoemance, micro-livestock, snail

Contact Address: Youssouf Toukourou, University of Parakou, Fac. of Agronomy, Dept. of Sciences and Techniques of Animal and Fisheries Production, B.P. 123, Parakou, Benin, e-mail: ytoukourou@gmail.com

Effects of three Plant Extracts on Growth Performance, Histopatholoy and Sensory Properties of Broiler Birds

Rosemary Ozioma Igwe[1], Jude T. Ogunnupebi[1], Isaac Ikechukwu Osakwe[2], Udo Herbert[3]

[1]*Ebonyi State University Abakaliki, Animal Science, Nigeria*

[2]*Alex-Ekwueme Federal University, Animal Science, Nigeria*

[3]*Michael Okpara University of Agriculture Umudike, Animal Breeding and Physiology, Nigeria*

The ban on the use of antibiotics due to their deleterious effects on consumers' necessitated the need for prophylactic use of plant extract. An experiment was conducted to evaluate the phytogenic effects of three plants extracts (Mango leaves, Pawpaw leaves and Guava leaves) on the performance, intestinal histopathology and sensory characteristics of broiler birds. A total of 128 day – old broiler chicks of Cobb strain were randomly divided into four treatment groups of 32 birds, each replicated four times with 8 birds per replicate in a completely randomised design (CRD). The groups were fed four diets. Diet 1(T1) is ordinary water, contained no extract and served as the control. Diets 2, 3, and 4 designated T2, T3 and T4, respectively, contained 100ml each of mango, pawpaw and guava leaves extract, respectively. The extract was obtained by squeezing 1000 g each of mango, guava and pawpaw fresh leaves in 1litre of water. There was no form of medication given to the birds outside routine immunisation. Data were collected on growth performance organ, histopathological parameters and sensory properties and analysed using Mini Tab Statistical software. Results showed that there were significant ($p < 0.05$) differences on the body weight gain, daily water intake and feed conversion ratio. Birds fed guava and pawpaw leaves extract significantly ($p < 0.05$) had higher body weight gain with low feed conversion ratio compared to the control and group fed mango. Group fed pawpaw leaves had higher final body weight of 2623 g followed by groups fed mango leaves, guava and control, 2400 g, 2232 g and 2098 g, respectively. There were no significant ($p > 0.05$) differences in the organ and intestinal histopathology but significant ($p < 0.05$) differences occurred in the sensory properties. Meat of birds on T2 and T3 were significantly ($p < 0.05$) more tender and juicy compared to those on T4 and T1. Results from this study indicate that the extracts had no deleterious effect on the wellbeing of the broiler birds but rather improved growth performance and welfare of the bird. Therefore, the use of plant extracts, especially pawpaw and guava leaves extracts, should be encouraged in broiler production.

Keywords: Broiler, guava, histopatholoy, leaves, mango, pawpaw

---

Contact Address: Rosemary Ozioma Igwe, Ebonyi State University, Abakaliki, Ebonyi State, Animal Science, Behind 7-Up Abakaliki-Enugu X-Press Way Ebonyi State Nigeria, Abakaliki, Nigeria, e-mail: alorosemaryozioma@gmail.com

Predicting Voluntary Dry Matter Intake of *Bos indicus* Cattle: A Case for Conceptual Models

Rodrigo Beltrame[1], Christian Adjogo Bateki[2], Uta Dickhoefer[2]

[1] *University Hohenheim, Inst. of Agricultural Engineering, Tropics and Subtropics Group, Germany*

[2] *University of Hohenheim, Animal Nutrition and Rangeland Management in the Tropics and Subtropics, Germany*

Productive performance of cattle in the (Sub-)Tropics is largely determined by their voluntary dry matter intake (VDMI). Conceptual mathematical models (CMM) have been suggested as a reliable option for predicting VDMI, although they have only been evaluated using meta-data. The present study assessed the reliability of three CMM to predict VDMI of 52 stall-fed (LW = 277.2 $\pm$ 74.1 kg) and 45 grazing cattle (LW = 237.7 $\pm$ 27.7 kg) from ICRISAT near Niamey, Niger, and the Station du Sahel in Niono, Mali, respectively. The CMM of Conrad et al. (1964) (C1), the modified Conrad et al. (1964) (C2), and modified Mertens (1987) (M4) were selected based on their reported adequacy to predict VDMI of cattle kept under tropical conditions. For C1 and C2, VDMI is estimated as 0.0107 * LW/(1-D) and as DDM/D, where 0.0107 (increased to 0.0116 for C2) is the daily fecal output (g/kg LW), LW liveweight (kg), D digestible fraction per unit of ingested dry matter, and DDM digestible DM intake (kg/d). For C1 and C2, the lower VDMI estimate of the two equations was retained as predicted VDMI, while for M4, DMI is the mean of equations, 0.0135 *LW/NDF and MEreq/MEd, where 0.0135 is the maximum intake of neutral detergent fiber (NDF) intake (g/kg LW), NDF the NDF concentration in the diet (g/kg DM), MEreq the animal's metabolisable energy (ME) requirements (MJ/d) and MEd the dietary ME concentrations (MJ/kg DM). Daily MEreq for maintenance was estimated as 0.631 MJ ME/kg $LW^{0.75}$ and for gain as 0.0243 MJ ME/g LW gain. The CMM were statistically evaluated through the mean bias (MB), root mean square error predicted (RMSEP), and relative prediction error (RPE).

The CMM predicted VDMI more accurately for grazing than stall-fed animals. For both datasets, the M4 had lower MB (-0.11 and -0.8 kg per animal and day and RPE (18.33) than C1 and C2 (MB $>$ 1.4 kg per animal and day and RPE $>$ 1.48). Therefore, M4 is the most reliable CMM however, further evaluation using better data is needed under stall-fed conditions.

Keywords: *Bos indicus*, conceptual models, voluntary dry matter intake

Contact Address: Rodrigo Beltrame, University Hohenheim, Inst. of Agricultural Engineering, Tropics and Subtropics Group, 70593 Stuttgart, Germany, e-mail: rodrigobeltrame@yahoo.com

Carcass Characteristics and Meat Quality of Finished Lambs Born to Supplemented Hamari Ewes under Range Conditions, Sudan

Omer Bushara[1], Babiker[2]

[1]*West Kordofan University, Dept. of Meat Production, Sudan*

[2]*University of Khartoum, Dept. of Meat Production, Sudan*

This study focused on carcass characteristics and meat quality of finished lambs to concentrate supplemented ewes on the open rage of Kordofan state, Sudan. The concentrate diet was offered to the dams on group at a rate of 500gper ewe and day. Ninety ewes of similar age and weight were divided into three groups (A), (B) and (C) with 30 animals each. (A) was given the supplement for 30 days before and after mating, (B) was given the supplement for 30 days before mating and (C) was left without supplement. Sixteen weaned lambs born to concentrate supplemented ewes and another equal number left to graze naturally was used for concentrate supplementation. Group (A) and (B) each contained eight lambs, while (C) contained sixteen lambs which were divided into two groups (C) and (D). All lambs were allowed to graze naturally. Lambs were given the diet for 60 day free choice, expect lambs of group (D) was allowed to graze on natural pasture only (control). Carcass characteristics, meat chemical composition and quality were studied on five lambs from each group. Statistical analysis indicated that there were no significant differences in slaughter, hot, cold and half carcass weights and empty body weight, among supplemented groups, but were significantly ($p < 0.05$) heavier compared to the control. No significant differences were found in dressing percentage and whole sales cuts between all groups. Only genital organs and fat depots were significantly ($p < 0.05$) heavier in supplemented groups than those on control. Carcass composition showed that concentrate supplementation resulted in significant ($p < 0.001$) increase in muscle, fat, and trim percentages than in control. Meat chemical composition showed higher protein and fat percentage and the reverse was true for moisture in supplemented groups than in the control. Meat quality attributes indicated that cooking loss decreased significantly ($p < 0.001$) while water holding capacity improved in meat of supplemented groups than in the control. It was concluded that concentrate supplementation of grazing lambs born to supplemented ewes enhanced lamb carcass weight and improved meat quality. Thus it is recommended to adopt concentrate supplementation of lambs to improve their carcass yield and their meat quality.

Keywords: Carcass, Hamari ewes, meat, range conditions

Contact Address: Omer Bushara, West Kordofan University, Dept. of Meat Production, Ommdadda Albuhyra Street, Omdurman, Sudan, e-mail: omerboshara@gmail.com

Fodder Quality Comparison in two Sorghum Populations under Drought

Vinutha K Somegowda[1], Prasad Kvsv[2], Jalaja Naravula[3], Anilkumar Vemula[1], Sivasubramani Selvanayagam[1], Abhishek Rathore[1], Chris S. Jones[4], Santosh P. Deshpande[1]

[1]*International Crops Research Institute for the Semi-Arid Tropics (ICRISAT), GG GTD, India*
[2]*International Livestock Research Institute (ILRI), India*
[3]*VFSTR, Biotechnology, India*
[4]*International Livestock Research Institute (ILRI), Kenya*

Digestibility and lignin content are very important in determining feed quality and plant fitness, with higher lignin content reducing digestibility and vice versa. A 5 % variation in fodder digestibility between poor and premium fodder is reported to result in a 20 % price variation. Sorghum is gaining importance as a food, feed, and fuel crop, it has similar feed quality to maize, that demands high nutrient and water availability. Additionally, drought in the semi-arid tropics is also affecting feed quality and sorghum is known to encounter drought mainly in the post rainy season. Therefore, two sorghum populations, a recombinant inbred line (RIL) population (n=320) and reference set (n=130) were evaluated under drought for fodder qualities. In this study, irrigation was withheld at the booting stage for the stress plots whereas the control plots were fully irrigated, all other crop management practices were performed equally. The dry weight (DW) was recorded at maturity, and the fodder was subjected to near infrared spectroscopy (NIRS) scanning to record: nitrogen content on dry matter basis (NDM); neutral and acid detergent fibre (NDF and ADF); acid detergent lignin (ADL); metabolisable energy (ME) and; *in vitro* organic matter digestibility (IVOMD). Significant variation was recorded across treatments and for all traits in the RIL population. Significant genotypic variation and genotype by treatment variation was recorded for the reference set in 2016 only. Pearson's correlation was pooled for across years and treatments for both populations and DW and IVOMD showed negative correlation with NDF and ADL, while positive correlations were observed between DW, ME and IVOMD in the RIL population. However, in reference set there was no strong positive or negative correlation between DW, ME and IVOMD. Genotype by sequence (GBS) analysis was used to perform quantitative traits loci (QTL) estimation for the RIL population while a genome wide association study (GWAS) was performed for the reference set. A total of 98 and 47 associated genes were extracted from Phytozome v12.1.6 for the RIL and reference populations respectively. Several genes belong to pathways that may help explain a causal functionality with the associated traits.

Keywords: Digestibility, fodder quality, GWAS, QTL, sorghum

Contact Address: Santosh P. Deshpande, International Crops Research Institute for the Semi-Arid Tropics (ICRISAT), GG GTD, 300 GG GTD ICRISAT, 502324 Hyderabad, India, e-mail: s.deshpande@cgiar.org

Evaluating Equations to Predict Organic Matter Digestibility Of tropical Ruminant Diets from Fecal Crude Protein Concentrations

Anil Kumar Gajaga Krishnappa, Joaquín Castro-Montoya, Uta Dickhoefer

University of Hohenheim, Animal Nutrition and Pasture Management in the Tropics and Subtropics (490i), Germany

The apparent total tract organic matter digestibility (OMD) is one of the primary parameters to evaluate nutritive value of feed consumed by ruminants. Estimating OMD *in vivo* requires abundant work and the need to collect the totality of faeces voided by the animal. To facilitate this estimation, equations based on the crude protein concentration in faeces have been developed: 1) Wang equation = 0.899–0.644 × exp (−0.5774 × faecal CP (g kg^{-1} OM) / 100)), 2) Peripolli equation = 0.7326 – 0.3598 exp[(−0.9052 CP (g / kg^{-1} OM)) / 100 and 3) Lukas equation = 79.76 – 107.7e (−0.01515 CP) have been developed to predict OMD in domestic ruminants fed forage-based diets.Since the equations were developed under specific experimental conditions, the potential of these equations to accurately predict OMD under other settings needs to be verified.Therefore, the potential of all the three equations was evaluated by comparing OMD estimated by these equations with the *in vivo* OMD measured in diets using tropical forages obtained from literature. In total, 224 *in vivo* measured OMD from different experiments conducted in the Tropics with cattle, sheep, and goats were included. Diets were selected so that the forage portion contained either 100 % legume, 50 % legume+50 % grass, or 100 %grass. Calculated OMD were regressed on the measured OMD and the slope, intercept, and R square were estimated to evaluate the performance of the equations using R-studio statistical software. A total of 12 comparisons were conducted, all R square were below 0.23 and slopes and intercepts ranged from 0.21 to 0.58 and from 200.7 to 546.7, respectively. Hence, the ability of all three equations to predict accurately OMD is very limited regardless of the characteristics of the forage in the diet. These results indicate that the equations developed to predict OMD from faecal N cannot be applied with confidence in a variety of tropical feeding conditions.

Keywords: Digestibility, ruminants ,tropical feedstuffs

Contact Address: Anil Kumar Gajaga Krishnappa, University of Hohenheim, Animal Nutrition and Pasture Management in the Tropics and Subtropics (490i), Egilostrasse 45 0213, 70599 Stuttgart, Germany, e-mail: anilgk5009@gmail.com

Willow and Panicum Silages Effects on Milk Yield and Quality of Black Mountain Goats

Sami Awabdeh, Rawad Swidan

National Agricultural Research Center, Livestock, Jordan

A study was conducted at al Khanasri station to evaluate the potential of willow and panicum silages as a forage source for lactating black mountain does and their nursing kids. Twenty-seven black mountain does and their kids were randomly assigned to one of the three dietary treatments (9 does per treatment); control group (CON.) were does fed wheat straw, a common relatively expensive forage source in Jordan, willow silage group (WS) were does fed willow silage, panicum silage group (PS) were does fed panicum silage as a source of forage of the diets. Concentrate were formulated accordingly to each diet for iso-caloric, iso-nitrogenous rations. Does were fed high concentrate diet ad libtum with 25:75 F:C ratio for 8 weeks of lactation. Intake and refusal were measured daily. Milk yield and milk component were measured biweekly. There were no differences in final body weight of does among groups. No significant different in milk yield between CON. and PS throughout the experiment and there were no significant different in milk yield between CON and WS group in the first 4 weeks of lactation. On the other hand, WS group were significantly different from other groups after 6 week of lactation (825 ml milk per day vs. 633 and 616 ml milk yield for CON. and PS, respectively) and the different continue to the end of the experiment. Total milk yield was significantly higher in WS compared to other groups for the entire experiment. No significant differences among treatment in total solids, fat, protein and lactose content. Cost/kg milk production (US$) was higher ($p < 0.05$) in PS and CON. groups compared with WS group. As a results, using willow silage in nursing doe's diets will increase milk production, with no changes in its components and reduce cost of milk production, which demonstrate a potential of willow silage to use as a forage source for black mountain goats.

Keywords: Black mountain goat, Panicum silage milk production, willow silage

Contact Address: Sami Awabdeh, National Agricultural Research Center, Livestock, Jarash Highway, 19381 Ain Basha, Jordan, e-mail: sami_awabdeh@yahoo.com

Determining the Factors Affecting the Adoption of Fodder Crops by Farmers in Ethiopia and Kenya

SAHIN KARAPACA[1], EVA SCHLECHT[1], NILS TEUFEL[2]

[1]*University of Kassel / University of Goettingen, Animal Husbandry in the Tropics and Subtropics, Germany*

[2]*International Livestock Research Institute (ILRI), Policies, Institutions & Livelihoods (PIL), Kenya*

Fodder crop production enables farmers to improve their livestock production and reach a higher level of self-sufficiency with livestock products. This study aimed to determine the level of adoption of fodder crops and assess factors that influence their adoption by farmers in selected regions of Ethiopia and Kenya. Study areas were selected purposively since forage production is limited to specific regions with intensifying dairy production in both countries. Village-level data was collected by the International Livestock Research Institute (ILRI) from Tigray, Amhara, Southern Nations Nationalities and Peoples Region (SNNPR), and Oromia regions of Ethiopia, as well as Upper Rift, Western, Nyanza, Central and Coast regions of Kenya, in 2015. Data was obtained from 180 villages or peasant-associations per country through group interviews based on a structured questionnaire. Descriptive statistics and a Tobit model were used to identify factors influencing the adoption of fodder crops.

Research findings show that fodder adoption intensity, expressed as the proportion of the total area allocated to fodder crop production, was 3.04 % and 10.66 % in Ethiopia and Kenya, respectively. The most commonly adopted species were Napier (*Cenchrus purpureus*, various varieties), Sesbania (*Sesbania sesban* (L.) Merr), and Rhodes grass (*Chloris gayana* Kunth) in Ethiopia; Napier, Calliandra (*Calliandra calothyrsus* Meisn.) and Rhodes grass in Kenya. The analysis revealed that altitude, mobile phone ownership, awareness about fodder, population size of dairy cows, the availability of a milk collection centre and milk marketing activities had a positive influence ($p < 0.05$) on the adoption of fodder crops, while distance to the nearest town, distance to the nearest all-weather road, arable land per farm household and population size of male (local) cattle had a negative effect ($p < 0.05$). In addition, the area of grazing land per farm household and female (local) cattle population of villages had positive and negative effects on fodder adoption in Ethiopia and Kenya, respectively.

Therefore, in regions where dairy cattle keeping predominates, the adoption of improved fodder crops should be enhanced by raising fodder awareness among farmers, providing infrastructure facilities such as all-weather roads, and promoting milk collection centres plus milk marketing activities.

Keywords: Dairy production, East Africa, extension, fodder cultivation

Contact Address: Nils Teufel, International Livestock Research Institute (ILRI), Policies, Institutions & Livelihoods (PIL), Kabete, Old Naivasha Road, PO Box 30709, 00100 Nairobi, Kenya, e-mail: n.teufel@cgiar.org

Dietary Sodium Diformate and Probiotic Yeast Improve Performance of Aging Laying Hens Against Positive Control

Christian Lückstädt, Anant Deshpande, Stevan Petrović

ADDCON, Germany

Annually, the layer industry in Asia suffers from poor bird health due to contamination with pathogenic bacteria, which often results in reduced performance and increased mortality. In this respect, organic acids' proven history against Gram-negative pathogenic bacteria in feeds, showing beneficial effects on health and performance, comes into effect. Sodium diformate (NDF) has been commercially used in layer production in tropical areas since more than a decade, but its impact on layers in South Asia was not yet thoroughly investigated, especially in combination with probiotics. A commercial trial was performed in India to measure the impact of 0.1 % dietary NDF and 0.05 % probiotic yeast on performance and health in laying hens from 50 weeks of age over a period of 30 weeks. The trial ran back to back: the first 15 weeks served as positive control, with the inclusion of antibiotic growth promoters. Treatment and control groups (approx. 25,500 Lohmann laying hens) each received a commercial diet throughout the trial. Feed intake was significantly lower in birds that received the NDF-yeast diet (106 v. 117 g/bird/day; $p < 0.001$), while hen-day egg production improved over the same period (88.5 v. 80.5 % in NDF and control groups, respectively; $p < 0.001$), despite the age of the birds. Hen-day production with the acidifier diet was >4 % above the standard given by the Lohmann management guide for birds of that age. Uniformity of lay was greatly improved, with a highly reduced standard deviation in the NDF-yeast fed birds. The lower feed intake, combined with increased laying performance, resulted in improved production efficiency (feed/egg) of the NDF-yeast birds (3.6 % - from 148 g to only 142 g feed per egg). Weekly mortality was significantly lower ($p < 0.001$) by almost 45 % in the group fed the acidified diet. These data show that sodium diformate (traded as Formi NDF), in combination with yeast, can improve performance and survival rates in layers under commercial tropical conditions in India and offer a viable alternative for antibiotic growth promoters.

Keywords: Layer, performance, sodium diformate, yeast

Contact Address: Christian Lückstädt, ADDCON, Parsevalstrasse 6, 06749 Bitterfeld-Wolfen, Germany, e-mail: christian.lueckstaedt@addcon.com

Energy Value and Crude Protein Fractions of Brewery Byproducts for Ruminants – Ethiopian Perspective

Ziade Abraha Kahsay, Christian Böttger, Karl-Heinz Südekum
University of Bonn, Inst. of Animal Science, Germany

Feed scarcity in terms of quantity and quality becomes a substantial issue in most tropical countries, and particularly in Ethiopia where more than two-thirds of the population own livestock. Brewery byproducts cost less than concentrate feeds and, as a supplement, could enhance protein and energy values of low quality cereal crop residues and herbage from pastures. In the current study, 21 samples of brewers grains (BG) and tella-atella (TA) – a byproduct of locally produced tella drink – obtained from Ethiopia were analysed using *in vitro* methods. Samples were subjected to analysis of chemical composition and crude protein (CP) fractionation. An *in vitro* rumen gas production technique was applied to estimate energy values. In BG, substantial contents of neutral detergent fibre (443–738 g/kg dry matter, DM), starch (43–255 g/kg DM), and ether extract (70–120 g/kg DM) contributed to mean estimated values of 8.4 and 6.1 MJ/kg DM for metabolisable energy and net energy for lactation, respectively, with a considerable variability between breweries. Samples of TA exhibited a comparable or higher energy value indicating the potential to be used as a supplement in ruminant feed. Particularly high starch concentrations (220–280 g/kg DM) in TA samples were most likely related to the brewing process and type of grains used. Both BG and TA contained high CP concentrations ranging from 187–299 and 160–219 g/kg DM, respectively. Moderately and slowly degradable CP fractions constitute together a mean proportion of more than three-fourth of CP. High content of cell wall associated CP fractions indicate low rumen protein degradation of BG and TA. These findings underline that, both BG and TA, can be utilised as supplements in ruminant diets and thus contribute to sustainable resource use. However, based on the variability observed in the current study, periodic evaluation of feeding value is vital for efficient utilisation of brewery byproducts from Ethiopian breweries due to inconsistent use of raw ingredients and unmalted cereal grains.

Keywords: Brewers grains, crude protein fractions, ruminant, tella-atella

Contact Address: Christian Böttger, University of Bonn, Inst. of Animal Science, Endenicher Allee 15, 53115 Bonn, Germany, e-mail: cboe@itw.uni-bonn.de

Actor Configuration, Constraints and Opportunities in the Forage Seed Value Chain in Kenya and Uganda

Kevin Maina, Ben Lukuyu, Isabelle Baltenweck
International Livestock Research Institute (ILRI), Kenya

Dairy production in East Africa is crucial for rural development, poverty reduction, food and nutrition security. Dairying has increased recently due to the high demand for milk and value-added milk products by a growing population and an expanding urban middle class. The sector contributes to more than 9 % of the gross domestic product (GDP) in Kenya and Uganda. However, sub-optimal feeding forms a major constraint for further growth and development of the dairy sector. As feeding represents 65 % of production costs, improved forage productivity will greatly increase milk production efficiency and thus reduce the production costs and price of milk. Currently, farmers mainly rely on grazing poor pastures, feeding crop residues and collected feeds. As a result of a poorly functioning forage seed value chain, promising and demanded species and varieties which provide for high quality forage for Kenya and Uganda, remain under-utilised. To promote forage production, a study was conducted to assess constraints and opportunities of forage seed production in Kenya and Uganda. The study used desk reviews and key informant interviews with sellers of forage planting materials and seed companies with a samples of 15 seed companies/entities to assess existing production and marketing business models for different forage species considering the biophysical and socio-economic contexts in Kenya and Uganda. Preliminary findings from the key informant interviews indicates that more than 50 % of seed transfer/sale to farmers is conducted through the informal sector. The most commonly traded propagation materials are grasses and leguminous forages. The seed quality certification standards are limited more to large-sized companies and thus, small-mediums sized companies often trade in uncertified seeds/planting materials. The study concludes that in order to create demand for improved forages, there is need to raise awareness and improve knowledge through innovative promotion pathways for the forages and extension among farmers. There is need to develop the nascent informal seed sector by supporting and developing quality declared seed standards. This willl increase seed availability and reduce cost of seed for smallholder farmers. There is also need to harmonise seed policies in Kenya and Uganda that allows smooth importation of forages seed.

Keywords: Business models, forage quality, forages, livestock productivity

Contact Address: Kevin Maina, International Livestock Research Institute (ILRI), P. O. Box 30709-00100, Nairobi, Kenya, e-mail: mainakevin.km@gmail.com

Effects of Nucleotides Supplementation on the Productive Performance of Rabbit

Guilherme Carvalho Rocha[1], Paola Cristina. Piza[1], Andressa Santana Natel[1], Ariane Flávia Nascimento[1], Lucas Alberto Teixeira de Rezende[2], Eloi Portugal[2]

[1]*University José Do Rosário Vellano (UNIFENAS), Agronomy, Brazil*

[2]*Federal Institute Sul de Minas Gerais, Veterinary Medicine, Brazil*

The ban of in feed-antibiotics in animal production has led to increasing interest in alternatives to overcome weaning-associated problems. Among others, dietary nucleotides are one group of bioactive agents which may have the potential to diminish challenges related to weaning, beside of exert positive effects on development animal. The objective of this study was to evaluate the effects of levels of nucleotides (NC) inclusion in diets to weaning rabbit on the productive performance. The experiment was developed at the Cuniculture Sector, Federal Institute of Education, Science and Technology of the South of Minas Gerais - Campus Muzambinho (CEUA - 5532061219). Thirty New Zealand White weaning rabbit with 30 day old and 0.575 kg (p = 0.32, SE = 0.117) were used, distributed in a randomised block design in four treatments according to the inclusion of nucleotides in the diet plus a witness: 0 (Control), 0.02%; 0.04%; 0.06 % NC inclusion; and 0.004 % commercial coccidiostat. Additives were included in the feed before pelleting. The rabbits were housed in individual cages and fed ad libitum during the 60-day experiment. NU supplementation did not affect the off total feed intake (6,51 kg, P=0,122, 37,7 % CV). A quadratic regression ($p < 0.05$) was observed for final weight (2.05, 1.73, 2.29 and 2.15), daily weight gain (23.96, 19.52, 27.28 and 26.42) and food convention (3.04, 6.06, 5.07 and 4.46), with the inclusion of NU in the diet (0, 0.02, 0.04 and 0.06 %). There was no difference ($p > 0.05$) by contrast orthogonal between Coccidiostatic and NC supplementation treatments (0.02, 0.04 and 0.06) for analysed variables. NU supplementation at 0.04 level resulted in a improvement in performance rabbits. Coccidiostatic can been replaced by NC, enabling the same results more safely.

Keywords: Additive, coccidiostatic, cuniculture, weight gain

Contact Address: Andressa Santana Natel, University José do Rosário Vellano (UNIFENAS), Dept. of Agronomy, Alfenas, Brazil, e-mail: andressa.natel@unifenas.br

Evaluation of On-Farm Goat Fattening Using Cowpea Hay with Concentrate in North-Western Dry Lands of Ethiopia

Belete Shimelash Abebe, Tikunesh Zelalem Abebe, Solomon Abegaz Guangul

Gondar Agricultural Research Center, Livestock Research Directorate, Ethiopia

An experiment was conducted in Gumara-Maksegnit watershed to evaluate fattening of 36 castrated yearling goats with initial weight of 28.2 kg using cowpea hay and concentrate supplementation with different proportions. The treatments were browsing alone (T1), browsing +100 % concentrate (T2), browsing +100 % cowpea hay (T3), browsing+ 50 % cowpea hay with 50 % concentrate(T4), browsing +25 % cowpea hay with 75 % concentrate (T5) and browsing +75 % cowpea hay with 25 % concentrate (T6) for 105 days. Each animal has free access to browsing and water. Average initial body weight was not different between treatment groups. The mean final weight (Kg), total body weight gain (Kg) and average daily weight gain (g) obtained were 33.88±4.00, 5.65±1.19 and 62.87±13.24, respectively. Total body weight gain and average daily body weight gain were significantly different ($p < 0.05$) between treatments. The group fed on browsing plus 75 % concentrate supplement +25 % cowpea and browsing +100 % concentrate had better final body weight, total body weight and average daily body weight gain. The groups fed on browsing +100 % concentrate and browsing +75 % concentrate with 25 % cowpea hay had higher daily feed conversion efficiency while those fed on other feed groups. The highest daily gain from T5 is related to a good nutrient balance available in the commercial concentrate supplement despite the relatively higher crude protein content supplemented with 25 % cowpea hay with 75 % concentrate. Average dry matter intake and feed conversion efficiency were calculated for indoor fed treatment groups (concentrate with cowpea hay fed treatments) only. There was significant difference in average daily dry matter intake (DM) between treatments. Increased supplementation of concentrate with cowpea hay forage increased total DM intake, that treatment groups fed on 75 cow pea hay, 25 concentrate and sole cowpea hay had higher DM intake than the other treatments but the average daily weight gain of the animal is higher for T2 and T5 that means goats fed on concentrate alone and 75 % concentrate with 25 % cowpea hay had better daily weight gain. Therefore based on the results T5 was recommended but as famers point of view T6 was preferred to reduce the concentrate level.

Keywords: Body weight, cowpea hay, dry land , fattening, goat

Contact Address: Belete Shimelash Abebe, Gondar Agricultural Research Center, Livestock Research Directorate, Kebele 16, Gondar, Ethiopia, e-mail: belyanst@yahoo.com

In vitro Gas Production and Short-Chain Fatty Acid Production from Tropical Forage Legumes Incubated with and without Polyethylene Glycol

Risma Rizkia Nurdianti[1], Alberto Jefferson Da Silva Macêdo[2], Odilon Gomes Pereira[2], Uta Dickhoefer[1], Joaquín Castro-Montoya[1]

[1]*University of Hohenheim, Animal Nutrition and Rangeland Management in the Tropics and Subtropics, Germany*

[2]*Federal University of Viçosa, Brazil,*

The aim was to evaluate the effects of polyethylene glycol (PEG) supplementation to young or old tropical forage legumes (TFL) that were either fresh or ensiled on *in vitro* gas production (IVGP) and short-chain fatty acid (SCFA) production.

Samples of tropical forage grasses (TFG) i.e. *Pennisetum purpureum* (PP) and *Setaria sphacelata* (SS) were collected as control from El Salvador and Indonesia, and samples of TFL, i.e. *Arachis pintoi* (AP) and *Glycine max* (GM) from Brazil. All TFL differed in their stage of maturity (i.e. young and old) and in the type of conservation (i.e. fresh and silage).

The Hohenheim gas test was used to determine IVGP after 8 and 24 h of incubation. The sample of each species was incubated individually without and with 750 mg PEG supplementation. All data were analysed using the mixed procedure of SAS with species, maturity, forage type of conservation, PEG supplementation, and their interactions as fixed effects and run as random effect. Least squares means were compared at a significance level of $p < 0.05$.

The IVGP after 8 h incubation was lowest for SS with PEG supplementation (average $\pm$ standard deviation; 4.2$\pm$0.68 ml/200 mg dry matter; DM) and highest for fresh young GM without PEG supplementation (15.2$\pm$1.33 ml/200 mg DM). Meanwhile, IVGP after 24 h incubation was lowest for SS (24.7$\pm$1.39 ml/200 mg DM) and highest for old AP silage (39.5$\pm$0.77 ml/200 mg DM) when both were supplemented with PEG. PEG supplementation increased IVGP after 8 and 24 h of incubation in fresh old AP and old AP silage ($p < 0.001$). Similarly, the estimated SCFA production was lowest in SS (0.49$\pm$0.03 mmol) and greatest in AP silage old without and with PEG inclusion (0.80$\pm$0.02 and 0.80$\pm$0.05 mmol, respectively). Similar to the increase in IVGP, estimated SCFA production increased for young and old fresh of AP.

The difference between the IVGP with and without PEG supplementation in each sample is an indicator of tannin effect. According to the findings, tannins are more active in old than young AP and more active in fresh than ensiled AP. Instead, tannin have no effect on *in vitro* fermentation in GM and TFG.

Keywords: Gas production, keywords: PEG, SCFA production, tropical forage grasses, tropical forage legumes

Contact Address: Risma Rizkia Nurdianti, University of Hohenheim, Animal Nutrition and Rangeland Management in the Tropics and Subtropics, Fruwirthstraße 31, 70599 Stuttgart, Germany, e-mail: aninutrop@uni-hohenheim.de

Ruminal *in vitro* Gas Production Kinetics of Above-Ground Maize Plant and Plant Fractions

Ismael Rubio-Cervantes[1], Christian Böttger[1], Katrin Gerlach[1], Thiago Bernardes[2]

[1] *Bonn University, Inst. of Animal Science, Germany*
[2] *University of Lavras, Department of Animal Science, Brazil*

Whole plant (WP) maize is a common feedstuff for ruminant animals in many parts of the world. However, the non-grain parts (NG) of the plant may be of particular interest in animal nutrition because a high overall digestibility of WP is the aim. Moreover, especially in tropical and subtropical regions maize grain is harvested for human consumption and only NG are used as feedstuff. The objective of the present study was to estimate ruminal degradation of maize WP, grain and NG of the plant using *in vitro* gas production (GP) kinetics.
Forty-five maize genotypes were sampled at silage harvest time over two different growing seasons in Brazil, dried at 55 °C and ground through a 1-mm screen. For each sample, WP and its grain counterpart were incubated in rumen fluid-buffer solution using the "Hohenheimer Futterwerttest" to determine *in vitro* GP at 0, 2, 4, 8, 12, 16, 24, 36, 48, 72 and 96 h. Gas production of NG of the plant was calculated using the curve subtraction method, that is subtracting the average GP of the grain part (weighted according to its proportional contribution to WP) at each time from the average GP of the WP. Average GP values were fitted to a mono-exponential model to describe GP kinetics.
On average, WP had the highest potential GP at 96 h with 66.1 ml/200 mg dry matter (DM) (standard deviation, SD 4.1), of which grain proportional fraction contributed 31.3 ml/200 mg of WP DM (SD 5.5) and NG contributed 36.0 ml/200 mg of WP DM (SD 5.3). The GP rates (/h) were: WP 0.059 (SD 0.004), grain proportional fraction 0.086 (SD 0.009), NG 0.039 (SD 0.007). The generally higher GP rate of maize grain compared to NG was likely due to a faster degradation of starch compared to plant fibre. The observed variability in parameters describing GP kinetics was related to actual differences in degradation between varieties, but also the proportion that grain and NG accounted for in WP.

Keywords: Animal nutrition, feedstuff, Brazil, ruminant

Contact Address: Christian Böttger, University of Bonn, Inst. of Animal Science, Endenicher Allee 15, 53115 Bonn, Germany, e-mail: cboe@itw.uni-bonn.de

Economic Evaluation of Broiler Chicken Supplemented with Fermented Mansanitas (*Muntingia calabura*) Leaves

Regie Lloren

Camiguin Polytechnic State College, Institute of Agriculture, Philippines

The study endeavoured to investigate the growth attributes as well as to provide economic analysis of broiler chicken supplemented with Fermented Mansanitas Leaves (FML). An experiment was carried out from October to November 2019 and employed a Completely Randomised Design. Four (4) treatments with three (3) replications with five (5) experimental chicks per treatment were prepared in the study. A total of sixty (60) day-old chicks were randomly selected and were distributed to different treatments. The treatments employed include the following: Treatment 1 (Control 1.5 gram per liter of water), Treatment 2 (10 milliliter of FML per liter of water), Treatment 3 (20 milliliter of FML per liter of water) and Treatment 4 (30 milliliter of FML per liter of water). Data such as the average chick cost and, average daily feed and water intake were collected to calculate the feed and supplementation cost per unit. Average dressed weight and average price of dressed chicken were also gathered to calculate the sales per unit. From the data obtained, gross margin per unit was calculated. Data gathered were analysed using Analysis of Variance, and Tukey's test was employed to compare significant differences among treatment means. Results revealed that supplement cost and gross margin per unit showed significant differences among treatment means. Other indicators such as chick cost, feed costs and sales per unit revealed no significant differences. It was concluded that a supplementation of 20 milliliter of Fermented Mansanitas Leaves per liter of water showed significant increase in gross margin per unit.

Keywords: Broiler, fermented Mansanitas leaves, gross margin per unit, supplement cost

Contact Address: Regie Lloren, Camiguin Polytechnic State College, Institute of Agriculture, Purok 2 Tangaro, 9104 Catarman, Philippines, e-mail: rlloren18@gmail.com

In vitro Fermentation of Lablab and Jack Bean with Polyphenols Affected by Ensiling Conditions

Temitope Aloba, Uta Dickhoefer, Joaquin Castro-Monotya
University of Hohenheim, Animal Nutrition and Rangeland Management in the Tropics and Subtropics, Germany

Many tropical forage legumes may contain substantial amounts of protein-binding polyphenols. Yet, ensiling may modify the concentration of polyphenols in forage legumes. Thus, it is expected that ensiling length will elevate polyphenol activities in forage legumes, thereby binding the proteins available and reducing rumen fermentation. An experiment was conducted to assess the *in vitro* fermentation of lablab and jack bean forage with and without polyethylene glycol (PEG) addition as affected by ensiling length and storage temperature. Samples of lablab and jack bean forages utilised consist of fresh forage, forages wilted for 6 h and forage that were ensiled indoor or outdoor for 75 and 180 days. Total phenols (TP) and tannins were analysed in all samples. Forage samples (375 mg) were incubated *in vitro* in buffered ruminal fluid with or without PEG (750 mg) in triplicate for 24 h in three runs. After 24 h of incubation, cumulative gas production (GP) was recorded, and short-chain fatty acid (SCFA) and ammonia-nitrogen (NH_3-N) concentrations were determined. Data were analysed using the GLIMMIX procedure of SAS in a 2 × 4 factorial design and their interactions. There was an interaction effect between ensiling length and storage temperature for TP and tannin concentrations of lablab ($p < 0.05$) and for TP concentration of jack bean ($p < 0.01$). Tannin concentration in jack bean increases with ensiling length ($p < 0.01$). There was an interaction between ensiling length and PEG for NH_3-N concentration for lablab ($p < 0.05$). Variable effects of ensiling length were observed for GP, NH_3-N, and SCFA for lablab and jack bean. For lablab, the GP with PEG was greater than without PEG ($p < 0.01$). Branched-chain fatty acid (BCFA) concentrations of lablab and iso-butyrate concentration of jack bean with PEG was lower than without PEG ($p < 0.01$). The results suggest that conservation conditions did not protect the biological reactivity of the tannin present in lablab as there is a weaker affinity for protein as seen in the BCFA concentrations. The amount of tannin present in both legumes and the type of tannin could be the reason for the weaker affinity.

Keywords: Forage conservation, legumes, polyphenol, ruminal fermentation, tannin

Contact Address: Temitope Aloba, University of Hohenheim, Animal Nutrition and Rangeland Management in the Tropics and Subtropics, 70593 Stuttgart, Germany, e-mail: lanrelex@gmail.com

Non-Structural Carbohydrate Profile of some Tropical Forage Grasses

Ermias Habte[1], Yonas Asmare[2], Asebe Abdena[2], Chen Dong[3], Liu Yinghao[4], Qian Zhang[4], Chris S. Jones[2]

[1]*International Livestock Research Institute (ILRI), Feed and Forage Development Program, Ethiopia*

[2]*International Livestock Research Institute (ILRI), Kenya*

[3]*Nanjing Agricultural University, China*

[4]*Institute of Grassland Research, China*

Forage crops are one of the important feed sources in sub-Saharan Africa. They provide organic nutrients, such as structural and non-structural carbohydrates (NSC), which are required for livestock production. In temperate forages, it has been noted that increased NSC levels are associated with rapid regrowth of forage crops, increased energy value of feed and tolerance to drought stress. However the information about NSC content and seasonal dynamics in tropical forages is limited. This study was conducted to examine NSC content of some forage grass species held in the ILRI field genebank in Ethiopia. NSC content of five perennial grass species was determined using the Anthrone colourimetric protocol. The plant samples of 60 accessions from 5 grass species were harvested after every three-months of regrowth, at the beginning of the day, and the harvested samples were fully dried and quantified for NSC at the ILRI animal nutrition laboratory. The results of the study indicated the existence of significant accession, species and seasonal variation in NSC content. The highest overall NSC content was observed in *Urochloa decumbens* followed by *Cenchrus purpureus* but the variation within species was, in most cases, as large as the variation between and NSC levels also varied with the harvest season. In addition significant accession difference has been obtained in each species which signifies accession variations are also important for improving NSC content within each grass species. Generally, the study provides an insight into opportunities for future development of feed resources with increased NSC content from grass species and the fluctuation of NSC content over the season for enhanced livestock production.

Keywords: Feed nutrition, forage grasses, NSC, tropical forage

Contact Address: Ermias Habte, International Livestock Research Institute (ILRI), Feed and Forage Development Program, Addis Ababa, Ethiopia, e-mail: e.habte@cgiar.org

Effect of Dietary Protein Restriction on the Growth of Snails of the Species *Archachatina marginata*

Chadrac Tsrigbidzi, Youssouf Toukourou

University of Parakou, Fac. of Agronomy, Dept. of Sciences and Techniques of Animal and Fisheries Production, Benin

The effect of food restriction on compensatory growth was studied on giant African snails of the species *Archachatina marginata* (Swainson 1821). The study aimed to determine the ability of captive-bred giant African snails to efficiently use food and nutritional resources, based on the phenomenon of compensatory growth. A total of 90 snails with an average live weight of 52.48 $\pm$ 9.03 g and an average shell length of 6.83 $\pm$ 3.38 cm, were randomly divided into three batches of 30 individuals in semi-buried pens made of cinder block and fine mesh screen. The snails were fed for an initial period of 70 days with rations containing 20.26 % crude protein for the control lot (I), 17.18 % and 14.43 % crude protein for lots II and III. At the end of each feeding period, eight snails from each batch were randomly selected and slaughtered. The results obtained showed that the snails displayed at the end of the feeding restriction period a mean shell length of 8.1 $\pm$ 0.54 cm, 8.11 $\pm$ 0.43 cm and 8.13 $\pm$0.5 cm ($p > 0.05$) for a mean live weight of 79.6 $\pm$ 7.3g, 68.9 $\pm$ 11.22g and 66 $\pm$ 10.66g ($p < 0.05$) for batches I, II and III respectively. The average daily dry matter intakes, as well as the average feed conversion ratios were 1.04 $\pm$ 0.12g, 1.09 $\pm$ 0.13g and 1.13 $\pm$ 0.14g ($p > 0.05$), as well as 58.28 $\pm$ 15.41, 103.30 $\pm$ 32.07 and 129.30 $\pm$ 30.86 ($p < 0.05$) for lots I, II and III respectively. The average carcass yields were 38.70 $\pm$ 3.12%; 30.35 $\pm$ 2.03 % and 28.30 $\pm$ 1.26 % for lots I, II and III. The feed conversion ratio and carcass yield were 56.24 $\pm$ 7.89, 36.32 $\pm$ 35.28 and 35.28 $\pm$ 3.21 ($p < 0.05$) and 40.44 $\pm$ 4.00 %, 37.48 $\pm$ 2.56 % and 36.55 $\pm$ 1.75 % respectively. It is apparent from the present study that African giant snails of the species *Archachatina marginata*, previously subjected to dietary restriction, exhibited a more accelerated growth rate that allowed them to partially compensate for a significant growth delay.

Keywords: Carcass yield, compensatory growth, feeding, micro-livestock, snail

Contact Address: Youssouf Toukourou, University of Parakou, Fac. of Agronomy, Dept. of Sciences and Techniques of Animal and Fisheries Production, B.P. 123, Parakou, Benin, e-mail: ytoukourou@gmail.com

Livestock and rangeland systems

A New Approach to Classify Livestock Farming Systems in Sub-Saharan Africa

Sarah Graf, Thomas Daum, Regina Birner

University of Hohenheim, Inst. of Agric. Sci. in the Tropics (Hans-Ruthenberg-Institute), Germany

The farming systems approach has emerged as unique tool to deal with the enormous diversity of smallholder farming in the tropics. Following the pioneering work of scholars such as Hans Ruthenberg and Pierre de Schlippé, the approach has been further developed - most notably under the leadership of FAO, but the focus has mostly been placed on the cropping component of farming systems. With regard to livestock, an influential book on livestock framing systems in Africa was published by Hans Jahnke in 1982. In 1996, Carlos Seré and Henning Steinfeld devised a global classification, which remains an important basis for classifying livestock farming systems, but its global scope results in few rather broad categories of farms that keep livestock, which are not necessarily similar regarding their size, resource base, enterprise patterns, household livelihoods and constraints. To address this limitation, we propose a new approach that combines a classification at two levels: the herd level and the farm level. In the first step, herd systems are defined according to classification criteria that we derived from primary studies about livestock in sub-Saharan Africa. We applied an iterative process of reviewing, coding and classifying these studies. The following main classification criteria were thus identified: main feed source, production goal, how animals are confined, and – if applicable – the mobility pattern. The second step of the classification refers to the farm level and identifies what type of cropping system a particular herd system is combined with. This combination of herd systems and cropping systems forms a livestock farming system. Key livestock farming systems are then defined as a typical combinations of specific herd systems and specific cropping systems. This modular approach addresses the need for meaningful descriptions of livestock and herd management practices on the one hand, and farm level analysis on the other. By geo-referencing the literature that was reviewed using GIS, a map of herd systems was produced based on relationships between geographic conditions and the classification criteria. The proposed new classification system can inform future research and development interventions by guiding topic choice, implementation strategies and transfer of results.

Keywords: Africa, farming systems, GIS, livestock production systems

Contact Address: Sarah Graf, University of Hohenheim, Inst. of Agric. Sci. in the Tropics (Hans-Ruthenberg-Institute), Wollgrasweg 43, 50599 Stuttgart, Germany, e-mail: sarah.graf@uni-hohenheim.de

Public Policies and Silvo-Pastoral Systems in Latin America: A Comparative Study

LEONARDO MORENO LERMA[1], MANUEL DIAZ[2], STEFAN BURKART[2]

[1]*Independent Consultant, Colombia*
[2]*International Center for Tropical Agriculture, Tropical Forages Program, Colombia*

The global projections of population and food demand increases by 2050 highlight the importance of Latin America as one of the future big food suppliers for our planet. The region has high agricultural potential and activities such as cattle farming can increase the global food supply, i.e. through the adoption of sustainable technologies such as silvo-pastoral systems. Despite the importance of this economic sector for the region, its negative environmental impacts (especially those of traditional extensive production systems) are numerous and the shift towards sustainability is perceived as slow and uncoordinated. This study aims at identifying both success stories and difficulties in the implementation of public policies for the development of sustainable cattle production systems in Colombia, Argentina and Costa Rica during the period 2010–2020. Based on literature review, media analysis and legal document reviews, a qualitative descriptive analysis was carried out, documenting and outlining the main political activities in the region. The results highlight the development and application of policies aimed at the use of sustainable production technologies, the adaptation of pastures to changing environmental conditions and the use of silvo-pastoral systems for cattle production. Although common successes are identified in the three countries, such as the existence of a large number of public policies aimed at promoting sustainable livestock - which is strengthened through e.g. national level development plans and legislative advances - they also coincide in difficulties, such as a minimal articulation between national and local policies and the lack of continuity of development programs. We conclude that, although the selected countries have different socioeconomic characteristics, as well as different levels of progress in the implementation of their policies, the general perception among the three countries is relatively similar to the extent that their efforts are still insufficient, i.e. when the commitments made during the COP21 are being considered. Although the advances made so far provide valuable contributions, it is necessary to treat them as a first stage in a long-range process towards sustainability, and support their continuity and further out-scaling, i.e. for reaching the ultimate goal of a broader adoption of silvo-pastoral systems.

Keywords: Climate change, livestock

Contact Address: Stefan Burkart, International Center for Tropical Agriculture (CIAT), Trop. Forages Program, Km 17 Recta Cali-Palmira, Cali, Colombia, e-mail: s.burkart@cgiar.org

Adapting the One Health Approach to Pastoral Contexts

Constanze Boenig, Wibke Crewett, Igor Pilawski

Vétérinaires Sans Frontières Germany, Germany

The One Health (OH) approach is currently widely acknowledged as an effective strategy towards reaching the SDGs. It emphasises the necessity to not only address human health, environmental health and animal health equally, but also to acknowledge complex interlinkages of those three pillars. OH seeks to create effective strategies to sustain the wellbeing of humans and animals and at the same time respect our planetary boundaries. However, in practice OH is currently not widely implemented. The existing examples are mainly found in the public health sector.

OH is a particularly powerful approach to create strategies to support the wellbeing of communities that live in close proximity to livestock, in climate sensitive environments and in marginalised locations, such as pastoralists and agro-pastoralists. Pastoral communities are particularly threatened by multiple challenges, such as zoonotic diseases, biodiversity loss and inaccessible human health services, which OH seeks to tackle. However, to date there is relatively few attempts to effectively implement holistic OH projects that tackle all three dimensions of OH and its linkages. One reason is that we still miss tailored concepts that guide the design of comprehensive projects and programs in the agricultural sphere. We therefore propose a OH framework tailored to the needs of agricultural pastoral and agro-pastoralist communities. The framework builds on an analysis of three pioneering OH projects in Ethiopia, Kenya and Uganda, which are currently being implemented by Vétérinaires sans Frontières Germany. We use secondary project documentation and expert interviews in order to identify implementation challenges of OH projects while enriching existing OH frameworks by specifications relevant for agricultural projects. A comparison of practical needs of OH design and existing OH concepts highlights some gaps in current OH concepts: unclear concepts of transdisciplinary integration and insufficient attention to institutional embeddedness of the OH approach. We deduct a comprehensive framework for OH project design for pastoral and agro-pastoral communities meant to support the community of practice that seeks to support that marginalised group. We hereby advance current knowledge in OH towards better applicability and demonstrate its use from an agricultural perspective.

Keywords: Concept, Eastern Africa, One Health, pastoralism, Vétérinaires sans Frontières

---

Contact Address: Wibke Crewett, Vétérinaires Sans Frontières Germany, Marienstraße 19-20, 10117 Berlin, Germany, e-mail: wibke.crewett@togev.de

Cut it or Keep it - Is Recovery Time the Key to Healthy Rangeland?

Sabine Baumgartner[1], Anna C. Treydte[2,1]
[1]*University of Hohenheim, Ecology of Tropical Agricultural Systems, Germany*
[2]*Stockholm University, Dept. of Physical Geography, Sweden*

Despite the importance for the livelihood of millions of people worldwide, many semi-arid African savannahs are prone to heavy degradation caused by overutilisation and increased climate variability. Rangeland management can strongly affect the rangeland condition and its productivity. The Maasai traditionally practice transhumance and seasonal exclosures, granting seasonal resting time between rangeland use. However, few studies have quantified the effect of resting under current overuse and climate change challenges.
We collected data on grass and forb biomass in experimental plots exposed to different resting time after cutting (monthly versus seasonal) and under different wildlife accessibility (fenced versus open). Measurements were done in different rangeland types (rainy season rangeland, dry season rangeland and seasonal exclosures). We conducted the experiment during a dry growing period (GP1) and a wet growing period (GP2).
Grass biomass during GP1 was highest in fenced plots with seasonal resting time. Wildlife grazing resulted in significantly lower grass biomass compared to fenced plots. During GP2, we found higher grass and forb biomass in plots with seasonal compared to monthly resting time and no effect of wildlife grazing. Mean forb biomass increased by five times during wet conditions in GP2 compared to dry conditions in GP1, whereas mean grass biomass doubled. These trends were similar for all three rangeland types.
Our results suggest that seasonal recovery phases between heavy grazing events can help to maintain forage provision of rangelands in our study region. Limited forage resources during dry conditions can, however lead to competition between livestock and wildlife. Forbs seem to profit more from increased rainfall than grasses, especially in already disturbed areas. We conclude that traditional Maasai rangeland management, based on seasonal recovery periods, can support rangeland productivity despite intense grazing and unpredictable rainfall pattern. To avoid the dominance of undesired plant species, additional restoration measures are necessary.

Keywords: Productivity, rainfall pattern, rangeland management, recovery time

---

Contact Address: Sabine Baumgartner, University of Hohenheim, Ecology of Tropical Agricultural Systems, Garbenstr. 13, 70599 Stuttgart, Germany, e-mail: sabine.baumgartner@uni-hohenheim.de

Tracking Livestock - How Do Different GPS Measurement Intervals Affect Movement Pattern Results?

Marie-Luise Schmitt[1], Lena Michler[1], Jane F. Ploechl[1], Anna C. Treydte[2,1]
[1]*University of Hohenheim, Ecology of Tropical Agricultural Systems, Germany*
[2]*Stockholm University, Dept. of Physical Geography, Sweden*

Animal tracking has been widely practiced in rangelands over the last years to understand livestock and wildlife spatial and temporal movement patterns. Globally, rangeland degradation is progressive and effective management strategies are crucial to maintain healthy pastureland for livestock and wildlife. GPS collars help to understand movement dynamics with minimal animal interference. However, little is understood how frequently GPS positions need to be recorded to receive sufficient detail on distances travelled and preferred grazing locations. To understand appropriate tracking intervals, we used GPS collars on livestock herds (goat/sheep and cattle) in the semi-arid savanna of Tanzania and in the desert-steppe of Mongolia. The Maasai herders of Tanzania tend their livestock in a tropical climate with rainy and dry seasons, whereas nomadic herders of Mongolia live in a continental climate with four seasons and high temperature differences throughout the year. Data of 6 livestock herds ($n^{goat}=6$) in south-western Mongolia and of 10 livestock herds ($n^{goat}=5$, $n^{cattle}=5$) in northern Tanzania were collected over a period of 2 months. We tested whether different GPS point intervals (3 minutes, 15 minutes, 30 minutes and 60 minutes) affected the outcomes on daily walking distance and maximum daily distance from herder camp in both ecosystems. We further tested influence of daily elevation, average altitude of livestock per day and average temperature per day on the daily walking distance of goats in both study sites. Our results suggest that mean values of daily walking distance decrease with reduced measurement density in each study area. Further, in both ecosystems, daily elevation had a significant influence on daily walking distance [Tanzania: $X^2 = 14.0983$, $P < 0.001$; Mongolia: $X^2 = 57.9977$, $P < 0.001$]. In Mongolia, average altitude per day was as well significant [X2 = 47.5285, p-value < 0.001]. Our study adds valuable information to optimize GPS tracking and, therefore, contributes to improving the understanding of animal movement patterns and the appropriate tools.

Keywords: GPS Tracking, mobility, Mongolia, pastoralism, Tanzania

---

Contact Address: Marie-Luise Schmitt, University of Hohenheim, Ecology of Tropical Agricultural Systems, Windhalmweg 40, 70599 Stuttgart, Germany, e-mail: marieluise_schmitt@web.de

COVID-19 and the Colombian Cattle Sector: Current and Potential Developments, Impacts and Mitigation Option

Stefan Burkart[1], Manuel Diaz[1], Karen Enciso Valencia[1], José Luís Urrea Benítez[1], Andrés Charry[2], Natalia Triana-Angel[1]

[1]*International Center for Tropical Agriculture (CIAT), Trop. Forages Program, Colombia*
[2]*Alliance of Bioversity International and CIAT, Colombia*

The COVID-19 pandemic is affecting the Colombian cattle sector. First impacts and short-term mitigation measures are already visible in all links of the beef and dairy value chains. The full magnitude of the crisis is not yet clear but most impacts already are or will be negative and will affect the beef and dairy value chain's performance in the near future. However, some positive trends are also occurring and could, at least to some extent, endure the crisis and help building a more resilient food system for the future. The objective of this study is to shed light on the current and potential developments, impacts and mitigation options of the COVID-19 pandemic on the Colombian cattle sector. Through literature, media and legal document review, we provide a thorough analysis for the different beef and dairy value chain links and framework conditions, and also include a perspective of how the pandemic affects more vulnerable parts of the population, rural education and on-going efforts towards sustainable intensification of the cattle sector. Our results show that consumer preferences will change towards more food safety, traceability, animal welfare and sustainability and the sector will need to understand this and push value chain formalisation and consumer communication. The transformation of the primary sector towards more sustainability and efficiency is becoming urgent, not only to increase resilience during times of crisis (as in the actual COVID-19 situation), but also to face the aggravating effects of climate change and combat inequality. Digitalisation and virtualisation have become important means during the crisis in all links of the value chains, creating opportunities for sustainably increasing sector efficiency. Research can play a fundamental role in analysing and understanding the impacts posed by the current crisis, providing technologies and recommendations for recovery, and developing solutions for building resilient food systems. Our results serve as guide for policy- and decision-makers to help understanding potential impacts of the pandemic and for the development of adequate mitigation measures in order to prepare the sector better for future crises.

Keywords: Consumer behaviour, food safety, food system, resilience, sustainable intensification, traceability, value chain formalisation

Contact Address: Stefan Burkart, International Center for Tropical Agriculture (CIAT), Trop. Forages Program, Km 17 Recta Cali-Palmira, Cali, Colombia, e-mail: s.burkart@cgiar.org

Impacts of COVID-19 Lockdown on Dairy Farms in and around the Megacity of Bengaluru, India

Md Shahin Alam, Eva Schlecht, Marion Reichenbach
University of Kassel / University of Goettingen, Animal Husbandry in the Tropics and Subtropics, Germany

The Indian dairy sector is the largest worldwide, although 80 % of dairy animals are kept in herds of two to five cows and a large share of milk is still marketed via informal channels. From March 2020 onwards, COVID-19 strongly impacted global food systems, disrupting supply chains, demand for food products, and the livelihood of millions involved in agriculture. Therefore, this study aimed to assess the impacts of COVID-19 on small (< 3 lactating and dry cows plus heifers (LDH)), medium (4–6 LDH), and large (> 6 LDH) size dairy farms in and around the Indian megacity of Bengaluru. A total of 129 dairy farms that had participated in a baseline survey conducted from January to March 2020, before a first lockdown was enforced in India, were surveyed again by phone from August to September 2020 regarding input supply for cows, milk marketing, and strategies adopted to cope with the impacts of COVID-19. Results show that the share of dairy farmers not providing concentrate feeds to their cows increased from 1 % before lockdown to 7 % afterwards ($p < 0.05$), and for dry forages from 20 % to 33 % ($p < 0.05$). Increased price of dry forage was the main reason for stopping to feed it. The average milk yield per cow per day was 8.2 liters before lockdown versus 7.6 liters after lockdown. An immediate impact of COVID-19 was a decreased consumer demand for milk. In consequence, overall milk yield decreased by 26 % after lockdown because a higher ($p < 0.05$) share (30 %) of the surveyed dairy producers sold at least one lactating cow during or after the first lockdown and switched to an alternative income-generating activity such as meat production from small ruminants, or engaged in an off-farm activity. Six percent of the dairy producers abandoned dairy production altogether. Despite the severe disturbance caused by the pandemic, and especially challenged by decreased milk demand and input availability, most dairy producers in and around Bengaluru were able to adapt their production strategies within a few weeks of time, demonstrating how resourceful and flexible smallholder farmers can react to a shock.

Keywords: Coping strategy, dairy production, pandemic, smallholder, survey

---

Contact Address: Md Shahin Alam, University of Kassel / University of Goettingen, Animal Husbandry in the Tropics and Subtropics, Steinstrasse 19, 37213 Witzenhausen, Germany, e-mail: shahindps@uni-kassel.de

Towards a Paradigm Shift in Livestock Production in Africa: Using the Potential of Neglected Animal Species

Maria Oguche, Regina Birner, Juliet Kariuki
University of Hohenheim, Inst. of Agric. Sci. in the Tropics (Hans-Ruthenberg-Institute), Germany

Population growth and changing consumption patterns have resulted in an increasing demand for animal protein, especially in sub-Saharan Africa (SSA). In meeting this demand, the production of conventional livestock species, namely cattle, sheep, goats, pigs and poultry, has been intensified. However, negative environmental, nutritional, gender-related and economic trade-offs associated with this increased production have made it expedient to explore the potentials of some neglected livestock species. Compared to neglected or "orphan" crops, neglected livestock species have not attracted much research attention, so far. To address this knowledge gap, this paper presents a systematic literature review of five neglected species: grasscutter (*Thryonomys swinderianus*), guinea fowl (*Numida meleagris*), guinea pig (*Cavia porcellus*), rabbit (*Oryctolagus cuniculus*) and donkey (*Equus asinus*), which could play a larger role in SSA. The review aimed to: (i) assess the sustainable production of these species in SSA; (ii) identify drivers and barriers to increasing the production of these species; and (iii) identify policy solutions for an increased and sustainable utilisation of these species. Applying the checklist for "Preferred Reporting Items for Systematic Review and Meta-Analysis" (PRISMA) and using Boolean search operators in academic search engines, 128 studies were selected for this review. The results show that the drivers for promoting neglected livestock species include their nutritional importance, high economic gross returns, environmental sustainability, and importance for women's empowerment. However, considerable barriers such as feed and nutrition problems, diseases and pests, exclusion from policies and development strategies, lack of research and extension, inadequate markets and animal welfare issues were identified. This study derives five policy recommendations: increased efforts to recognise and integrate the role of neglected livestock species in national livestock policies, dedicated research and development towards genetic improvement, awareness creation, increased marketing, and technology application which if implemented could enhance the sustainable adoption of, and benefits from these neglected livestock species. We conclude that, compared to conventional livestock species, these neglected livestock have fewer trade-offs and a high and unexploited economic and nutritional potential which if given attention could yield more benefits.

Keywords: Neglected livestock, nutrition, sub-Saharan Africa

Contact Address: Maria Oguche, University of Hohenheim, Inst. of Agric. Sci. in the Tropics (Hans-Ruthenberg-Institute), 70593 Stuttgart, Germany, e-mail: maria.oguche@uni-hohenheim.de

Household Livelihood Strategies and Livestock Dependence in Rural Tanzania: Implications for Poverty Reduction

Sirak Bahta, Francis Wanyoike, Isabelle Baltenweck, Nils Teufel, Mark van Wijk

International Livestock Research Institute (ILRI), Kenya

The majority of the Tanzanian population (about 67 %) reside in rural areas and depend mostly on agriculture as a source of livelihood. While the literature on sources of livelihoods in rural Tanzania abounds, these studies have commonly relied on simple descriptive statistics to investigate the implications of livelihood strategies on welfare in households. This study used the sustainable livelihood framework to evaluate the effect of livelihood strategies on the welfare of Tanzanian livestock farm households. We have used data from 824 farm households to apply a rigorous methodology that follows a three-stage procedure, which ranked the outcomes from different livelihood strategies, identified factors that constrain households' entry into high-income earning livelihood strategies, and finally explored the relationship between the identified livelihood strategies and household characteristics with poverty (poverty probability index (PPI)).
We present sub-groups of households so as to better contrast types of livelihoods and compare a number of the sub-groups' characteristics and actions. We provide commentary and explanation regarding livelihood strategy and its direct and indirect connections to poverty. In general, the results suggest that households that rely on livestock or mixed farm activities are relatively better off than households that fall into only or mostly crop activities. This is particularly true for income per adult equivalent per day levels within the 0.1–1$ range, accounting for about 66 % of sample households. The multinomial (MNL) regression analysis results show that market orientation, education level, land size, low rainfall, and non-farm income determine the household's choice of livelihood strategies. Finally, the generalised linear model results showed that compared to the base livelihood strategy (slightly livestock dependent), the moderate livestock dependent (MLD) livelihood strategy, or diversified or crop-livestock mixed livelihood strategy, significantly reduces poverty. Recommendations include partnerships and facilitating actions that support crop-livestock mix livelihood strategy in association with improving market orientation and efficiency in Tanzanian smallholder farm households.

Keywords: Livestock, poverty, sustainable livelihoods approach, Tanzania

Contact Address: Sirak Bahta, International Livestock Research Institute (ILRI), Sirak Bahta, 00100, Nairobi, Kenya, e-mail: s.bahta@cgiar.org

Characterisation of the Variability in Monthly Rainfall and Temperature in Grazing Ecosystem Supporting Sahiwal National Stud Herd in Naivasha, Kenya

Macdonald Gichuru Githinji[1], Thomas Muasya[2], Evans Ilatsia[1], Bockline Bebe[2]

[1]*Kenya Agricultural and Livestock Research Organisation, Kenya*

[2]*Egerton University, Dept. of Animal Science, Kenya*

The present study evaluated the trends of rainfall and temperature at the national Sahiwal stud, Nakuru, kenya. Data on climatic variables comprising monthly rainfall (mm), minimum and maximum temperatures were obtained from the meteorological weather station within the Stud. The coefficient of variation (CV), percentage departure from the mean (Anomalies), Precipitation Concentration Index (PCI) and moving average were computed to evaluate the variability of rainfall and temperature. The detection of trends and their analysis were performed using parametric and non-parametric tests. The Mann-Kendall (MK) trend test was used to detect trends while the Sen's Slope test was used to compute the slope using Sen's method. The mean annual rainfall was 578.5 $\pm$ 151.3 mm and a CV of 24.2 %. The PCI revealed that the study area has had rainfall with moderate concentration over the years, with about 34 % of the years having high rainfall concentration. The rainfall anomalies found in the current study depict inter-annual variability with the trend in the anomalies being more varied in recent years. Mean annual and long season rain decreased by 36.5 and 25.5 mm per decade while short season rain increased by 69 mm per decade. The short season rain had higher CV (59.2 %) than long season rain (49.2 %). MK trend analysis test revealed found as statistically significant decreasing trend for long and annual rainfall and a significant increase for short rains. The mean temperature for the study area ranged from 10.4 to 26.5 °C. The rate of change of minimum, maximum and mean monthly temperature was found to be was 0.017 °C, -0.156 °C and -0.09 °C per decade. The overall anomalies of mean annual temperature showed inter annual variability. The MK trend analysis revealed non-significant increase and decline for minimum and mean temperature, respectively. The months of April and May showed significant increase while the months of February and September had significant decline, indicating inter-annual variability in minimum temperature. The results of the current study point towards the need to adjust farming activities with the variability occurrence and design mitigation strategies to enhance adaptive capacity and resilience to climate change for livestock production systems.

Keywords: Mann-Kendall trend test, rainfall, Sen's Slope estimator, trends

---

Contact Address: Thomas Muasya, Egerton University, Dept. of Animal Science, Njoro, Kenya, e-mail: thomas.muasya@egerton.ac.ke

Influences on Livestock Weight Gain by Grazing Patterns and Herd-management Strategies in the Gobi, Mongolia

Lena Michler[1], Petra Kaczensky[2], Anna C. Treydte[3,1]

[1]*University of Hohenheim, Ecology of Tropical Agricultural Systems, Germany*
[2]*Inland Norway University of Applied Sciences, Dept. of Forestry and Wildlife, Norway*
[3]*Stockholm University, Dept. of Physical Geography, Sweden*

Pastoralism is a widely distributed livelihood strategy in semi-arid and arid regions, which are often high in environmental variability but low in overall biomass production. Globally, pastoralists adapt their herding strategies to external factors, and mobility has been the key for sustainable utilisation of pastureland for centuries. In Mongolia, nomadic pastoralism is still practised by around 1/3 of the human population, and pastoralists have to cope with harsh climatic conditions. To select high quality pasture resources, Mongolian pastoralists frequently travel between and around camp sites. Pastoralists of the Dzungarian Gobi actively tend their small livestock (sheep and goats) throughout the day while following them on horseback, motorbike or by foot. Livestock has to gain enough weight during the short vegetation period in summer to get through the hard winter months. To understand whether herd management strategies influence weight gain of sheep and goats, we equipped 19 livestock herds with GPS collars over a period of 20 months and combined movement patterns with weight gain measurements of 320 animals. Animals were weighed in two spring seasons, and once in autumn to determine weight gain during summer and weight loss during winter season and associate those factors to location. Furthermore, we assessed socio-economic drivers of 20 local pastoralists through individual interviews on their rangeland management. Our preliminary results suggest that more pro-active herd management strategies, mostly practised by younger herders, correlate with longer daily walking distances of livestock. The majority of herders (67 %) followed their livestock on horseback, in line with traditional herding practices. Besides, our results indicate that a longer daily grazing time influences weight gain positively. We conclude that herding management practices are directly influencing animal weight gain which is important for the livestock wellbeing and, hence, herders' livelihoods.

Keywords: GPS collars, movement patterns, pastoralism

Contact Address: Lena Michler, University of Hohenheim, Ecology of Tropical Agricultural Systems, Stuttgart, Hohenheim, Germany, e-mail: lena.m.michler@gmail.com

Spatial Rangeland Utilisation by Livestock of Maasai Pastoralists in Northern Tanzania

Jane F. Ploechl[1], Anna C. Treydte[2,1]
[1]*University of Hohenheim, Ecology of Tropical Agricultural Systems, Germany*
[2]*Stockholm University, Dept. of Physical Geography, Sweden*

Pastoral mobility is an important strategy adopted by millions of pastoralists around the globe in arid and semi-arid environments to meet the nutritional demands of their livestock. Mobility promotes rangeland sustainability by distributing the grazing pressure throughout the landscape. Maasai pastoralists of Tanzania traditionally optimise their livestock foraging by moving their herds in response to seasonal rainfall patterns. However, increasing human population and livestock numbers together with climatic changes result in an increasing grazing pressure at local and landscape level. The Enduimet Wildlife Management area (E-WMA) is a community-run conservation area, where wildlife, livestock, and humans coexist. To minimise the negative interactions on the grazing land, it is of great importance to understand the livestock utilisation and movement patterns. Therefore, we equipped seven cattle, five goats and four sheep with GPS-collars in November 2019 in E-WMA over a period of 4 months. Furthermore, we assessed herding strategies and grazing management practices through key-informative interviews with 10 elders (above 60 years of age) and two participatory mapping sessions with each 12 participants. Our preliminary results suggest that there is a higher utilisation of the low-quality grazing areas close to the permanent settlements during the rainy season, with highest densities within 1 km around settlements and water holes. High-quality grazing areas at the village land boundaries are used during the dry season, with low cattle grazing pressure around settlements. In contrast, the utilisation of rangelands by goat and sheep herds is highest within 500 meters around settlements and overall shorter travel distances, regardless of the season. Furthermore, we found a correlation between the age of the herder and distance travelled by the livestock. Our results are of great importance as they then can be used by authorities to develop participatory and demand-driven development approaches and sustainable resource use plans.

Keywords: GPS collars, livestock movements, Maasai, pastoralists, rangeland degradation

Contact Address: Jane F. Ploechl, University of Hohenheim, Ecology of Tropical Agricultural Systems, Pflaumheimer Weg 4, 63785 Obernburg, Germany, e-mail: jane.ploechl@gmail.com

Camel Milk Market Structure in the Arid and Semi-Arid North Eastern Kenya

Simon Gicheha, Ernst-August Nuppenau

Justus-liebig University Giessen, Institute for Agricultural Policy and Market Research, Germany

Camel milk enterprise is crucial to cash incomes, rural employment and food and nutrition security among pastoral communities in the arid and semi-arid North-Eastern Kenya. Despite evidence for the growing demand for camel milk outside the pastoral population and interventions to promote the value chain through trader cooperatives, the enterprise is still characterised by low quality and marketed volumes, losses and inadequate access to the end market and price. Although the intermediaries provide a crucial market linkage due to the nature of the production system, the significant variation in size and scale can be interpreted as a source of inefficiency. This study investigated the features that characterise and determine competition between traders operating in the same market by computing the Gini coefficient and Lorenz curve for the milk business in the year 2019. Random cross-sectional data on trader and trading enterprise characteristics, operating and marketing costs were collected from 135 camel milk producers and 193 camel milk traders in Isiolo County Kenya. The Gini coefficient showed a concentration ratio of 0.49 indicating an uneven spread of market share. The Lorenz curve of the income of traders under cooperative dominated and lied above individual operation indicating less inequality in income distribution compared to individual trade. Generalized Lorenz dominance also showed that cooperative trade is superior from a welfare perspective. The presence of large and small traders in the market is therefore a source of market inefficiency. An analysis of prospective policies is therefore necessary to minimise the economic effects of non-cooperation.

Keywords: Cooperative, inefficiency, intermediation, Lorenz, market, welfare

Contact Address: Simon Gicheha, Justus-liebig University Giessen, Institute for Agricultural Policy and Market Research, Unterhof, 35390 Giessen, Germany, e-mail: simon.kariuki-gicheha-2@agrar.uni-giessen.de

Sustainability in Small Ruminants Systems: Integrated Assessment in an Indigenous Community of La Guajira, Colombia

Martha Lilia Del Río Duque[1], Clara Viviana Rúa Bustamante[2], Michelle Bonatti[1], Juan Sebastian Valencia Sanchez[2], Katharina Löhr[1], Stefan Sieber[1]

[1]*Leibniz Centre for Agricultural Landscape Research (ZALF), Sustainable Land Use in Developing Countries (SusLAND), Germany*

[2]*The Colombian Agricultural Research Corporation (Agrosavia), Colombia*

In Colombia the Tropical Dry Forest (TDF) is one of the ecosystems most affected by deforestation and the expansion of livestock farming. Only 720.000 hectares of the 8.8 million hectares of this type of forest remain today. Given the depletion of the TDF resources, it is essential to design sustainable land use strategies in communities that have a direct influence on these remnants. One of these communities are the Wayuu indigenous people in the area of influence of the Macuira National Natural Park in the Upper Guajira, whose main productive activity is the free-grazing of small ruminants (sheep and goats). The lack of grazing management causes a negative impact on the natural values conservation, threatens natural regeneration processes, contributes to soil compaction and erosion, and pollutes water sources. Therefore, to promote sustainable systems for small ruminants, The Colombian Agricultural Research Corporation –Agrosavia– designed a silvopastoral arrangement with native forage species of the TDF for the feeding of sheep and goats, taking into account the traditional knowledge of the indigenous communities, and fostering agroecological techniques.

This study aimed to do an *ex-ante* impact assessment of these silvopastoral arrangements. The integrated assessment was done using SCALA-PB tool among different stakeholders of the project. ScalA-PB is a standardised survey questionnaire compounded of a total of nine different steps with multiple questions related to sustainability, climate change mitigation, adaptive capacities, peacebuilding and scaling-up potential. The results showed that the major constraints for scaling up silvopastoral arrangements in these kind of indigenous communities were the economic conditions at the local/regional level, since there is no infrastructure such as access to roads, irrigation, electricity and tap water available, a lack of support for the spread of this kind of initiatives by other economic actors, a high level of financial capital (initial and maintenance costs) for the implementation is required. Finally, it can be seen that SCALA-PB is a useful tool to do a sustainability impact assessment, even when markets are not the main reason to implement projects, as it is the case of silvopastoral systems for small ruminants in indigenous communities.

Keywords: *Ex-ante* impact assessment, goat and sheep farming, indigenous people, scaling up, silvopastoral arrangements

Contact Address: Martha Lilia Del Río Duque, Leibniz Centre for Agricultural Landscape Research (ZALF), Sustainable Land Use in Developing Countries (SusLAND), Eberswalder Straße 84, 15374 Müncheberg, Germany, e-mail: martha_lilia.del_rio_duque@zalf.de

Scaling-Out Knowledge: How the Pandemic Helped Spreading the Voice for a More Sustainable Cattle Sector

José Luis Urrea-Benítez, Stefan Burkart

International Center for Tropical Agriculture (CIAT), Trop. Forages Program, Colombia

The adaptation to the restrictions caused by the COVID-19 pandemic included considering new approaches, such as a broader use of information and communication technologies (ICTs). The Colombian Roundtable for Sustainable Cattle (MGS in Spanish) is a multi-actor platform supporting the transformation towards a sustainable cattle sector. The MGS mainly focused on face-to-face activites, such as meetings for information exchange or capacity building events in the field, which had to be stopped once the pandemic had started in Colombia. In order to mitigate the related negative impacts during the strict lockdown, four MGS member institutions (FAO, National University of Colombia, CATIE and Alliance Bioversity-CIAT) decided to organise a virtual seminar series on sustainable beef and dairy systems and value chains. The objective of this article is to describe this effort, its reach and importance for the sector, its viral spread, as well as the lessons learned. The event was comprised by four modules, each with a weekly seminar during four weeks: i) greenhouse gases; ii) biodiversity, landscapes, and ecosystem services; iii) markets and consumers, and iv) agricultural extension. Each module invited experts from different institutions/sectors, to present their (scientific) advances and approaches. Initially aimed at the 53 MGS member institutions in order to keep the exchange active and build capacities, the initiative had an outstanding record of assistance, by reaching an average of 1,300 $^{\text{views}}/_{\text{week}}$. The importance of the topics and the viralisation of the initiative gathered people from 23 countries. The feedback received was very positive, and the attendees rated the presenters regarding clarity, level of expertise, time management, and if the goals were reached. About 30 % of the audience were cattle producers, reaching a key stakeholder and contributing to strengthening their capacities regarding the implementation of sustainable production technologies. Other participants came from academia, NGOs, governmental institutions, and private companies. With these promising results and a big database of people interested in sustainable cattle production, the MGS could strengthen its role as actor for a regional dialogue, and contribute to the strengthening of the public policy framework and the formulation of high impact projects for sustainable intensification.

Keywords: Capacity building, knowledge sharing, partnerships, webinars

Contact Address: José Luis Urrea-Benítez, International Center for Tropical Agriculture (CIAT), Tropical Forages Program, Km 17 Recta Cali-Palmira, 763537 Palmira, Colombia, e-mail: j.l.urrea@cgiar.org

Transformation of Traditional Livestock Systems under Land Use Changes from the 1970s to the 2018/2019 in Ladakh, India

Maximilian Ibing[1], Martin Wiehle[2], Thanh Thi Nguyen[3], Imke Hellwig[3], Eva Schlecht[1], Andreas Buerkert[3]

[1]*University of Kassel / University of Goettingen, Animal Husbandry in the Tropics and Subtropics, Germany*

[2]*University of Kassel, Tropenzentrum / Organic Plant Production and Agroecosystems Research in the Tropics and Subtropics (OPATS), Germany*

[3]*University of Kassel, Organic Plant Production and Agroecosystems Research in the Tropics and Subtropics, Germany*

Ladakh in N-India is an example for remote mountain areas that are still largely dedicated to subsistence agriculture while facing major rural-urban transformation processes. It is home to traditional agriculturists, agro-pastoralists, and nomads that are highly adapted to a unique and demanding high-mountain environment. Since the 1970s the local agricultural communities are subject to a profound structural change which accelerated with the recent advent of rapidly increasing tourism. This study aimed to analyse the (1) drivers of change, (2) adaptation mechanisms of the agro-pastoral subsistence communities to a changing market for labour and agricultural products, and (3) rampant land use change processes and urbanisation. To this end, four study sites, divided into two pairs, one for agro-pastoralists (Leh and Diskit), and one for pastoralists (Kharnak and Kharnakling) were selected. This allowed to compare sites with different levels of transformation and exposure to tourism. We conducted 98 semi-structured household interviews to determine agricultural practices, cropping patterns, livestock keeping, changes in herd composition and product marketing and motivation. Additionally, socio-economic parameters such as household expenditure, (off-farm) occupation, and impact of tourism on household revenue were surveyed. Besides, a remote sensing and GIS approach allowed classifying land use maps in the 1970s, the 2000s, and 2018/2019. The land use classification revealed an overall expansion of urban areas, particularly in Leh with an eightfold increase in built-up land since 1974, while in Diskit the urban area doubled. In Leh, 41.7 % of the agricultural area transformed to urban compared with only 1.7 % in Diskit. Households adapted to this process by incorporating new modes of cropping and livestock keeping in their traditional systems, resulting in a higher specialisation. Socio-economic data indicated increasing off-farm income generating activities while the available

Contact Address: Andreas Buerkert, University of Kassel, Organic Plant Production and Agroecosystems Research in the Tropics and Subtropics, Steinstraße 19, 37213 Witzenhausen, Germany, e-mail: buerkert@uni-kassel.de

workforce dedicated to agriculture declined by 30 % since the 1970s. Results showed that main drivers of transformation are tourism (direct) and border tensions (indirect). There is evidence that the local change processes may result in a complete abandonment of traditional systems in the long run, even if religious and cultural reasons may slow down this process.

Keywords: Agriculture, GIS and remote sensing, pastoralists, rural-urban transformation, tourism

Economic Contribution of Cart Donkeys to the Livelihoods of Donkey Cart Families in the Rural and Peri-Urban Areas of Punjab, Pakistan

Muhammad Tariq

University of Agriculture Faisalabad, Dept. of Livestock Management, Pakistan

Among livestock, donkeys are one of the major draught/carriage animals, source of transportation, livelihood means and security for the family crisis in Pakistan. Donkeys support at great extent to integrated agriculture in rural-areas and are efficiently used to pull wooden carts carrying variety of goods e.g fruits, vegetables, cereals, wood, bricks etc. and earn livelihood for many poor families in peri-urban areas. A baseline survey was conducted using semi-structured pre-tested questionnaire, 60 each from districts Faisalabad (Metropolitan) and Toba Tek Singh (Rural) donkey-cart owners (n=120) during January-February 2019, using snow-ball sampling to explore different transportation services by donkeys and economic contribution cart-donkeys and its impact on livelihoods of donkey-cart families in peri-urban and rural-areas of Punjab, Pakistan.

Results showed that interviewed cart-donkey owners had age averaged 38.5 ± 11.90 years and mostly were uneducated (n=66, 55 %) and 33 % had 5years education. Most (78 %) were married and supporting family, averaged size of 6.0±2.08HH members. The donkey-carts keeping experience averaged 12.9 ± 4.78 years. Overall, donkey-cart price averaged 29816.7 ± 5768.58PKR (≈177.3 Euro), was mostly made from Dalbergia sissoo wood (97 %). Donkeys age averaged 2–4 years (36 %), 4–7 (59 %) and 8–10years 5 %. All donkeys were exclusively males and adult-donkey price averaged 35229.2 ± 19051.45PKR (≈209.69 Euro) and Lassi breed price was significantly ($p < 0.001$) higher than Spurki and non-descript. Daily working hours averaged 9.0±2.50 hours/day. Normally, most donkey-carts (97.5 %) were used around the year with averaged use 6.1±1.11 days/week. All donkey-carts (98 %) were used exclusively for earning livelihoods with a few for household activities. Daily income averaged 853.8±361.96 PKR/day (≈5.08 Euro/day) and was significantly ($p < 0.05$) different among two cities. Daily feeding expenditures averaged 185.9 ± 56.89 PKR/day (≈1.11 Euro/day). All the interviewed owners were keeping donkey-carts as their prime source of livelihood and earning exclusively being used to fulfil daily household needs (98.5 %) plus donkey-keeping expenses. In general, it is concluded that, donkey ownership and participation of donkey-carts in transportation of goods both in peri-urban and rural-areas has a positive impact livelihoods by increasing their income and livestock ownership. But despite of valuable contribution in providing livelihoods to the poor families, still it is extremely neglected species in terms of facilitation, conservation and resource allocation in Pakistan.

Keywords: Donkey, draught power, Pakistan, poverty, Punjab, subsistence, wooden carts

Contact Address: Muhammad Tariq, University of Agriculture Faisalabad, Dept. of Livestock Management, 38000 Faisalabad, Pakistan, e-mail: tariqlm@uaf.edu.pk

Factors Affecting Livestock Profitability in Egyptian Rural Farming Systems

Helmy Metawi, Eman El- Bassiouny

Animal Production Research Institute, Sheep and Goats Research, Egypt

Several studies have attempted to understand why livestock production on some farms is more profitable than others. This paper examined the factors affecting the profitability of livestock production in rural farming systems in two agro-ecological zones in Egypt: the irrigated areas in the Nile Valley (Damietta Governorate) and the oases area in the Western Desert (New Valley Governorate). The two regions represent different integrative agriculture-livestock systems with different levels of intensification.A total of 180 farmers were randomly selected from farm households engaged in agricultural production Structured questionnaire was used to collect data during the period from March 2020 to January 2021. Data collected were analysed using descriptive such as frequency and percentages. Gross margin and regression models were also used. Results showed different contributions of animal species to household livelihood according to asset endowments, societal and agro-ecological environments. Livestock herds differed in size and composition between the different areas. The variability of herd size is more important in the New Valley according to land tenure and water source. Socio-economic characteristics such as farmer age, family size, education level, herd size, income and breeding experience were statistically significant at different levels and were the main factors affecting the profitability of animal husbandry. The recommended policy measures should be directed towards establishing small ruminant breeding centres in order to increase herd size and transfer knowledge by providing extension services to educate farmers about modern management techniques. Farmers should be encouraged to form cooperative societies that can help them increase their access to credit facilities that provide them with adequate inputs .

Keywords: Livestock, profitability, smallholder, socio-economic characteristics

Contact Address: Helmy Metawi, Animal Production Research Institute, Sheep and Goats Research, 5 Nadi Elsaid Str.Dokki, Giza, 16128 Dokki Cairo, Egypt, e-mail: hrmmetawi@hotmail.com

Goat Production and Distribution Pattern in the Derived Savannah Area of Oyo State, Nigeria

Adedayo Sosina[1], Olaniyi Babayemi[2], Philips Adewuyi[3]

[1]*Oyo State Agri-Business Development Agency (OYSADA), Research and Technical Services Department, Nigeria*

[2]*University of Ibadan, Department of Animal Science, Nigeria*

[3]*University of the Gambia, School of Agriculture and Environmental Sciences, Gambia*

The location, population, and distribution pattern of goat for efficient policy formulation for improvement are challenges to the livestock industry. There is a great disconnect between the goat farmers' needs and policy framework towards ameliorating these production constraints. The study tries to investigate the population and distribution pattern of goat in Oyo state, Nigeria.

A multi-stage sampling technique was used to elicit information from two hundred and twenty five purposively selected respondents from Egbeda, Oluyole, Ona-ara, Akinyele, Ido, and Ibarapa East LGA's. The herd sizes were classified less than 8 (smallholders), 9–25 (medium), 26–50 (large), and above 50 (commercial). The goat production systems were mapped with the Participatory Rural Approach method. The location of the farm households were recorded using a GPS device and the average goat herd size estimated in Tropical Livestock Units (TLU). The GPS data were transferred into the ARC-GIS software and processed with the ARC-GIS model 10.0. Samples of the selected feed resources (FR) fed to goat were collected and analysed for chemical compositions: crude protein (CP), neutral detergent fibre (NDF) and ME (MJ kg^{-1} DM) using near infra-red reflectance spectroscopy (NIRS). Data were analysed using descriptive statistics. The average goat herd size (TLU) for Egbeda, Ibarapa East, Ona ara, Akinyele, and Oluyole was 37, 25, 8, 7, 5 and 4, respectively. The majority (45 %) of farmers were smallholders. The FR CP (%) ranged from 5.81 ± 0.26 (cassava leaf) to 24.91 ± 0.91 (*Amaranthus spinosus*), NDF (%) ranged from 22.38 ± 4.43 (*Amaranthus spinosus*) to 67.96 ± 2.58 (*Althemanthe dedentata*) while ME ranged from 7.88 ± 0.24 (*Althemanthe dedentata*) to 10.68 ± 0.18 (cassava leaf). The goat farmers were evenly distributed across rural areas due to the availability of abundant feed resources. Most feed resources available were below goat protein requirement level, hence supplementation necessary for productivity. The bio-informatics can provide relevant information for goat production for policy framework and intervention strategies.

Keywords: Bio-informatics, ecological zone, feed resources, goat production systems, interventions, policy

---

Contact Address: Adedayo Sosina, Oyo State Agri-Business Development Agency (OYSADA), Research and Technical Services Department, 1 Secretariat Road, Ibadan, Nigeria, e-mail: dayososina@gmail.com

The role of wildlife

Why Bushmeat Should Not Be Banned

Louw Hoffman

University of Queensland, Centre for Nutrition and Food Sciences, Australia

With the initial belief that the SARS-CoV^{-2}, (COVID-19 virus) came from exotic animals whose meat was being sold at a wet market in Whuan, China, there were calls all over the world that bushmeat should be banned from all markets worldwide. During May 2017 to November 2019, 47,381 animals from 38 species were sold at this specific market. Numerous scientists and organisations believed that prohibiting the sale and consumption of wild meat would protect public health and biodiversity. Other scientists argued that this sudden removal of wild meat could impact negatively on people and wildlife; up to 15 countries have been identified as losing food security if such a ban were to be implemented. It has been calculated that if the protein supplied by wild meat had to be replaced by traditionally farmed livestock, an additional 124,000 km^2 of land would need to be transformed into agricultural land which could drive >260 species towards extinction. This presentation discusses these issues and suggests a more holistic approach to address these matters due to the complex nature of the interconnectivity of global food systems and nature around the background of vulnerable people, particularly the silent masses in developing countries.

Keywords: Holistic approach, public health, wildlife

Contact Address: Louw Hoffman, University of Queensland, Centre for Nutrition and Food Sciences, QLD 4072 Brisbane St Lucia, Australia, e-mail: louwrens.hoffman@uq.edu.au

Wild Mammal Dung Abundance in Lake Mburo National Park Is Lower Than in Adjacent Ranchlands

Antonia Nyamukuru
Makerere University, Environmental Management, Uganda

The establishment of livestock ranchlands adjacent to protected areas in savannah ecosystems is believed to threaten wild animals. Intensive competition for vegetative resources, water and poaching are considered to be immediate factors that reduce the capacity of protected areas to sustain wild mammals. The coexistence of wild mammals and ranchlands is common in Southern Africa but has rarely been suggested as a viable conservation option in East Africa. To assess the importance of ranchlands in conserving wild mammals, 36 plots of 20 × 20 m dimension were positioned along a 7240 m stretch from the boundary in Lake Mburo National Park (LMNP) and 36 plots of similar dimension were set within the ranchlands adjacent to the Park. The dung counts of different species recorded in the plots were used as a relative index of mammal abundance in the ranchlands and in LMNP. The results reveal 18 wild mammal species recorded in both sampled areas, 12 within LMNP and 17 in the adjacent ranchlands. The topi Damaliscus lunatus was only found in the park. Total dung count estimated in both ranchlands and LMNP was 2,586 with LMNP accounting for 29 % and ranchlands 71 %. In terms of wild mammal dung, ranchlands had a higher wild mammal dung count than LMNP (30 % higher). The study points to the compatibility of the two land uses in conserving wild mammals and biodiversity in general, negating the common belief of competition and exclusion. Future research is needed on the compatibility of ranchlands with protected areas on biodiversity status of other species.

Keywords: Abundance, cattle, land use, livestock, protected area, ranches, savannah, Uganda, wild mammals

---

Contact Address: Antonia Nyamukuru, Makerere University, Environmental Management, Kampala Road, Kampala, Uganda, e-mail: nyamukuru@gmail.com

Human-Wildlife Conflict and Household Livelihoods in Communal Conservancies, Kunene Region, Namibia

Esther Simaneka Ndinelago Nambahu, Meed Mbidzo, Thinah Moyo
Namibia University of Science and Technology, Agriculture and Natural Resources Sciences, Namibia

Human Wildlife Conflict (HWC) incidents threaten the livelihoods of households whose income is largely dependent on livestock farming. This study aimed at determining the impact of HWC on the livelihoods of households in communal conservancies with a particular focus on livestock predation. Household surveys were used to assess the impact of livestock predation on the livelihoods of nomadic pastoralists in four conservancies in Kunene Region, namely, Marienfluss, Orupembe, Okanguati, and Epupa. Descriptive statistics, cross-tabulations, Chi-square test and one-way ANOVA were used to analyse the data. Livestock selling is important for food security in the four conservancies. Livestock selling is the paramount source of income in the study areas, with few households having members employed by the conservancy as game guards or in joint venture lodges and campsites. The dependence of households on livestock threatens food security because of livestock predation, which takes away livestock that would be sold for food and sustenance of the family. Thus, making livestock predation a problem, especially for poor households with few livestock numbers. The study found that households experience livestock predation caused by different predators. There is a lack of offset payments for livestock losses to farmers even though the HWC-SRS was established in the conservancies. Additionally, even though conservancies generated income, little or no benefits reached the household level. An evaluation of the cost of HWC incidents of livestock predation found that HWC is costly to household wellbeing affecting food security, particularly of poor households. Households in conservancies, therefore, need assistance to mitigate the effects of drought and HWC incidents.

Keywords: Benefits, conservancies, households, HWC, HWCSRS, livelihoods, livestock predation, pastoralists

---

Contact Address: Esther Simaneka Ndinelago Nambahu, Namibia University of Science and Technology, Agriculture and Natural Resources Sciences, No. 13 Jackson Kaujeua Street, 9000 Windhoek, Namibia, e-mail: esthernambahu@gmail.com

The Lessepsian Immigrant Fish Species, Is it a National Disaster or a National Wealth

Sahar Mehanna

National Institute of Oceanography and Fisheries (NIOF), Fish Population Dynamics, Egypt

There are probably no ecosystems on earth which may resist the introduction of an alien species. Success of an introduced species in the new environment generally depends on a combination of several bio-ecological factors. The migration of Red Sea fishes through the Suez Canal, which is known as "Lessepsian migration has impacted the ecosystem of the Eastern Mediterranean. Not all the invasion through the Suez Canal is of negative impacts, the richness of Red Sea species introduced through the Suez Canal (Lessepsian species) to the eastern Mediterranean coastline, reaching a maximum of 129 species per 100 km^2, as well as many Lessepsian species have positive impacts on the ecosystem and biodiversity and securing food for millions of people. In Egypt, more than 50 % of the Mediterranean catch is of Red Sea origin. Out of 43 Lessepsian fish species recorded in the Egyptian Mediterranean waters, about 20 fish species are became of commercial importance and compose the main food for the coastal communities on the Egyptian Mediterranean. Most of Lessepsian species distributed in the region have begun to be caught in increasing quantities in recent years and became one of the economically important species landed in many Mediterranean countries like Turkey, Greece, Cyprus, Lebanon, ...etc. On the other hand, few numbers of species have caused some health problems but this could be mitigated by some regulations. This study is prepared to discuss and demonstrate the economic importance of most Lessepsian immigrant fish species and their contribution to the eastern Mediterranean economy and food security as well as the negative impacts of some Lessepsian immigrant and how to mitigate it. So, this study will answer a very important question is the Lessepsian immigrant fish species a national disaster or a national wealth.

Keywords: Alien species, commercial importance, lessepsian migration, positive impacts

Contact Address: Sahar Mehanna, National Institute of Oceanography and Fisheries (NIOF), Fish Population Dynamics, P.O. Box 182, Suez, Egypt, e-mail: sahar_mehanna@yahoo.com

Parasites, pathogens and animal health

Posters

Effect of Communication Towers on the Performance and Behaviour of *Apis mellifera* (Internal Activities) in Iraq

Faiza Salah[1], Alaa Alnaimee[2]
[1]*University of Gezira, Dept. of Crop Protection, Sudan*
[2]*University of Baghdad, Plant Protection, Iraq*

Beekeeping of *Apis mellifera* isthe most important branches of agricultural investments. Bees are most active pollinators. Many factors affect their activities, one of them is electromagnetic radiation. The aim of this study was to investigate the effect of the radiation emitted by communication towers on the behaviour of honey bee communities internally. The experiments were conducted in Iraq. The first location was 500, the second was 150 meters from the telecommunication tower and the third transaction was placed directly under the tower. The height of the tower was 30 m and the amount of radiation emitted from it was 925 MHZ. The results of the internal activity, measuring the activity of the queen in laying eggs, showed that the highest rate of activity in the first treatment, followed by second treatment while the lowest rate was recorded on the third treatment. The area of the closed brood was highest in the first treatment followed by the second treatment and the lowest rate in the third treatment .The results showed that the highest drone brood of the third treatment was 6.76 inches, and the first and second treatments recorded 4.48 and 14.1 inch, respectively. As for the area of sealed honey, there were no significant differences between treatments. As for the results of pollen area, there were significant differences for the first and second treatments, while the third treatment recorded the lowest rate. The results of the measurement of the speed of achievement of the wax foundation showed that there were significant differences between the two treatments, where the second treatment surpassed the rest of the transactions within four days, followed by the first treatment and the third transaction recorded the lowest. The highest rate of density of bees was recorded in the first treatment and the lowest rate for the third treatment. It could be concluded that communication towers have negative effect on the internal activities of *Apis mellifera.*

Keywords: *Apis mellifera*, communication tower, Iraq

Contact Address: Faiza Salah, University of Gezira, Dept. of Crop Protection, P.O. Box20 Nishishiba, 111111 Wad Madeni, Sudan, e-mail: faizaruba2@gmail.com

Endoparasite Infections in Dairy Cattle Along a Rural-Urban Gradient in the Megacity of Bangalore

Ana Pinto[1], Katharina May[1], Tong Yin[1], Marion Reichenbach[2], Pradeep K. Malik[3], Regina Rössler[2], Eva Schlecht[2], Sven König[1]

[1]*Justus-Liebig-University Giessen, Inst. of Animal Breeding and Genetics, Germany*

[2]*University of Kassel / Georg-August-Universität Göttingen, Animal Husbandry in the Tropics and Subtropics, Germany*

[3]*National Institue of Animal Nutrition and Physiology, Bangalore, India*

With the urban share of the world population projected to increase to more than 65 %, the need to pursue a healthy and sustainable urban livestock system becomes fundamental. Endoparasite infections can lead to lower performance, impair health and welfare status of the animals. Environmental and host-related factors such as the close human-animal cohabitation and changes in animal housing conditions contribute to endoparasite infection intensity and probability. The aim of the present study was to investigate social-ecological effects on gastrointestinal nematode (GIN) and *Eimeria* spp. infections in dairy cattle along a rural-urban gradient in the emerging Indian megacity Bangalore. In this regard, 726 fecal samples from 441 dairy cattle of different ages and physiological stages were collected from 101 farms and examined at three visits between June 2017 and April 2018. Based on a survey stratification index (SSI) comprising built-up density and distance to the city centre, we assigned the farms to urban, mixed and rural areas. GIN eggs were identified in the faeces of 243 cattle (33.5 %), and *Eimeria* spp. oocysts in the faeces of 151 cattle (20.8 %). Co infection rates of GIN and *Eimeria* spp. were 8.5 to 12.2 % higher in rural compared to urban and mixed areas. The SSI effect significantly influenced *Eimeria* spp. infection probability and oocyst per gram of faeces (OpG; $p < 0.001$) with an infection probability and OpG higher than 26 % and 40 % for cattle kept in rural areas compared to cattle from urban areas. However, the SSI effect was not significant for the infection probability of GIN and for GIN eggs per gram of faeces (EpG). The variations in endoparasite infection intensity and probability observed along the rural-urban gradient of Bangalore reflect the variability in dairy husbandry systems governed by the social-ecological context.

Keywords: Dairy cattle, *Eimeria* spp., gastrointestinal nematodes, husbandry system, India, social-ecological effect, urbanisation

Contact Address: Ana Pinto, Justus-Liebig-University Giessen, Inst. of Animal Breeding and Genetics, Ludwigstr. 21b, 35390 Giessen, Germany, e-mail: ana.pinto-garcia@agrar.uni-giessen.de

A Qualitative Assessment of the Context and Enabling Environment for the Control of *Taenia solium* Infections in Endemic Settings

Nicholas Ngwili[1,3], Nancy Johnson[2], Raphael Wahome[3], Samuel Githigia[3], Kristina Roesel[1], Thomas Lian[1]

[1]*International Livestock Research Institute (ILRI), Animal and Human Health, Kenya*
[2]*IFPRI, CGIAR Research Program on Agriculture for Nutrition and Health, United States*
[3]*University of Nairobi, College of Agriculture and Veterinary Sciences, Kenya*

Background

Taenia solium is a zoonotic helminth causing three diseases namely, taeniasis (in humans), neurocysticercosis (NCC, in humans) and porcine cysticercosis (PCC, in pigs) and is one of the major foodborne diseases by burden. The success or failure of control options against this parasite in terms of reduced prevalence or incidence of the diseases may be attributed to the contextual factors which underpin the design, implementation, and evaluation of control programmes.

Methodology

The study used a mixed method approach combining systematic literature review (SLR) and key informant interviews (KII). The SLR focused on studies which implemented T. solium control programmes and was used to identify the contextual factors and enabling environment relevant to successful inception, planning and implementation of the interventions. The SLR used a protocol pre-registered at the International prospective register of systematic reviews (PROSPERO) number CRD42019138107 and followed PRISMA guidelines on reporting of SLR. To further highlight the importance and interlinkage of these contextual factors, KII were conducted with researchers / implementers of the studies included in the SLR.

Results

The SLR identified 41 publications that had considerations of the contextual factors. They were grouped into efficacy (10), effectiveness (28) and scale up or implementation (3) research studies. The identified contextual factors included epidemiological, socioeconomic, cultural, geographical and environmental, service and organisational, historical and financial factors. The enabling environment was mainly defined by policy and strategies supporting *T. solium* control.

Conclusion / Significance

Failure to consider the contextual factors operating in target study sites was shown to later present challenges in project implementation and evaluation

Contact Address: Nicholas Ngwili, International Livestock Research Institute (ILRI), Animal and Human Health, PO Box 30709, Nairobi, Kenya, e-mail: n.ngwili@cgiar.org

that negatively affected expected outcomes. This study highlights the importance of fully considering the various domains of the context and integrating these explicitly into the plan for implementation and evaluation of control programmes. Explicit reporting of these aspects in the resultant publication is also important to guide future work. The contextual factors highlighted in this study may be useful to guide future research and scale up of disease control programmes and demonstrates the importance of close multi-sectoral collaboration in a One Health approach.

Keywords: Contextual factors, enabling environment, *Taenia solium*

Trade-Offs and Synergies between Livestock Intensification and Emergence of Zoonotic Diseases in Sub-Saharan Africa

Sarah Schmidt, Christoph Reiber, Regina Birner, Mizeck Chagunda
University of Hohenheim, Inst. of Agric. Sci. in the Tropics (Hans-Ruthenberg-Institute), Germany

Understanding the link between agriculture and zoonotic diseases is important for a healthy and sustainable future. Zoonotic diseases are infectious diseases that are transmitted between humans and animals. Subject to type and intensity, livestock-human interactions significantly contribute to emergence and re-emergence of zoonotic diseases. The objective of the current study was to determine the trade-offs and synergies between intensifying livestock production and cases of zoonotic diseases in Sub-Saharan Africa. Specifically, the study focused on the factors that contribute to the development, spread and persistence of zoonotic diseases. Further, strategies for preventing and controlling zoonotic diseases were investigated. To address these objectives, a systematic literature review was conducted. A defined search criteria resulted in 237 peer-reviewed scientific articles and other publications covering the period between 1990 and 2021. Of these 122 articles were utilised. Results showed some direct zoonotic trade-offs with increased livestock intensification. Increased livestock population due to intensification leads to increased transmission of infectious diseases. In this regard, the general spread of zoonoses is significantly influenced by changes in land use (such as deforestation). Further, intensification usually results in homogenous populations. Such homogenous livestock population have increased vulnerability to zoonotic epidemics spread. Intensification also results in increased international trade and human travel which in itself increases disease transmission risk. In most African countries like in most Low- and Middle-Income Countries, the problem is often exasperated by lack of infrastructure, poor human medical care, and education. One synergistic aspect that comes with intensification is that of the use of technology for improved productivity. This has a positive effect in that the same technology can also be used to monitor and mitigate any outbreaks of zoonotic diseases. In order to mitigate the trade-offs and capitalize on the synergies, it is more important than ever to rely on interdisciplinary cooperation and to combine knowledge from different sectors in a 'One Health' approach., which links the health of animals, humans and the environment and thus provides the basis for future action .

Keywords: Intensification, one-health, synergies, trade-offs, zoonosis

Contact Address: Mizeck Chagunda, University of Hohenheim, Inst. of Agric. Sci. in the Tropics (Hans-Ruthenberg-Institute), Stuttgart, Germany, e-mail: mizeck.chagunda@uni-hohenheim.de

Mapping of *Borrelia* in Exotic Farm Animals of Czech Republic

Johana Hrnková[1], Jiří Černý[1], Natasha Rudenko[2], Francisco Ceacero Herrador[1], Maryna Golovchenko[2]

[1]*Czech University of Life Sciences Prague, Department of Animal Science and Food Processing, Czech Republic*

[2]*Institute of Parasitology, Laboratory of Molecular Ecology of Vectors and Pathogens, Czech Republic*

Borreliosis is a disease caused by bacteria of the *Borrelia burgdorferi sensu lato* complex. It can cause severe disease in humans, and usually milder, chronic disease in some animals. In central Europe *Borrelia* is predominantly spread by the tick *Ixodes ricinus*. The effect of *Borrelia* infection on animals exotic to Europe has been previously studied on various zoo animals. Many positive samples were detected during wide-scale samplings in Germany and Czech Republic. Susceptibility of the zoo animals to the pathogen has varied between species and individuals but the studies pointed out the necessity to further investigate the impact of *Borrelia* on exotic animals. We have collected samples from several smaller-scale farm facilities keeping exotic animals either for meat or milk production, as a tourist attraction, or for research purposes. The sampled species included ostriches, Carpathian and milk buffalo, dromedary and Bactrian camels, alpaca llamas and finally common eland antelopes. From these samples we have discovered an unexpectedly high prevalence of *Borrelia* bacteria in the blood serum of the exotic animals. The giant elands were displaying the highest overall prevalence. This fact was surprising since only 2 live ticks were found during collections in surrounding areas and on the pasture. This preliminary result leaves us with a question about the mode of transmission of this bacterium between the antelopes. Transovarial transmission has been only scarcely suggested in the scientific community, but it might explain the paradox of low tick numbers and high *Borrelia* prevalence on the common eland farm. The research continues and more tick and serum samples are being collected to further deepen the reliability of outcoming data. We hope to shed more light on the paradigm of the mode of transmission of *Borrelia* between animals and to provide the data for the monitoring of overall *Borrelia* prevalence in Czech Republic.

Keywords: Borreliosis, exotic animals, *Ixodes ricinus*, ticks

Contact Address: Johana Hrnková, Czech University of Life Sciences Prague, Department of Animal Science and Food Processing, Kamýcká 129, 16000 Praha 6, Czech Republic, e-mail: hrnkova@ftz.czu.cz

Selfmedicative Behaviour in Gastrointestinal Parasite Infected Goats: Shift in Preferences for Tanniferous Fodder Plants

Marvin Heuduck[1], Christina Strube[2], Katharina Raue[2], Eva Schlecht[3], Martina Gerken[1]

[1]*Georg-August-University Goettingen, Dept. of Animal Science, Germany*

[2]*University of Veterinary Medicine Hannover, Inst. for Parasitology, Germany*

[3]*University of Kassel / University of Goettingen, Animal Husbandry in the Tropics and Subtropics, Germany*

Gastrointestinal nematode infections are a worldwide major cause of economic losses and a thread to pasture-based livestock due to their adverse effects on animal health and productivity. Especially regarding the trichostrongyles, the most prevalent nematodes in ruminants, the excessive usage of conventional anthelmintic drugs over the last decades has led to an emergence of resistant nematode populations, and decreased the efficacy of available anthelmintics. This underlines the relevance of a paradigm shift towards a more sustainable and unconventional approach to control nematode infections.

This study evaluated a possible change in feed preferences of GIN-infected Boer goats and hypothesised that infected goats increase their intake of tannin-containing fodder plants.

Feed preferences of eighteen juvenile male Boer goats (3–4months) were evaluated over a period of 12 weeks. The individually fed animals were assigned to three treatment groups: I) Non-infected + feed choice II) Infected + feed choice, and III) Infected without feed choice. At the beginning of the trial, all goats were free of nematodes. After four weeks, group II) and III) were experimentally infected with third-stage nematode larvae representing the local species spectrum. Group I and II animals were offered a free choice cafeteria trial for 30 min prior to the usual daily feeding time. Four fodder plants (pelleted leaves of sainfoin, willow, walnut, blackberry) of varying tannin contents (low to high) and tannin-free hay pellets were simultaneously offered. The position of the pellets in the trough was randomised each day.

During the cafeteria trial plant preferences were recorded by video surveillance and amounts of ingested pellets were measured. In addition, blood parameters, saliva composition and feces were analysed on a weekly basis in order to evaluate the course of infection and potential shifts in feed preferences due to changes in taste perception. The cafeteria trial revealed a shift from tannin-free (hay) and low tannin-containing feed (sainfoin) to plants with higher tannin-contents (walnut, blackberry) with proceeding nematode infection.

Keywords: Anthelmintic drug agents, condensed tannins, goats, nematodes, selfmedicative behaviour

---

Contact Address: Marvin Heuduck, Georg-August-University Goettingen, Dept. of Animal Science, Albrecht Thaer Weg 3, 37075 Goettingen, Germany, e-mail: marvin.heuduck@uni-goettingen.de

Are There Gender Differences in Access to and Demand for East Coast Fever Vaccine? Empirical Evidence from Rural Smallholder Dairy Farmers in Kenya

Humphrey Jumba[1], Henry Kiara[1], George Owuor[2], Nils Teufel[1]

[1] *International Livestock Research Institute (ILRI), Policies Institutions and Livelihoods, Kenya*

[2] *Egerton University, Agricultural Economics and Agribusiness Management, Kenya*

Women lag in the adoption of agricultural innovations compared to men, mainly due to gender inequalities in access to complementary inputs, capital, and knowledge / information. The Infection-and-Treatment-Method (ITM) is considered a safe and effective method of controlling East Coast fever. However, since its commercialisation in Kenya differences in demand for this vaccine among smallholder men and women dairy cattle keepers have not been assessed. Using a sample of 448 respondents, we used an Average-Treatment-Effect framework to estimate ITM adoption rates under awareness constraints and the determinants of adoption among smallholder male-headed (MHHs) and female-headed (FHHs) households. We found some difference in ITM awareness between MHHs (57 per cent) and FHHs (46 per cent). However, gender adoption gaps in the actual and potential adoption rates were considerable, with actual adoption rates of 41 per cent and potential adoption rate of 62 per cent among MHHs, compared to 19 per cent actual and 31 per cent potential adoption for FHHs. The smaller adoption gap for FHHs indicates that only increasing awareness amongst FHHs will not reduce inequities. ITM adoption in both household headships was mainly determined by education, extension interventions, access to financial services, and social capital. In addition to this, ITM adoption in FHHs was positively influenced by age, landsize, and group membership. To realise adoption beyond the current potential and to reduce inequities at the scale-up stage, gender-specific interventions targeting resource-poor women cattle keepers would be effective, in addition to ensuring that women have access to extension and financial services.

Keywords: Adoption, awareness, East Coast fever, gender, infection and treatment method, Kenya

Contact Address: Humphrey Jumba, International Livestock Research Institute (ILRI), Policies Institutions and Livelihoods, Po Box 30709, Nairobi, Kenya, Kenya, e-mail: h.jumba@cgiar.org

Consumers and food

Pathways to improved food and nutrition security

Oral Presentations

Posters

Identifying Nutrition-Sensitive Development Options in Madagascar through a Positive Deviance Approach

Arielle Sandrine Rafanomezantsoa[1], Jonathan Steinke[2], Christoph Kubitza[2], Sarah Tojo Mandaharisoa[1], Narilala Randrianarison[1], Denis Randriamampionona[1], Harilala Andriamaniraka[1], Stefan Sieber[3,2]

[1]*University of Antananarivo, Trop.l Agric. and Sustainable Develop., Madagascar*

[2]*Humboldt-Universität zu Berlin, Thaer-Institute of Agricultural and Horticultural Sciences, Germany*

[3]*Leibniz Centre for Agric. Landscape Res. (ZALF), Sustainable Land Use in Developing Countries (SusLAND), Germany*

In Madagascar, food insecurity is a major public health concern. Atsimo Antsinanana region is one of the most affected regions with its 6-months lean period in the year. To identify local strategies to improve household food security, we adopt the positive deviance approach. It is based on the concept that in all societies, there are positive deviant individuals who are more successful while having the same resources as others. The approach has already proved its worth when applied to child malnutrition at individual level. Here, we extend the analysis to food and nutrition security at household level. Thus, the objective of this study is to identify the particular practices and behaviours adopted by households with superior food and nutrition security. Household performance was assessed on four dimensions: Household food insecurity access scale (HFIAS), minimum acceptable diet (MAD) of children under two years, women's dietary diversity score (WDDS), and child diarrhea frequency (as a proxy for nutrition-related hygiene behaviour). Subjects of the study were 413 households located in 3 districts of the Atsimo Antsinanana region. The methodology used combines qualitative and quantitative methods. The data analysis started with removing the influence of household resources on food security performance, then the identification of positive deviant households by using Pareto frontier of the four performance dimensions. As results, 20 positive deviant households were identified. We find that positive deviance is strongest for HFIES, and weakest for child diarrhea frequency. To identify 'deviant' practices, 16 positive deviant households were interviewed, and 8 focus group discussions with nutrition experts in the region were conducted to evaluate the findings on possible innovative practices. This methodology led to the identification of several strategies used by positive deviant households, including the strategy of buying rice during the harvest season and selling in the off-season, adoption of new cultivation techniques and new agricultural channels, the renewal of fruit plantations, and cultivation of cloves. These findings can contribute to the life improvement of Atsimo Antsinanana population especially on their food security so that they can lead a healthy life.

Keywords: Atsimo Antsinanana, food security, household, positive deviance

---

Contact Address: Arielle Sandrine Rafanomezantsoa, University of Antananarivo, Trop.l Agric. and Sustainable Develop., Antananarivo, Madagascar, e-mail: arielsandrine1@gmail.com

Enhancing Rural households' Food Security through Diversification: A Food System Analysis in Tanzania

Mwanga Ronald[1], Million Seleshi[2], Stefan Sieber[3], Constance Rybak[4]

[1]*Humboldt University Berlin, Faculty of Life Sciences, Germany*
[2]*Himaraya University, Rural Development and Agricultural Extension, Thailand*
[3]*Leibniz Centre for Agric. Landscape Res. (ZALF), Sustainable Land Use in Developing Countries (SusLAND), Germany*
[4]*Leibniz Centre for Agric. Landscape Res. (ZALF), Inst. of Socio-Economics, Germany*

Inadequate access to nutritionally rich food is a global challenge with particular salience in developing countries. Thus, sustainable food-based approaches are vital for reducing malnutrition due to insufficient dietary intakes and nutrient-deficient diets. The diversification of diets through increased production and utilisation of underexploited but nutrient-dense crops like indigenous vegetables (IVs) and legumes is an ideal way to reduce hidden hunger and food insecurity, especially, among the vulnerable and lower-income rural population. This paper analyses the potential of crop diversification to enhance food security in rural households using an integrated food system analytical framework. The effect of diversified on-farm production, including the inclusion of IVs and legumes in farm systems, on nutrition outcomes is analysed using Household Dietary Diversity Score (HDDS) as a proxy. The study uses cross-sectional data collected from 667 rural households by the Vegi-Leg Project based in Lindi region, Tanzania. Employing an Order logit model, it identifies the effect of production diversification (indigenous vegetables and pigeon peas) on household food security status and other determinants of food security. The results indicated that the mean HDDS was 8.29 while the production diversity was 3.07. The order logit regression model result reveals that production diversity positively affects HDDS. ($p < 0.05$). The result also shows that HDDS is negatively influenced by age of household head ($p < 0.01$), while positively influence by sex of household head ($p < 0.05$), indigenous vegetable production ($p < 0.1$), and districts ($p < 0.01$). Therefore, the uptake of resilient agro-biodiversity enhancing crops like pigeon pea and IVs should be promoted and considered as a strategy to ensure sustainable food and nutritional security, especially in low-income rural households.

Keywords: Food security, households dietary diversity score (HDDS), production diversification

Contact Address: Mwanga Ronald, Humboldt University Berlin, Faculty of Life Sciences, Oberfeldstrasse 132, 724-01-02-01, 12683 Berlin, Germany, e-mail: rnldmwanga@yahoo.com

Ethnobotany of Wartime: Wild Plants for Human Nutrition During the Conflict in Syria

Naji Sulaiman[1], Zbynek Polesny[1], Cory Whitney[2]

[1]*Czech University of Life Sciences Prague, Faculty of Tropical AgriSciences, Department of Crop Sciences and Agroforestry, Czech Republic*

[2]*University of Bonn, Inst. Crop Sci. and Res. Conserv. (INRES), Germany*

Wild edible plants have been an important source for human nutrition since ancient times, in particular when access to food is constrained in emergency situations such as natural disasters and conflicts. The war in Syria has now been going on for 10 years since it started in 2011. In addition to the many consequences of this conflict, 6.5 million people are unable to meet their food needs and a further 2.5 million people are at risk of food insecurity. Wild food plants are already culturally important in the region and may be supplementing local diets during this conflict. Our study aimed to uncover which wild plant species are used by local people, and what role they play in human nutrition during the current crisis. The fieldwork was carried out between March 2020 and March 2021 in the Tartus governorate located in the coastal region of Syria. We used semi-structured interviews with 50 participants in 26 villages about their use of wild plant species for food and drink. We recorded the vernacular names, uses, plant parts used, modes of preparation and consumption and frequency of use. We documented 75 wild food plant species used for food and drink. Asteraceae, Fabaceae and Lamiaceae were the most represented botanical families, whereas *Origanum syriacum*, *Rhus coriaria*, *Eryngium creticum*, and *Cichorium intybus* were among the species most quoted by informants. *Sleeq*, *Zaatar*, and *Louf* were the most popular wild plant-based dishes. The young aerial part was the most common plant part to be used. However, the nutritional composition of these wild food plants is not well documented. More research will be helpful in determining the role that these wild food plants play in supplementing local diets during the conflict.

Keywords: Eastern Mediterranean, emergency human behaviour, middle East, traditional food, wild vegetables

---

Contact Address: Zbynek Polesny, Czech University of Life Sciences Prague, Faculty of Tropical AgriSciences, Department of Crop Sciences and Agroforestry, Kamýcká 129, 16500 Praha - Suchdol, Czech Republic, e-mail: polesny@ftz.czu.cz

Importance of Social Capital in Enhancing Child Feeding and Nutrition in Drylands in Northern Benin

Kouété Paul Jimmy[1], Ange Honorat Edja[1], Georges Djohy[2], Brigitte Kaufmann[3]

[1]*University of Parakou, Dept. of Rural Economics and Sociology, Benin*

[2]*University of Parakou, National School of Statistics, Planning and Demography (EN-SPD), Benin*

[3]*German Institute for Tropical and Subtropical Agriculture (DITSL), Germany*

In Benin as in many developing countries, stunting remains a public health problem, touching 32 % among children under five years at national level, and reaching 41 % among children from poor households. Infant and young child feeding and nutrition (IYCFN) practices depend on many factors including social factors and cultural norms. Social capital with trust, norm and network as attributes, has been recognised to stimulate positive nutrition outcomes. However, little is known about the role that different types of social capital play in generating of and adopting locally appropriate IYCFN practices. This paper aims at filling this knowledge gap by exploring the social capital- based factors influencing mothers' IYCFN decisions and practices. Within a transdisciplinary collaboration, qualitative research following an ethnographical approach was conducted. Data was collected from mothers of children under five years in Banikoara and Nikki Districts in the drylands of Northern Benin. Participant observations, in-depth narrative interviews and focus-group discussions were conducted with regard to mothers' social capital, i.e., relationships, roles, information and resources flow, norms, and their influences on mothers' decisions and practices in the domain of IYCFN. Social capital was conceptualised in threefold terms: bonding social capital describes the links between people with similar objectives, such as mothers' groups; bridging describes the capacity of such groups to make links with others that may have different views; and linking describes the ability of groups to engage with external agencies.

Results revealed the importance of bonding capital (e.g. through women self-help groups) to increase household food availability for more diversity in complementary feeding through existing collective action and food exchanges practices. Also, linking social capital showed direct positive effects when external actors developed trust-based relations with child mothers by: i) valorizing local food resources and knowledge, and ii) helping to develop and add value to local complementary foods that fit to mothers bonding capital (i.e., family capacities and capitals). Therefore, taking bonding social capital and trust into account would be a promising perspective to support locally appropriate IYCFN practices, and consequently improving child nutrition status in households with low levels of financial and physical capital.

Keywords: Benin, child mothers, feeding and nutrition practices, local food resources, social capital

---

Contact Address: Kouété Paul Jimmy, University of Parakou, Rural Economics and Sociology, Okedama, Parakou, Benin, e-mail: jimmykouetepaul@yahoo.fr

Sun-Drying Vegetables amongst Smallholder Farmers in Western Kenya – A Matter for Food Insecure Households Only?

Irmgard Jordan[1], Eleonore C. Kretz[2], Annet Itaru[3], M. Gracia Glas[1], Lydiah Maruti Waswa[3]

[1]*Justus-Liebig University Giessen, Center for International Development and Environmental Research, Germany*
[2]*Justus-Liebig University Giessen, Inst. of Nutritional Sciences, Germany*
[3]*Egerton University, Dept. of Human Nutrition, Kenya*

Background: A year-round availability of food is a challenge in many sub-Saharan African countries. Vegetable preservation methods can play a role in contributing to food security and is considered to be an easy option to enhance vegetable intake. This study aimed at investigating predictors of vegetable preservation by sun drying and possible associations with food security.
Methods: Cross-sectional household surveys were conducted in December 2018 (baseline) and January 2021 (endline) targeting 190 households in Teso-South Sub-County, Kenya. Among others, data on socio-economic status and the Food Insecurity Experience Scale (FIES) were captured. The endline survey followed up on experiences gained during Trials of Improved Practices (TIPs) which were carried out among a subgroup of 53 households in 2019, testing among others recommendations on home-scale sun-drying of vegetables. Analysis of the survey data included estimation of food insecurity, and predictors for sun-drying of vegetables.
Results: The majority of the respondents who sun-dried vegetables in 2020 had been part of the TIPs-group (78 %). There was no significant but positive correlation between sun-drying and education level of the respondent or spouse. The prevalence of households who remained food insecure (FI) from 12/2018 to 1/2020 was 4 % compare to 16.5 % among control households, while about 12 % of the households were no/mild FI in both groups stayed to be no/mild FI. The increase in FI level was generally higher in the control group compare to TIPs households. Drying of vegetables was more likely practised in households who were food insecure at endline.
Conclusion: The preservation of vegetables by home-scale sun-drying is not commonly practised in Teso-South. Although sun-drying was more likely to be implemented with increasing wealth and education level, it remains associated with food insecurity only. A promotion campaign is needed to enhance the (cultural) acceptability and utilisation of dried vegetables as easy option to improve vegetable intake.
Funding: The study was conducted within the EaTSANE-project financially supported by BMEL/ptble (Germany) and MOEST (Kenya) within the LEAP-Agri initiative.

Keywords: Dried vegetables, food intake, food security, sun-drying, vegetable

Contact Address: Irmgard Jordan, Justus-Liebig University Giessen, Center for International Development and Environmental Research, Senckenbergstr. 3, 35390 Gießen, Germany, e-mail: Irmgard.Jordan@ernaehrung.uni-giessen.de

Homegardens: Commercialisation and Contribution to Food Security in the Upper East Region of Ghana

Harriet Tweneboah, Vladimir Verner

Czech University of Life Sciences Prague, Fac. of Tropical AgriSciences, Czech Republic

Homegardens have become a center for development initiatives by governments, non-governmental organisations as well as development agencies to help in the global challenge of food production and food insecurity especially in developing countries. They embark on this through providing aid and building the local capacity to promote sustainable intensification of homegardens to improve food and cash security of households. This study therefore aimed at determining homegarden commercialisation and its impact on the food security of smallholder households in the Upper East Region of Ghana. Mixed sampling technique was used to select 120 farmers who were running homegardens. Commercialisation was calculated using commercialisation index, which served to categorise the homegardens into more and less market oriented. Food security was measured via Hfias Score. Ordered probit regression was used to analyse the contribution of homegarden commercialisation on the household food security. The results showed that homegarden commercialisation significantly contributes to food security of farmers in the Upper East region. Also, outcomes from qualitative analysis confirmed that farmers highlighted important role of homegardens in terms of food security and gaining additional income. Major constraints in homegardening were high initial capital of investment, lack of water and lack of agricultural extension service. We, therefore, argue that governments and development agencies should include and support homegardens in agricultural and rural development policies in Ghana.

Keywords: Commercialisation index, Hfias score, multiple linear regression, ordered probit model

Contact Address: Vladimir Verner, Czech University of Life Sciences Prague, Fac. of Tropical AgriSciences, 129 Kamycka str., 16500 Prague - Suchdol, Czech Republic, e-mail: vernerv@ftz.czu.cz

Home Gardening for Climate Change-Resilience: Co-Designing a Nutrition-Sensitive Intervention in Rural Kenya

Lea-Sophie Hansen[1], Isabel Mank[1], Grace W. Kihagi[1], Erick Agure[1], Erick M O Muok[2], Ina Danquah[1], Raissa Sorgho[1]

[1]*Heidelberg University, Heidelberg Institute of Global Health (HIGH), Germany*

[2]*Kenya Medical Research Institute, Centre for Global Health Research (CGHR), Kenya*

Climate change will slow down the reduction of child undernutrition in rural sub-Saharan Africa, where households largely rely on rainfed subsistence agriculture. An integrated programme of nutrition-sensitive (home gardening) and nutrition-specific (nutrition counselling) interventions offers a promising adaptation strategy. Co-design with the local community and stakeholders is key for the successful implementation of such a program. In this qualitative study, we aimed at identifying perceptions and experiences of farming families and local stakeholders with regard to home gardening in rural Kenya. Between September and November 2020, we conducted semi-structured in-person interviews with 30 caregivers of young children living in Siaya County and online interviews with 28 local stakeholders working in the agriculture and nutrition sectors. The interviews focused on experiences with home gardens, including the preferred set-up, maintenance strategy and crop choice, as well as barriers. The audio records were transcribed and analysed by content analysis using the software Nvivo. Among these rural communities in western Kenya, home gardening was a well-established practice, and perceived as a means of diversifying the households' nutrition and livelihood. However, the participants acknowledged that home gardening was rarely high yielding, only conducted seasonally, and partly depended on the use of chemical fertilisers and pesticides. Barriers to a prospering home garden were severe shortage of seeds and saplings, access to water and land, particularly for women, as well as destruction by straying livestock. For an improved garden set-up, the interviewees mentioned the relevance of capacity building for resilient designs, self-sufficient resource management and a strong community collaboration. Stakeholders stressed the importance of co-designing the project, together with the community and local leadership, and enabling an independent replication of techniques and resources in the community. Thus, home gardening remains a promising approach to combat the impacts of climate change on food self-sufficiency in rural Kenya. The exact set-up should be tailored closely to individual needs and resources to facilitate caregivers' empowerment to contributing to household food security.

Keywords: Adaptation, food security, home gardens, undernutrition

Contact Address: Lea-Sophie Hansen, Heidelberg University, Heidelberg Institute of Global Health (HIGH), Im Neuenheimerfeld 130.3, 69120 Heidelberg, Germany, e-mail: lea-sophie.hansen@uni-heidelberg.de

The Potential of Neglected and Underutilised Species for Enhancing Food and Nutrition Security in Northern Ghana

Lara Elena Thiele[1], Rashida Chantima Ziblila[2], Lydia Madintin Konlan[3], Fred Kizito[4], Margareta Lelea[5], Oliver Hensel[1]

[1]*University Kassel, Agricultural and Biosystems Engineering, Germany*
[2]*University for Development Studies, Dept. of Agricultural Extension, Rural Development and Gender Studies, Ghana*
[3]*University of Kassel / University of Goettingen, Dept. of Agricultural Econ. and Rural Developm., Germany*
[4]*International Insititute of Tropical Agriculture (IITA), Natural Resource Management, Ghana*
[5]*German Institute for Tropical and Subtropical Agriculture (DITSL), Germany*

Food and nutrition security remains an issue and while some crops supply a significant amount of calories worldwide, other neglected and underutilised species (NUS) are not considered for commercial agriculture despite that many have nutritional benefits. In Ghana, anemia, and vitamin A deficiency are prevalent among women and children.

The aim of the study is to assess the nutritional status of selected NUS and their availability, and accessibility, in relation to their potential to increase food and nutrition security in northern Ghana. The location in northern Ghana includes 27 women's groups around Tamale. Publicly available databases were utilised to assess the nutritional composition of selected NUS. Then two specific examples are contrasted – one which is introduced - orange-fleshed sweet potato (*Ipomoea batatas*) and – one that is indigenous – African Locust Bean (*Parkia biglobosa*). Transcripts from group activities such as a seasonal calendar and other focus group discussions as well as semi-structured interviews are coded for access, availability, volume, and utilisation among other themes. This analysis is complemented by IITA data of biofortified OFSP released in Ghana.

The analysis of the databases shows the high potential of selected NUS. For example, ALB contains a high amount of folate necessary for women during pregnancy. The semi-structured interviews revealed that the situation of the ALB trees is dire with diminishing availability due to agricultural mechanisation, firewood pressures, and other factors arising from access rights and land governance. Multiple approaches are needed to ensure the propagation of young trees. OFSP has been introduced and promoted due to its high Vitamin A content. Currently, it is distributed by international aid organisations

Contact Address: Lara Elena Thiele, University Kassel, Agricultural and Biosystems Engineering, Witzenhausen, Germany, e-mail: letmino85@gmail.com

and agricultural extension, but it is in need of a functioning supply chain and increased market demand.
The analysis revealed that data on NUS is scarce and the availability of NUS is a key factor for farmer's uptake and therefore using the potential that the species offers. *Parkia biglobosa* is not sufficiently available anymore and *Ipomoea batatas* L. is not sufficiently available yet. Furthermore, biofortification of OFSP can bring the vitamin A content to an unprecedented amount compared to already released varieties.

Keywords: African locust bean, Ghana, neglected and underutilised species, nutrition security, orange-fleshed sweet potato

Factors Explaining Purchase Choices of Packaged Child Food in Kenya and Benin

Ina Cramer[1], Iris Schröter[1], Diba Tabi Roba[2], Georges Djohy[3], Hussein Wario[2], Marcus Mergenthaler[1]

[1] *South Westphalia University of Applied Science, Agricultural Economics, Germany*

[2] *Center for Research and Development in Drylands, Kenya*

[3] *University of Parakou, National School of Statistics, Planning and Demography (ENSPD), Benin*

Child malnutrition is high in drylands and is, among other factors, influenced by choice of child food. Particularly in peri-urban areas, diverse packaged child foods are commercially available and increasingly used in child feeding. Shifting paradigms towards healthier and more sustainable product choices, e.g. through nudging or other interventions, therefore presupposes a comprehensive understanding of parents' motives underlying food choice for their children. Since data on behavioural routines related to choices of child food from Sub-Sahara Africa is rare, the present study aims to reveal Kenyan and Beninese parent's food choices and points out similarities and divergences between both countries. Shops and supermarkets in the town of Marsabit, northern Kenya and in Nikki, Banikoara and Parakou in northern Benin offering a larger than average range of child food products were selected for data collection. Shop keepers in theses shops were interviewed (computer assisted personal interviews) and asked to identify various categories of child feeding products offered in their shops and the top and slow selling products in each category. In addition, they shared their assessment, why customers like or dislike the mentioned products. An online form was used to collect the data and participants could choose predefined reason and add additional ones. The categories chosen by shopkeepers in both countries were similar as were the main reasons and frequency of reasons given to explain the purchase choices: taste, nutritional value and price. A combination of reasons suggest that purchase decisions are in many cases influenced by visual cues. Subconscious believes caused by the packaging design, brand and presentation of products are assumed to impact purchase choices. The importance of the influence of packaging design and presentation on parents' food choices suggests that nudging might break up existing behavioural routines. As nudges are not universal but depend on culture specific features, learning about attractive attributes will help to develop a more tailored way of presenting child food. This would help to promote more sustainable and healthier food choices, thereby making locally produced and processed foods more attractive and breaking the dominance of standard child food produced by multinationals.

Keywords: Child feeding, malnutrition, nudging

Contact Address: Ina Cramer, South Westphalia University of Applied Science, Agricultural Economics, Lübecker Ring 2, 59494 Soest, Germany, e-mail: cramer.ina@fh-swf.de

Identifying Farm-Level Pathways to Food Security in West Africa: A Qualitative Case Study

Lucille Gallifa, Filippo Lechthaler

Bern University of Applied Sciences (BFH), School of Agricultural, Forest and Food Sciences (HAFL), Switzerland

Background: While the essential interlinkage between agriculture and food security has been widely recognised, there is little evidence that analyses this relationship at farm-level. Understanding these multiple linkages is needed to identify policy levers that reduce malnutrition especially for small-scale farming households in rural areas of West Africa.
Objective: This research aimed at embracing livelihoods complexity by identifying farm-level pathways to household food security, using a people centred and qualitative approach. Assessing how key assets and contextual elements influence pathways to food security. Finally, this research investigated whether farm-level pathways to food security can be identified discussing farmers livelihoods.
Methods: The Sustainable Livelihood Framework was used to operationalize the research questions and develop data collection tools. Data was obtained from semi-structured interviews, focus group discussions, key informant interviews, dietary diversity questionnaires and structured interviews in four villages of the department of Taabo, Côte d'Ivoire. Participants were farmers from two different ethnic groups as well as representatives of the health and rural development sector. Interviews where recorded, transcribed, and coded based on content analysis.
Results: Interviewed farmers are largely food self-sufficient but rely on the market on a regular basis. Two main farming systems coexist in the region, based on farmer's ethnic group. Income and food stocks depend strongly on the latter, which implies different perceptions of the lean period and translates in a different diet.
Women are responsible for subsistence crops and use the income from selling surplus to buy additional foods. Men rely on cocoa for their income, which is partly used to cover non-food expenditure. While support to small-scale farmers exists for growing cocoa, subsistence crops under women responsibility are completely neglected .Although cocoa represents an important share of household's income, women's activities are more closely linked to household food security.
Conclusion: Agriculture as main source of food and income plays an undisputable role in food availability, access, and stability at household level. Women's role is central in each identified pathway therefore their responsibilities and constraints must be carefully considered to achieve food security. Local food and health systems shall be sensitive and act upon these specific needs.

Keywords: Agriculture, food security, livelihoods, smallholders, West Africa

Contact Address: Lucille Gallifa, Bern University of Applied Sciences (BFH), School of Agricultural, Forest and Food Sciences (HAFL), Sandrainstrasse 83, 3007 Bern, Switzerland, e-mail: lucille.gallifa@bfh.ch

Examining the Impact of Linking School Feeding Program on Smallholder Farmer Income and Household Food Security Status

Bulus Barnabas, Miroslava Bavorova, Mustapha Yakubu Madaki

Czech University of Life Sciences Prague, Fac. of Tropical AgriSciences - Dept. of Economics and Development, Czech Republic

The study determines the food security status of smallholder farmers and analyses the factors influencing food security status in northeastern Nigeria. The school feeding programs have been identified as highly effective in reducing childhood hunger and malnutrition while also increasing local food production. This vehicle stimulates local economies by providing a market and source of income for local smallholder farmers by buying food from local suppliers. A purposeful selection of three states, namely Adamawa, Bauchi, and Gombe, was made for the study, and twelve local government areas, four from each state, and sixty wards with four farmers each were randomly selected to form 240 smallholder farmers. Data collected for the study include the farmers' socioeconomic characteristics, institutional factors affecting their productivity, and their seven-day food consumption recall. To determine factors influencing smallholder farmer food security status, descriptive analysis, as well as a binary probit regression, was used. The farmers' food consumption score revealed that 2.5 % of the respondents were in the poor, 67.1 % in the borderline, and 30.4 % acceptable category. The results of the regression show that access to extension services, market information, and an increase in income positively affect farmers' food security ($p < 0.1$, 0.1, and 0.5, respectively), whereas an increase in age and a poor source of household diversification negatively affects it ($p < 0.5$). Based on our findings, emphasis should be placed on improved extension services, market information, concern for elderly farmers, low-income households, and farmers encouraged to diversify their sources of income in order to improve the food security of the farming households.

Keywords: Food security, school feeding Nigeria, smallholder farmers

Contact Address: Miroslava Bavorova, Czech University of Life Sciences Prague, Fac. of Tropical AgriSciences - Dept. of Economics and Development, Kamýcká 129, 16521 Prague - Suchdol, Czech Republic, e-mail: bavorova@ftz.czu.cz

A Disaggregated Analysis of Fish Demand in Myanmar

Yee Mon Aung, Manfred Zeller, Ling Yee Khor, Nhuong Tran
University of Hohenheim, Inst. of Agric. Sci. in the Tropics (Hans-Ruthenberg-Institute), Germany

The rapidly growing aquaculture sector and concurrent stagnation of capture fishery production are observed globally. Myanmar is one of the major consumers of fish worldwide and its fish demand has been increasing rapidly over the years, but no study has investigated its fish demand parameters at the household level in particular. We estimate demand elasticities for fish in Myanmar by fish supply sources (aquaculture, freshwater capture, marine capture, and dried fish) and household groups (wealth group and household location). A multi-stage budgeting framework combined with the Quadratic Almost Ideal Demand System (QUAIDS) is applied to provide the micro-level evidence of fish demand in Myanmar using household survey data from 2015. Our findings show that fish demand from all sources and household groups has increased with income, but less than unity in all cases, showing that all sources of fish in Myanmar are normal goods. A substantial share of increasing demand for all fish groups is likely to come from poor and rural households because the income elasticity of demand for all fish groups is higher for poor (0.40) and rural households (0.32) than non-poor (0.26) and urban households (0.29). Aquaculture fish consumption is the most income-responsive in all household groups. Compensated own-price elasticities by all household groups reveal a downward-sloping demand curve for all sources of fish. Demand for fish tends to be less price elastic for poor and rural households in most cases because fish is their cheapest animal protein source, and substitutes are limited. Effective management policies and new technologies are essential to sustain fish supply from capture fisheries and aquaculture to meet the increasing fish demand in Myanmar. Interventions that increase aquaculture production will have the most effective and significant effects on household's food and nutrition security.

Keywords: Fish demand elasticities, Myanmar, QUAIDS model, three-stage budgeting framework

---

Contact Address: Yee Mon Aung, University of Hohenheim, Inst. of Agric. Sci. in the Tropics (Hans-Ruthenberg-Institute), 70593 Stuttgart, Germany, e-mail: yeemonyau@gmail.com

Implementation of Kawasan Mandiri Pangan (KMP) Program in the Border Area Indonesia, and its Impact on Local Food Security

Endar Purnawan[1], Gianluca Brunori[1], Paolo Prosperi[2]

[1]*University of Pisa, Agricultural Food and Environment, Italy*

[2]*CIHEAM-IAMM, Montpellier Interdisciplinary center on Sustainable Agri-food systems, France*

This paper addresses the implementation of the Kawasan Mandiri Pangan (KMP) programme as a microfinance programme for farmer groups, whether the programme affects farmers' decisions of production, marketing, and consumption or not, and its impacts on household food security along three dimensions: food availability, food access, and food utilisation. Based on a qualitative and theory of change mix method analysis derives from interviews and focus group discussions (FGDs), this research sheds light on the program's success between two farmer groups. Both groups have improved productivity and increased food availability, but only one group sustains the program. The result indicates no intervention of specific crop commercialisation from the program, where best crops selling price, relation, and commitment are factors affecting farmer marketing-decision. Other findings are food access at the household level increased when crop selling price was reasonable, and food utilisation is much influenced by local wisdom. Taken together, the research findings highlight the importance of capability of the management, commitment of the members, and supervision of agricultural extension agents. There is a need for a local-owned enterprise to absorb agricultural products to maintain crops selling price, which is the primary driver of food accessibility and utilisation at the household level.

Keywords: Food accessibility, food availability, food security, food utilisation, KMP Program

Contact Address: Endar Purnawan, University of Pisa, Agricultural Food and Environment, Via del Borghetto, 80, 56124 Pisa, Italy, e-mail: e.purnawan@studenti.unipi.it

Pengembangan Usaha Pangan Masyarakat (PUPM) Program for Food Security: Investment for Small Family Farms in Inter-Country Border Area, Indonesia

Endar Purnawan, Gianluca Brunori
University of Pisa, Agricultural Food and Environment, Italy

This paper addresses PUPM programme implementation and its contribution to local food security. Based on a qualitative method analysis derives from the farmer, agricultural extension worker, and management of Toko Tani Indonesia (TTI) interviews, this research sheds light on the local rice farming condition, the efficiency of TTI in the rice supply chain, and the programme contribution in food security: food availability and food access. The result indicates that the programme had definite success in shortening the supply chain, even though it still could not influence the stabilisation of supply and market price. The programme benefits farmers because now they can get income from rice farming and help the community at the same time because people can buy or access rice at a low price compared to the market. Taken together, the research findings highlight the importance of standardisation of the crop and provide more TTI numbers with more volume and design the criteria for the consumer to contribute to poverty alleviation and local food security at the same time.

Keywords: Food security, Indonesia, PUPM program, rice, TTI

Contact Address: Endar Purnawan, University of Pisa, Agricultural Food and Environment, Via del Borghetto, 80, 56124 Pisa, Italy, e-mail: e.purnawan@studenti.unipi.it

Factors Affecting Food Security in the Kurdistan Region of Iraq

Niga Abdalla, Miroslava Bavorova
Czech University of Life Sciences Prague, Fac. of Tropical Agrisciences - Dept. of Economics and Development, Czech Republic

In Iraq, 4,7 % of the population do not have sufficient food consumption (food insecure) and 23,6 % have chronic hunger; 15–25 % are undernourished; two-thirds of food insecure people located in the rural areas. Among the main drivers of food insecurity in the area are conflict; socio-economic and demographic characteristics; and ineffective safety net. This is a proposed research that will be conducted in Sulaymaniyah Governorate of the Kurdistan Region of Iraq. The main aims of the study are i. describing the food security of households in the study area; ii. investigating factors influencing food security; iii. finding household access to safe water and availability of water; iv. analysing the effect of public distribution system on the household food security. We will in particular investigate, if there is a difference in households' food security based on rural and urban areas and different heights from seal level, and how the army conflict affects household food security. The household food security will be measured using two indicators: i. Food Consumption Score (FCS) to capture the availability and variety of food; ii. Household Food Insecurity and Access Scale (HFIAS) to measure the household access to food. The data will be collected in Sulaymaniyah Governorate during May and June 2021 with an estimated sample size of 400 households. A combination of purposive and random sampling will be applied to collect data in three different zones (rural-with conflict, rural - without conflict, and urban areas) and three different heights from sea level (very lowlands "lower than 300 meters height from sea level", lowlands "300 to 1500 meters height from sea level", and high lands "higher than 1500 meters height from sea level"). Data will be analysed using regression model and student's t-test and chi squared test. The results will document the geographic distribution and characteristics of food insecure population and hence, can be used as a tool by the aid organizations or government for better targeting the credit or food aid distribution. Moreover, the findings regarding the importance of the Public Distribution System for household food security will help the policymakers in the ongoing reforming process of the system to better target the groups in need.

Keywords: Army conflict, public distribution system, water consumption

Contact Address: Niga Abdalla, Czech University of Life Sciences Prague, Fac. of Tropical Agrisciences - Dept. of Economics and Development, Prague, Czech Republic, e-mail: nega.rzgar@gmail.com

Consumer perception - challenges to increase diet diversity

Role of Trust in Consumer's Willingness to Eat Reared Crickets in Myanmar

Myint Thu Thu Aung, Jochen Dürr
University of Bonn, Center for Development Research (ZEF), Germany

Insects are generally considered a valuable source of proteins, fats, and micronutrients. Cricket is one of the most popular insect to be reared. It needs less investment than traditional livestock production and requires less land and water to provide food, jobs, and income opportunities. Cricket rearing can also help to achieve different sustainable development goals. Despite crickets being the most popular edible insect in Myanmar, only very few farmers are rearing them and the rearing business is not thriving yet as in other countries, mainly due to a lack of consumers which are still used to only eat crickets collected in the wild. Thus, this study aimed to identify the effect of the role of trust on the attitude and willingness to consume reared crickets in Myanmar. A sample of 224 respondents from Yangon and Mandalay who have recently eaten wild-harvested crickets was used. Data were collected through telephone interviews. Confirmatory factor analysis was applied for validation, and data were analysed using structural equation modelling. The result revealed that trust in cricket producers significantly influences attitude and consumption intention, whereas trust in retailers did not show a significant effect on both attitude and willingness to eat reared crickets. At the same time, the attitude had a significant relationship with the willingness to consume reared crickets. According to these results, producers should try to build the trust of the public by improving transparency and honesty in the cricket production process such as openly sharing information about how to rear the cricket, how to control and maintain food safety to increase consumers' interest in reared crickets and thereby to improve acceptance and increase consumption.

Keywords: Acceptance, attitude, confirmatory factor analysis, insect rearing, structural equation modelling, wild harvesting

Contact Address: Myint Thu Thu Aung, University of Bonn, Center for Development Research (ZEF), Genscherallee 3, 53113 Bonn, Germany, e-mail: myintthuthuaung@gmail.com

Consumer Preferences for Safe Pork Products in Rural Kenya

Cianjo Gichuyia[1], Nadhem Mtimet[2], Faical Akaichi[3], Joshua Onono[4], Eric Fèvre[5,1], Lian Thomas[1,5]

[1]*International Livestock Research Institute (ILRI), Kenya*
[2]*International Fund for Agricultural Development (IFAD), Italy*
[3]*Scotland's Rural College (SRUC), United Kingdom*
[4]*University of Nairobi, Public Health, Pharmacology and Toxicology, Kenya*
[5]*University of Liverpool, Inst. of Infection, Veterinary and Ecological Sci., United Kingdom*

Food-safety is classified as a basic human right and is therefore controlled and regulated by governments. Some governments however, due to resource constraints and the diverse nature of the food in their countries, fail to enforce appropriate legislation. To remedy such situations, studies have suggested alternatives to government intervention which leverage market incentives to achieve greater efficiency in food safety management. These incentives are largely determined by consumers preference and their willingness to pay for safer food.

Rural pork value chains in Kenya are mostly low input systems with several public health inadequacies which pose a threat to the health of consumers as well as other livelihood systems dependent on pork production. This is a common characteristic of livestock production in developing countries where smallholder production and informal marketing systems are quite dominant. This study analyses the preferences and willingness to pay for safe and high quality pork products by consumers in such a system. Results indicate that consumers in Kenya state a preference for safer pork products and a willingness to pay more for them. These highlight a potential opportunity to exploit market based incentives such as a 'safe pork' premium to encourage a certain degree of self-regulation of meat value chains. Additionally, investing in increasing consumer awareness about food safety issues should be considered in order to generate an effective market demand especially in rural areas with relatively lower literacy levels.

Keywords: Consumer preferences, food safety, pork, value chains, willingness to pay

Contact Address: Cianjo Gichuyia, International Livestock Research Institute (ILRI), Old Naivasha Road, Nairobi, Kenya, e-mail: m.gichuyia@cgiar.org

Consumer Preference/Acceptance of Intrinsic/Extrinsic Attributes of Dried Mango in Germany and Kenya

Götz Uckert[1], Jil Soika[1], Salama Simon Lerantilei[2], Stefan Sieber[1], Turoop Losenge[3]

[1]*Leibniz Centre for Agric. Landscape Res. (ZALF), Sustainable Land Use in Developing Countries (SusLAND), Germany*

[2]*Jomo Kenyatta University of Agriculture and Technology (JKUAT), Dept. of Food and Nutritional Sciences, Kenya*

[3]*Jomo Kenyatta University of Science and Agriculture, Dept. of Horticulture, Kenya*

The international research project STEP-UP had linked stakeholder within the mango value chain in joint events. Here processing, value adding and production upgrading strategies were participatory identified to enforce the mango production. Training of sustainable intensification and market linkage strategies enabled small farm enterprises to step up to-wards food and nutrition security, sustainable development and income generation at farm and community levels. After two seasons of solar drying of mangoes researchers tested sensory attributes of the products. Intrinsic product attributes of dried indigenous mangos from Kitui County have been assessed in Kenya and in Germany. Additionally, quality characteristics in Kenya and extrinsic product attributes in Germany have been tested to draw implications for value-adding marketing strategies. Indigenous mango varieties (Kikamba, Boribo) as well as grafted straints from Kenya were tested against a market reference. Data of sensory evaluation from 200 panelist was achieved by hedonic scaling. Additionally, consumers were asked to indicate their willingness to pay for various mango products. The panellist received an online questionnaire together with instructions on the conduction of the sensory test on a number of test samples of dried mango slices. It was shown that colour and flavour were identified as the main drivers for positive consumer preferences of the samples. Moreover, panellists buying behaviour has an impact on overall liking and assessment of sensory attributes. The quantitative analysis exemplified the preferences for extrinsic attributes. Labels, country of origin and additives allow for market segmentation and pricing. The studies results indicate relevance of tailored marketing strategies towards meeting extrinsic attributes and sustainability preferences. Packaging and labelling of produced dried mango from Kenya, should envelope aspects of sustainability like indicating product origin from distinct farms, using indigenous trees species, rescued from felling. This is intended to appeal to consumers who value extrinsic attributes and

Contact Address: Götz Uckert, Leibniz Centre for Agric. Landscape Res. (ZALF), Sustainable Land Use in Developing Countries (SusLAND), Eberswalder Strasse 84, 15374 Müncheberg, Germany, e-mail: uckert@zalf.de

are willing to pay the prices necessary to generate sufficiently high incomes for livelihoods and farmers. Solar drying of tropical fruits, is suitable to reduce post-harvest losses, especially at peak volumes during the fruit harvest season, while creating income opportunities for smallholder farmers, women and youth producing healthy and nutritious products.

Keywords: Consumer preference analysis, Dried mango quality, health aspects, nutritious facts, Small farm enterprises, Solar drying, sustainability attributes, value addition

Processing and Consumption of Fruits and Vegetables: Knowledge, Attitude, and Practices of Rural Households in East Africa

Jacob Sarfo, Elke Pawelzik, Gudrun B. Keding
University of Goettingen, Dept. of Crop Sciences, Division Quality of Plant Products, Germany

Despite the potentials of fruits and vegetables (FVs) for good human health, consumption has been below recommended intake amounts, notably in Sub-Saharan Africa. High post-harvest losses of FVs and limiting availability account for low consumption, among others. Addressing this challenge includes innovations to process more FVs directly at the heart of areas of food production, mostly rural communities. Therefore, the "Fruits and Vegetables for all seasons" project, in one work component gauges the knowledge, attitudes, and practices (KAPs) of rural households in East Africa on the processing and consumption of FVs. Household surveys with women between 15–49 years were conducted in six study sites – three fruit and three vegetable production areas – in Kenya, Tanzania, and Uganda. Two 24-hour dietary and 7-day FVs recalls were carried out in plenty and lean seasons. Closed and open-ended questions were constructed to assess knowledge, attitudes, and practices on FVs processing and consumption. Demographic and socioeconomic data were collected in one season. The final analysis consisted of 586 and 734 households from the fruit and vegetable areas, respectively. Mean intakes for vegetables were 122–146 g day^{-1} and 88–180 g day^{-1} for plenty and lean seasons, respectively; intakes for fruits were 27–134 g day^{-1} in plenty and 3–45 g day^{-1} in lean season, being far below the recommended 200 g day^{-1} for fruits or vegetables. More than 60 % of the women obtained high knowledge scores on FV processing and consumption. Equally, a large proportion (>85 %) had positive attitudes towards FV consumption. However, FV processing among participants was abysmally low, largely attributable to a lack of technical knowhow and processing equipment. A small number of households (85) who process FVs rely on traditional drying methods for reasons being simple, readily available, or best method. A positive attitude towards FV consumption was positively associated with vegetable consumption in Kenya and negatively in Uganda. High knowledge scores on FV consumption positively correlated with fruit intake in Tanzania. The positive attitude towards FV consumption, yet lack of processing knowledge should be a basis for local and regional interventions to increase FV processing and thus, FV availability year-round.

Keywords: East Africa, fruits, KAPs, processing, vegetables

Contact Address: Jacob Sarfo, University of Goettingen, Dept. of Crop Sciences, Division Quality of Plant Products, Carl-Sprengel-Weg 1, 37075 Goettingen, Germany, e-mail: jsarfo@gwdg.de

The Social Role of Food Vendors in Nutrition Security and Information Dissemination in Turkana County, Kenya

Marisa Nowicki, Irene Induli, Francis Odhiambo Oduor, Céline Termote

Alliance of Bioversity International and CIAT, Food Environment and Consumer Behaviour, Kenya

This Food environment (FE) study was conducted in Turkana County, Kenya, which is characterised by a harsh climate, high poverty levels, and remoteness. A vendor analysis was conducted across Lodwar town, and 10 rural community health units (CHU) randomly selected in Loima and Turkana South sub-counties. The vendor analysis contained three components. First, geocoding of approximately 380 vendors throughout the area of interest. The distribution of geocoded vendors was: Kiosks/Retail Shops (53 %), Roadside Vendors (14 %), Wholesalers (11 %), Restaurants (10 %), Mobile Vendors (4 %), Open Air Markets (3 %), Street Hawkers (2 %), and Supermarkets (1 %). Second, a stratified sample was taken to select four vendors per vendor type. With this sample, enumerators conducted a full inventory of the food and drink items they sold. In part 3, enumerators conducted in-depth interviews regarding the vendors' businesses. Data suggests food vendors play a key social role in their communities. For example, 84 % of vendors reported a willingness to sell food items on credit to community members in need, including the poor, disabled, and elderly. These vendors may act as a small safety net for their community. Vendors also expressed interest in disseminating nutrition information, with 45 % reporting they at least "sometimes" give nutrition advice to customers. When asked about the "healthiest" foods, most vendors cited vegetables (72 %), fruits (65 %), and pulses (58 %). When asked about the "least healthy" foods, most vendors cited alcohol (72 %), and Soda (35 %). However, when discussing unhealth food items, vendors cited grains/ white roots/ tubers/ plantains (16 %) and nuts/ seeds (14 %), as often as they cited candy (14 %), and crisps/ cookies/ crackers (14%). Although food vendors are rarely included in nutrition programs and policy discussions related to food security, they play a key role within the food environment. Their willingness to help community members through credit and nutrition advice could be utilised to enhance food and nutrition security. Organisations working in LMICs should 1) engage vendors in nutrition and agriculture programming, and 2) highlight the importance of diverse, nutritious diets, which include traditional staples. This will improve the accuracy of vendors' nutrition advice, as well as empower vendors to take further actions to ensure food and nutrition security in their communities.

Keywords: Food environment, Kenya, LMICs, nutrition security, vendor analysis

Contact Address: Marisa Nowicki, Alliance of Bioversity International and CIAT, Food Environment and Consumer Behaviour, Nairobi, Kenya, e-mail: Nowicki.marisa@gmail.com

Understanding Local Needs and Barriers for an Integrated Nutrition Intervention Towards Sustainable Diets, Burkina Faso

Louise-Caroline Büttner[1], Fanta Zerbo[2], Moubassira Kagoné[2], Ali Sié[2], Ina Danquah[1], Isabel Mank[1]

[1]*Heidelberg University, Heidelberg Institute of Global Health (HIGH), Germany*
[2]*Institut National de Santé Publique (INSP), Centre de Recherche en Santé de Nouna (CRSN), Burkina Faso*

Undernutrition among children <5 years continues to be a challenge in West Africa. Specifically, seasonal food insecurity and climate change-related impacts on food production accelerate the prevalence of child undernutrition. A combination of nutrition-related interventions such as home gardens combined with nutrition counseling may reduce child undernutrition and increase the resilience of rural population to a changing climate. Home gardens are defined here as small-scale horticultural production sites at the household-level. The present study aimed to understand needs and barriers to implement such a nutrition-related intervention in rural Burkina Faso.
In preparation of an integrated intervention project, a qualitative study was carried out in Burkina Faso in 2020. A semi-structured guide led the interviews, which were recorded, transcribed and translated. Directed content analysis was applied. 32 in-depth interviews were carried out with mothers (24 to 54 years old) of children <5 years living in rural Burkina Faso, and 11 in-depth interviews were conducted online with experts from Burkina Faso. 20 out of the 32 mothers did never practice any gardening and eight mothers reported to have never received nutritional advice. Among the 11 stakeholders, most had either experience with gardening or with nutrition counseling projects, but rarely combined.
A reoccurring communicated challenge to the implementation of gardens was a lack of water. As gardens were often only cultivated during the rainy season (May-Oct), the mothers expressed interest to learn more about irrigation and planting vegetables during the dry season (Nov-Apr). The women also explained that the men tend to decided what the family eats, wherefore a holistic family approach should be considered in order to increase acceptance and behavioural change towards a sustainable and nutritious meal for children. The mothers acknowledged the importance of nutritional advice to improve the feeding practices of their children and the overall health of their family. In conclusion, the mothers of children <5 years showed great interest in participating in the intervention project. These results support the design of an integrated intervention to improve the nutritional status of children and increase the resilience of small-scale agricultural households to climate change.

Keywords: Child undernutrition, home gardens, nutrition counseling, qualitative research, West Africa

Contact Address: Isabel Mank, Heidelberg University, Heidelberg Institute of Global Health (HIGH), Im Neuenheimer Feld 324, 69120 Heidelberg, Germany, e-mail: isabel.mank@uni-heidelberg.de

Jackfruit-But-Bars - A Promising Product to Enhance Jackfruit Utilisation in Uganda?

JOHANNA TEPE, DOMINIC LEMKEN

University of Göttingen, Agricultural Economics and Rural Development, Germany

The nutritional value of jackfruits in Uganda has long been acknowledged. However, due to its large size and sticky insides it is difficult to handle and currently being underutilised. Also, once the fruit is opened the edible bulbs must be consumed within a few days as they are easily perishable. This causes large amounts of jackfruits to go to waste. One solution is to process the fruits into easily accessible products with longer shelf-lives. As processing of jackfruits is still rare there are limited insights on consumer demand for jackfruit products. Therefore, one work-package of the "Fruits and Vegetables for all seasons" project analyses the potential of jackfruit-nut-bars (JNBs) as a channel to enhance utilisation of jackfruit.

JNBs are mainly made up of jackfruit bulb, mango, peanuts, and lemon. The ingredients were mixed and oven dried. The dried product is a healthy snack that can be consumed away from home. It fits into the ongoing shift towards more processed and easily accessible products in Uganda. To analyse consumer demand for the products, we first assessed sensory perception of four slightly different jackfruit-nut-bars and then used Van Westendrop's price sensitivity metre to elicit consumers' willingness to pay (WTP). The survey was conducted at Makerere University in Kampala, Uganda. Very few study participants had eaten processed jackfruit products before. While fresh jackfruit was mainly consumed because of its nutritional value and sweet taste, it was mostly disliked because of its stickiness.

Results of the sensory analysis of the JNBs show that most characteristics (colour, aroma, texture in the mouth, taste, and general appearance) of all four JNBs were rated by over half of the participants as "like it" or "like it very much". WTP was predominantly influenced by their current attitudes and practices towards food and snack consumption. Frequency of snack consumption, as well as consumers neophobia and consumption of sugared snacks because they are cheap had a positive and significant effect on consumers WTP. Based on our findings we conclude that JNBs are a promising product to enhance utilisation of jackfruits. More comprehensive research is necessary to identify further consumer groups.

Keywords: Consumer demand, jackfruits, Processing, Snacks, Uganda

---

Contact Address: Johanna Tepe, University of Göttingen, Agricultural Economics and Rural Development, Platz der Göttinger Sieben 5, 37073 Göttingen, Germany, e-mail: johanna.tepe@uni-goettingen.de

Exploring Intervention Options for Small-Scale Family Poultry Development in the Atsimo Atsinanana Region, Madagascar

Barbara Kurz[1], Stefan Sieber[2], Jonathan Steinke[1], Sarah Tojo Mandaharisoa[3], Arielle Sandrine Rafanomezantsoa[3]

[1]*Humboldt-Universität zu Berlin, Thaer-Institute of Agricultural and Horticultural Sciences, Dept. of Agricultural Economics, Germany*

[2]*Leibniz Centre for Agric. Landscape Res. (ZALF), Sustainable Land Use in Developing Countries (SusLAND), Germany*

[3]*University of Antananarivo, Tropical Agriculture and Sustainable Development Dept., Madagascar*

Poultry farming occupies an important role in the Malagasy rural economy and the livelihoods of its people. Research in the Atsimo Atsinanana region (South East Madagascar) shows that consumption of eggs is uncommon, despite proven health benefits and the existence of large numbers of poultry. This study tries to contribute to improving the local nutritional situation and health of the rural population, especially for women of reproductive age and children in the 1000-day window. The main objective of this study was to describe current challenges and opportunities to small-scale family poultry rearing and egg consumption in the rural area of the Atsimo Atsinanana region and to explore viable options for the advancement of poultry production and egg consumption via development interventions. A multi-tiered survey approach was used, including online international and local expert interviews as well as on-site interviews with beneficiaries in the study area. Findings show that critical constraints for poultry production include high mortality, poor husbandry, cultural constraints, low productivity of local chickens, and poor marketing. The major health constraint that hinders poultry farming is Newcastle disease. Small-scale family poultry production offers opportunities for nutrition and income, as poultry provides animal protein and can be sold at local markets. It can promote gender equality and empowerment since women tend to have more control over poultry production than other activities. Birds are slaughtered for socio-cultural purposes and sold to meet other needs like the purchase of seeds, school supplies, or medicine. Only small numbers of eggs are consumed because people prefer hatching instead of eating them due to the low egg-laying potential of local chickens. Possible solutions to this situation include improvements in breeding and production. Interventions stressed by international experts, local experts and rural peo-

Contact Address: Barbara Kurz, Humboldt-Universität zu Berlin, Thaer-Institute of Agricultural and Horticultural Sciences, Dept. of Agricultural Economics, Mariendorfer Weg 51, 12051 Berlin, Germany, e-mail: kurzbarb@hu-berlin.de

ple from the research region include health interventions (vaccinations and biosecurity) and interventions with a focus on animal husbandry or nutrition. Furthermore, training for farmers on poultry management practices, marketing, and poultry health were suggested. Another focus should be put on creating awareness about the nutritional benefits of poultry products and the potential benefits of making poultry farming a primary activity for farmers.

Keywords: Development interventions, Madagascar, poultry, small-scale family poultry development, smallholder farmers

Rural Turkana Food Environment: Consumers' Perspectives

Irene Induli, Marisa Nowicki, Francis Odhiambo Oduor, Céline Termote

Alliance of Bioversity International and CIAT, Food Environment and Consumer Behaviour, Kenya

The Food Environment (FE) study was conducted in Turkana county, Kenya, one of the arid regions in the country. Loima and Turkana South sub-counties were purposively selected because of security and access. To our knowledge, no FE studies have been previously conducted in these areas. 10 community healthy units (CHUs) were randomly selected. With the help of Community Health Volunteers, 2 households were purposively selected from the 10 CHUs to ensure interviewees were conversant with their FE. Target respondents were people who regularly acquire food in households, mainly men and their wives. In-depth interviews were used for data collection. The interviews were conducted in the local Turkana language following a predetermined question guide. 18 interviews were conducted in 10 health units. The interviews were recorded and later transcribed for analysis. From preliminary findings, the main source of food was shops/markets (89 %). More than 50 % said foods produced from their own farms were safer to consume and of better quality. 72 % regularly utilised barter trade to acquire food.

High food pricing was attributed to high transport costs, government taxes and reduced productivity due to unpredictable weather. All households equated food quality to nutrient content of food. Packaged foods were deemed safe and less contaminated; but their preference was divided, with 50 % preferring locally prepared/produced foods over packaged foods.

Only 22 % of households utilised wild foods (WFs), although they all indicated their neighbours consumed. All participants cited several wild edible fruits and vegetables- the sources and preparation methods. This may suggest there is stigma associated with utilising wild foods; despite them substituting domesticated foods during scarcity. From the findings, more research should be done on nutrient content of WFs, and the information disseminated to communities to encourage consumption and domestication of these foods for food and nutrition security.

Informal food sources like the wild and barter trade are important in rural populations, hence should be acknowledged in food security policies and programs. Data analysis is ongoing. More will be presented in the conference.

Keywords: Food quality and safety, rural food environment, wild foods

Contact Address: Irene Induli, Alliance of Bioversity International and CIAT, Food Environment and Consumer Behaviour, 00200 Nairobi, Kenya, e-mail: i.induli@cgiar.org

Nutritional Potential of Traditional Food Products for the Nutrition of Young Children of Arid Zones

Adrien Dogo[1], Georges Djohy[2], Franck Hongbete[1], Ange Honorat Edja[3]

[1]*University of Parakou, Foods Sciences Laboratory, Benin*

[2]*University of Parakou, National School of Statistics, Planning and Demography (EN-SPD), Benin*

[3]*University of Parakou, Dept. of Rural Economics and Sociology, Benin*

Global change processes induce complex transformations in African societies in general and particularly in dryland which already face environmental challenges for food production. This strongly affects the food environment of the inhabitants and makes it difficult for them to adopt a quality, balanced and healthy diet; which further increases the prevalence of chronic malnutrition and the various micronutrient deficiencies in drylands in general and North Benin in particular. Although these areas have traditional foods with high nutritional potential that can be used as local solutions to improve children's diets and nutrition, these foods are losing importance and particularly in peri-urban areas face competition from imported children's food. This study aims to assess the nutritional potential of traditional children's foods with a view to identify those that are highly nutritions and can counter micronutrient deficiencies. To do this, traditional children's foods were identified through focus group discussions and individual interviews with mothers of different generations in seven (07) villages of the municipalities of Nikki and Banikoara in North Benin. The identified children's food were ranked by mothers for their value and from the 12 most preferred child food samples were prepared according to the traditional recipes. Nutitional composition including the mineral and vitamin profile of the samples of food preparations were determined in duplicate using standard methods of food analysis. The results show a variety of traditional foods with high levels of protein, iron, calcium, zinc, vitamins A and C which are nutrients of interest to enhance nutrition of young children. In the next steps local processors will be involved on how these foods can be improved to make them more widely available and accessible to rural and peri-urban areas in order to improve the nutrition of young children in the dynamic context of the drylands of North Benin.

Keywords: Food and nutrition for young children, food environment, micronutrients, traditional food products

Contact Address: Adrien Dogo, University of Parakou, Foods Sciences Laboratory, Banikanni Maison Agonssaclounou, Parakou, Benin, e-mail: adriendogo@gmail.com

Fruit and Vegetable Intake: Knowledge, Attitude and Practices among Rural School-Aged Children in East Africa

Luisa Alves, Jacob Sarfo, Gudrun B. Keding

University of Goettingen, Division Quality of Plant Products, Department of Crop Sciences, Germany

Fruit and vegetable (FV) intakes are very low in Sub-Saharan Africa, leading to different forms of malnutrition. As there is only little data on FV consumption of school-aged children, the "Fruits and Vegetables for all seasons" project, in one work component, measured FV intakes, and the knowledge, attitudes, and practices (KAPs) of school-aged children on the processing and consumption of FVs in rural households in East Africa.

Data was collected from households in six study sites – three fruit and three vegetable production areas – in Kenya, Tanzania, and Uganda. While the main interviewee was the mother or caregiver, available children between 6-13 years were interviewed applying a 24-hour dietary recall and 7-day FV recall during two seasons. KAPs were assessed using closed and open-ended questions. In total 251 children, 68 from the fruit and 183 from the vegetable areas were considered for the analysis.

All 68 children from the fruit areas reported that they consume fruits in general and like to eat them. Top three favourite fruits and mostly known were mango (50 %), jackfruit (38.2 %) and orange (36.8 %). From the vegetable areas 97.8 % of children consumed vegetables in general and 98.4 % liked to eat them with cabbage (60.7 %), kales (54.6 %) and cowpea leaves (39.9 %) being the three most favourite vegetables and also the most known. Only 4.4 % of the children from the fruit and 4.9 % from the vegetable areas disagreed on the statement, that fresh fruits/vegetables are good for the body. Contrariwise, there were 86.8 % of children from the fruit and 99.5 % from the vegetable areas that did not know what a processed or preserved fruit/vegetable is. Moreover 66.2 % and 57.9 % of the children stated that they are unable to eat fruits or vegetables, respectively, year-round. Only 42.6 % of children eat fruits and only 4.4 % eat vegetables while in school.

First results show that FV consumption among school-aged children is seasonally insufficient and diversity of most preferred FVs is limited to few mainly exotic species. In addition, intake of FVs will be analysed and related to the household demographic, socioeconomic and KAPs data among all three countries.

Keywords: East Africa, fruits, KAPs, rural, school-aged children, vegetables

Contact Address: Luisa Alves, University of Goettingen, Division Quality of Plant Products, Department of Crop Sciences, Carl-Sprengel-Weg 1, 37075 Goettingen, Germany, e-mail: luisa.alves@stud.uni-goettingen.de

The Impact of Moral Opposition to Tampering with Nature on Consumers' Preference for Genetically Biofortified Food in Sub-Saharan Africa

Titilayo Akinwehinmi[1], Ramona Teuber[1], Oluwagbenga Akinwehinmi[2]

[1] *Justus Liebig University Giessen, Dept. of Agricultural Policy and Market Research, Germany*

[2] *University of Goettingen, Dept. of Agricultural Economics and Rural Development, Germany*

It remains unclear whether moral opposition to tampering with nature will significantly drive resistance from consumers as scientists push for the up scaling of genetically modified (GM) food in sub-Saharan Africa (SSA). This issue has been extensively investigated in literature focusing on developed countries. Yet, this information is largely missing in previous studies that focused on SSA. There is a consensus in certain literature that consumers are more positively disposed to GM foods that provide benefits to them justifying scientists' optimism that consumers in SSA will favor genetic biofortification of staple foods in a bid to address the problem of micronutrient deficiency. In contrast, another body of literature argued that much of consumers' resistance to GM food can be explained in human's absolute moral opposition to tampering with nature evoking insensitivity to benefits of GM food. In this study, we investigate the validity of these hypotheses in the context of SSA. Employing data from a discrete choice experiment, we modelled causality between absolute moral opposition to tampering with nature and consumers' preference for a GM beans with improved nutrient content and yield. The mediating role of consumers' objective knowledge of GM was also investigated. For our sample, we purposively target a scientific literate population of a university community in Nigeria. We estimated a latent class logit model using a total of 1512 observations generated from 168 respondents. First, we found that only 9.8 % of the sample manifested absolute moral opposition towards GM food. Further, the results reveal three segments of respondents with heterogeneous utility weights for nutrient content and yield attributes. Absolute moral opposition and objective knowledge towards GM food predicted consumers' probability of class assignment. Overall, the results show that, in this population, consumers' preference for GM food is mainly driven by knowledge of the objective purpose of GM and whether the GM application significantly improved nutrient density in a cost-effective way. Since our aim here was to provide evidence for what might be expected in a larger population of consumers, we call for further testing of these hypotheses among a less literate population of consumers.

Keywords: Biofortification, choice experiment, genetic modification, moral opposition, preference

Contact Address: Titilayo Akinwehinmi, Justus Liebig University Giessen, Dept. of Agricultural Policy and Market Research, Senckenbergstrasse 3, 35390 Giessen, Germany, e-mail: titilayo.akinwehinmi@agrar.uni-giessen.de

Drying Vegetables - A Sustainable Solution to Enhance Vegetable Intake and Culturally Acceptable for Everyone?

Irmgard Jordan[1], Annet Itaru[2], Eleonore C. Kretz[3], Paulina M. Kossmann[1], Daisy Alum[4], M. Gracia Glas[1], Lydiah Maruti Waswa[2]

[1]*Justus-Liebig University Giessen, Center for International Development and Environmental Research, Germany*
[2]*Egerton University, Dept. of Human Nutrition, Kenya*
[3]*Justus-Liebig University Giessen, Inst. of Nutritional Sciences, Germany*
[4]*Makerere University, Dept. of Food Technology and Nutrition, Uganda*

Background: A year-round availability of vegetable is a challenge globally. Vegetable preservation methods can play an important role in contributing to food security and is considered to be an easy option to enhance vegetable intake and thus dietary diversity. This study aimed at investigating on how best to promote vegetable drying and to introduce them into diets in two regions in East Africa.

Methods: Women from about 100 farming households with children below eight years of age in Kapchorwa District, Uganda, and Teso-South, Kenya took part in trials of improved practices (TIPs) to test locally adapted nutrition education messages in 2019. Qualitative data collection was carried out through private counselling with the primary caregiver at their homes. Part of the programme was a solar-dryer construction session together with the women and their husbands participating in the TIPs. Group discussions and tastings with and without dried vegetables were held with women and men separately. No incentives were provided for joining the solar-dryer construction sessions, while a travel allowance was paid to all participants for joining the group discussions. An evaluation survey was conducted in January 2021.

Results: Not all TIPs households joined the solar-dryer session which was positioned at a central place in the villages. Those who used the dryer did this despite unfavourable weather conditions, and were positive surprised about the opportunity to preserve vegetables even in times of rains. Challenges occurred in offering solutions how best to store home-scaled dried vegetables at the homesteads. Participatory cooking trials to process dried vegetables for consumption confirmed that they are an innovative and palatable option to enhance vegetable intake for all family members. A year later, drying of vegetables was associated with being a "TIPs-household" but not commonly practised.

Contact Address: Irmgard Jordan, Justus-Liebig University Giessen, Center for International Development and Environmental Research, Senckenbergstr. 3, 35390 Gießen, Germany, e-mail: Irmgard.Jordan@ernaehrung.uni-giessen.de

Conclusion: Innovative packaging is needed to store home-scale dried vegetables in an appropriate and sustainable manner. Dried vegetables are tasty if prepared in an appropriate manner which needs adaptation of existing recipes and promotion in participatory cooking trials to make them culturally acceptable for everyone.
The study was conducted within the EaTSANE-project funded by BMEL/ptble within the joint AU/EU-LEAP-Agri initiative.

Keywords: Dietary diversity, food packaging, food security, social behaviour change communication, solar drying, vegetables

Consumers' Sensory Perception of Fermented African Nightshades in Tanzania

Frank Sangija, Haikael Martin, Athanasia Matemu, Marynurce Kazosi
Nelson Mandela African Institution of Science and Technology, Food and Nutritional Sciences, Tanzania

African nightshades (ANS) are underutilised and neglected African indigenous vegetables regardless of their nutritional quality and high availability in Tanzania. This study seeks to assess the consumer acceptability and willingness to buy spontaneously fermented (SF) and controlled fermented (CF) *Solanum scabrum* (SS) and *S. villosum* (SV) relish. A 9-point hedonic scale was used to evaluate the fermented products' sensory attributes and overall consumer acceptability. In Kilimanjaro and Morogoro regions, 370 untrained consumers tested the relish. From the results, fermented relishes from SS-CF exhibited superior overall acceptability of 7.96 ($p < 0.05$). Also, it was liked much in other parameters such as colour, taste, texture, aroma, odor, saltiness, spiciness, sourness, and bitterness $p > 0.05$. Generally, the consumers liked other fermented relishes, SS-SF, SV- SF, and SV- CF, in all sensory parameters. Most consumers detected salt and sourness in the relish (85–90 %); about 50 % noticed bitterness. On the other hand, ugali (84–87 %) and rice (62–67 %) were the preferable staples to accompany the relish. At the same time, the amaranth (77–81 %) and spinach (58–64 %) were suitable vegetables to mix with the relish. Also, the majority preferred to mix relish with meat (80 %) and fish (70 %). The majority were willing to buy the relish (97–99 %), and only 53–59 % could pay 500 Tanzania shilling per 120 g of relish. Supermarkets (71–75 %), kiosks (72–69 %), markets (64–67 %), open market/magulio (49–53 %) and hotel (49–50 %) were the proposed outlets for the relish. Both CF and SF relish can be recommended to be eaten alone, accompany foods, and mix with other foods. Therefore, relish can improve nutritional value, utilisation, livelihood and ensure the availability of ANS year-round.

Keywords: African nightshade, consumer perception, controlled fermentation, relish, sensory parameters, spontaneous fermentation, willingness to buy

Contact Address: Frank Sangija, Nelson Mandela African Institution of Science and Technology, Food and Nutritional Sciences, Tengeru, Arusha, Tanzania, e-mail: sangijaf@nm-aist.ac.tz

Trends and Constraints in the Utilisation of African Nightshade (*Solanum nigrum* complex) in Tanzania

Frank Sangija, Haikael Martin, Athanasia Matemu, Marynurce Kazosi
Nelson Mandela African Institution of Science and Technology, Food and Nutritional Sciences, Tanzania

African nightshade (ANS, *Solanum nigrum* complex) is among the most widely distributed and consumed indigenous vegetables in Tanzania. Several challenges hamper the utilization of ANS. This study sought to assess trends and constraints to ANS utilization in Kilimanjaro and Morogoro regions, Tanzania. About 627 farmers participated in the interview. Both qualitative and quantitative methods were employed to collect information on ANS production, processing, and postharvest handling. The results showed that 72.1 % of farmers grow different ANS species, with *Solanum scabrum* being vastly cultivated. Also, 79.4% of ANS farmers use irrigation, handheld hoe (97.6 %) use pesticides (70.7 %), and 64.8 % use fertiliser in production. African nightshade is mainly used as food (97.9 %), animal feed (41.3 %), and medicine (38 %). On average, only 5 % of ANS sales contributed to family income. Findings show that main constraints to ANS utilisation include; pests and diseases (92.9 %), lack of knowledge (58 %), shortages fertiliser (51 %), shortages pesticides (50 %), inadequate means of transport (50.4%), lack of extension services (48%), improper postharvest handling (41.4 %) and inadequate storage facilities (34%). Postharvest losses accounted for 78.4 % loss of ANS. Mitigation measures were; harvesting in small quantities (54.5 %) and instant selling (61.9 %) of fresh ANS. There was minimal value addition on ANS, e.g., drying (5.3 %) and fermentation (1.1 %). Moreover, boiling (63.0 %) and frying (45.4 %) and (98.9 %) were the typical methods of cooking ANS. More emphasis should be placed on good agricultural practices, providing knowledge to farmers, and supporting inputs such as pesticides, fertilisers, and quality seeds. Furthermore, knowledge on the processing and preservation of ANS should be given to improve utilization, reducing losses, and ensuring ANS availability. Also, research should focus on breeding local cultivar, which is resistant to pests and diseases.

Keywords: African nightshade, cultivation, preservation, postharvest handling, processing, *Solanum nigrum* complex, utilisation

Contact Address: Frank Sangija, Nelson Mandela African Institution of Science and Technology, Food and Nutritional Sciences, Tengeru, Arusha, Tanzania, e-mail: sangijaf@nm-aist.ac.tz

Factors Influencing Cowpea Consumption among Household in Gombe State Nigeria

Isah Bakoji[1], Jibril Ayemi Salawu[2], Murtala Nasiru[2]
[1]*Nigerian Correctional Service Bauchi, Department of Agriculture, Nigeria*
[2]*Abubakar Tafawa Balewa University Bauchi, Dept. of Agricultural Economics, Nigeria*

The study analysed factors influencing cowpea consumption among households in Gombe state, Nigeria. Two hundred and eighty (267) consumers were randomly selected for the study. Data were collected using structured questionnaire and were analysed using descriptive and inferential statistics .The result reveals that age, sex, household size and education were the significant determinants of cowpea consumption among households in the study area. The opinion of cowpea consumers seems to be divided on the issue of post harvest protection of cowpea in Gombe State. Those who support the use of post harvest treatment of cowpea are mostly farmers and cowpea retailers while those who oppose the use of post harvest treatment seem to be mainly processors and household consumers. Additionally, the opinion of consumers about level of cowpea likeness seems to be the same between farmers and retailers as a group and processor and household consumers in the study area. However, the findings discovered that insect infestation (37.04 %), inadequate capital (65.38 %), high cost of cowpea (60.24 %) and abdominal discomfort (26.13 %) were the major constraints to consumers of cowpea in the study area. It is concluded that cowpea was consider as a food security crop in the area. It is therefore, recommended that public health strategies, effective and easy method for controlling cowpea infestation should be put in place to fight against indiscriminate treatment of cowpea in the study area. Government could partner with the private sector to invest in making Purdue triple hematic bag and be made readily available and affordable to the farmers and other cowpea actors in the study area.

Keywords: Abdominal discomfort, post harvest treatment, purdue triple hematic bag

Contact Address: Isah Bakoji, Nigerian Correctional Service Bauchi, Department of Agriculture, Tsohon Kamfanibauchi, +234 Bauchi, Nigeria, e-mail: bakojiisah@gmail.com

Do Mothers in Southwest Ethiopia Use Fermentation and Malting for Processing Complementary Foods?

Berhane Kebede[1], Dessalegn Tamiru[1], Nurezemen Gali[1], Tefera Belachew[2], Sirawdink Fikreyesus Forsido[3], Melese Sinaga Teshome[1]

[1] *Jimma University, Nutrition and Dietetics, Ethiopia*

[2] *Jimma University, Population and Family Health, Nutrition Unit, Ethiopia*

[3] *Jimma University, Post-harvest Management, Ethiopia*

Background: Poor complementary feeding is a significant driver of malnutrition, micronutrient deficiency and infant and child mortality, especially in developing countries. Traditional food processing methods such as fermentation and malting can improve the nutritional qualities of complementary foods. Information is lacking regarding their utilisation in the preparation of complementary foods in Southwest Ethiopia.

Objective:- To assess the use of fermentation and malting in the processing of complementary food and associated factors among index mothers of children aged 6–23 months in Jimma zone, Southwest Ethiopia.

Methods: A community-based cross-sectional study was conducted in Jimma zone among 636 index mothers of children aged 6–23 month, from April 1–30, 2018. Study participants were selected using a multistage stratified sampling technique. Data were collected using a pre-tested semi-structured interviewer-administered questionnaire and analysed using SPSS version 20. Bivariate and multivariable logistic regression analyses were used to isolate independent factors at 95 % confidence interval and $p < 0.05$.

Results: Three hundred sixty two (56.9 %) and 3.9 % of mothers used fermentation and malting in processing of complementary foods, respectively. Those mothers who considered fermentation to improve digestibility [AOR=16.5 ,95%CI=(7.12, 38.4), p-value<0.001], those mother who considered fermentation to improve taste [AOR=6.15, 95%CI= (4.2,10.12), $p < 0.001$], child age [AOR=1.12, 95 % CI= (1.08, 1.16), $p < 0.001$], who do not attended formal education [AOR=1.95, 95 % CI= (1.33,2.86), $p = 0.001$], who had no diarrheal morbidity in a year [AOR=1.88, 95%C I= (1.28,2.75), $p = 0.001$] and not got complementary feeding advice [AOR=2, 95 % CI= (1.34,2.99), p-value=0.001] were found to be significantly associated with the use of fermentation and malting.

Conclusion: In this community, most participants used fermentation but not malting as a food processing method for processing complementary foods. Age of the child, formal education, complementary feeding advice, and diarrheal morbidity in a year, perceptions such as fermentation improves di-

Contact Address: Sirawdink Fikreyesus Forsido, Jimma University, Post-harvest Management, Jimma University College of Agriculture, 307 Jimma, Ethiopia, e-mail: sirawdink@gmail.com

gestibility, and fermentation improves taste were independent predictors of using fermentation and malting. Therefore there is a need to give nutrition education to mothers on the benefits of malting and fermentation.

Keywords: Complementary feeding, fermented food, traditional food processing techniques

Consumers' Perception and Willingness to Pay for Vitamin A Fortified Gari in Ibadan Metropolis, Oyo State, Nigeria

Olalekan Taiwo, Oluwafunmiso Adeola Olajide

University of Ibadan, Agricultural Economics, Nigeria

The study was carried out to examine the consumers' perception and willingness to pay for vitamin A fortified gari in Ibadan metropolis, Oyo State. This was carried out by determining the awareness of consumers of the vitamin A fortified gari and estimating their willingness to pay using a contingent valuation method. A multistage random sampling technique was employed to select three local governments out of the 11 local governments in Ibadan and 200 respondents were randomly selected based on proportionate to size principle. Primary data were collected on socioeconomic variables, awareness level, and willingness to pay for vitamin A fortified gari from the respondents using a well-structured questionnaire. Methods of data analyses involved the use of descriptive statistics, and a logistic regression for the respondents sampled. 46.5 % of respondents sampled were between 21 and 30 years while the average age was 35 years. The result revealed that 66 % of the respondents were not aware of Vitamin A fortified gari. 64 % of the respondents has post-secondary school education, 84 % of the respondents were willing to pay for vitamin A fortified gari. 33.1 % were willing to pay between N150 to $N_2$00 for vitamin A fortified gari. The logistic regression estimates show that amount willing to pay and minimum price willing to pay by the consumers significantly impacted on the probability of being willing to pay for vitamin fortified gari. It is recommended that policy that will increase awareness of vitamin fortified gari and enhance purchasing power of the consumers should be promoted.

Keywords: Fortification, perception, regression , vitamin A, willingness to pay

Contact Address: Olalekan Taiwo, University of Ibadan, Agricultural Economics, No & Royal Close, 234 Ibadan, Nigeria, e-mail: taiwoolalekana@gmail.com

Consumer Bevaviour of Regional Cheese from Family Farmers: A Theoretical Model and Measurement Scale

Rejane Carmo Rezende Dias[1], Jose Elenilson Cruz[2]
[1]*Federal University of Goiás (UFG) / State University of Goiás (UEG), Postgraduate Program in Agribusiness (UFG) / Senador Canedo University Unit (UEG), Brazil*
[2]*Federal Institute of Education, Science and Technology of Brasília (IFB), Campus Gama, Brazil*

There is growing interest around the world in the consumption of unprocessed natural foods, including raw milk and dairy products. The production of cow's milk in Brazil is relevant in the social and economic context as it constitutes one of the main sources of income for the Brazilian family farmer. However, most family farmers are vulnerable to the competitive market of raw milk, have low income, and have a certain specialization of this production. Therefore, the production of regional cheese from cow's milk can be an alternative for these family farms to increase their income, mitigate the effects of the drop in the price of raw milk and improve economic stability. Despite the importance of knowing the consumer of regional cheese, there is insufficient evidence in scientific research of consumer behavioural effects of specific regional food and more specifically in consumer behavioural effects of regional cheese. So, this work aims to present a theoretical model with research propositions and a measurement scale for the study of behaviour consumers of regional cheese. The bibliographical research was carried out under the perspectives of the Theory of Planned Behaviour and Theory of Perceived Risk. The measurement items of each construct were adopted from scales already validated in previous works in English, and later were translated into Portuguese and adapted for Brazilian culture. For theoretical validation, the scale was sent to three researchers specializing in the construction of consumption scales and a specialist in the regional food market to evaluate the items in the scale. After the literature review and theoretical validation, the result obtained was a theoretical model with seven research propositions and a scale that measures the effects that cognitive attitudes, affective attitudes, personal norms, subjective norms, perceived behavioural control and perceived risk have on the intention of purchasing regional cheese consumers. This result will be useful for future empirical studies that wish to understand the influence of consumer behaviour on the intention to buy regional cheese and thus contribute with relevant information about this consumer for family farmers to promote and sell their regional cheese.

Keywords: Consumer behaviour, family farmers, regional cheese

Contact Address: Rejane Carmo Rezende Dias, Federal University of Goiás (UFG) / State University of Goiás (UEG), Postgraduate Program in Agribusiness (UFG) / Senador Canedo University Unit (UEG), Rua C-24 Quadra 22 Lote 16 Casa 1, 74265140 Goiânia, Brazil, e-mail: prof.rejane.adm@gmail.com

Food composition - processing - nutritional quality

Oral Presentations

Posters

Effect of Lactic Acid Fermentation on Nutritional and Antinutritional Compounds in African Nightshade

Frank Sangija, Haikael Martin, Marynurce Kazosi, Athanasia Matemu

Nelson Mandela African Institution of Science and Technology, Food and Nutritional Sciences, Tanzania

African nightshade (ANS) is among many underexploited and neglected indigenous vegetables. If adequately utilized, the crop can improve nutritional, sensory, and keeping quality. This study assessed the effect of fermentation on the nutrients and antinutrients composition of fermented *Solanum villosum* (SV) and *S. scabrum* (SS). Both spontaneous fermentation (SF) and controlled fermentation (CF) (*Lactobacillus plantarum* and *Leuconostoc mesenteroides*) were employed. Fermented relish was prepared from the pickle with the addition of cooking oil, onions, pepper, ginger, turmeric, and cinnamon. A significant reduction in pH from 7.4–8.2 to 3.1–3.2 and an increase in titratable acidity (TTA) from 0.045-0.072% to 0.85-1.42 % was observed in SF after day 30. On the other hand, a significant ($p < 0.05$) decrease in pH from 7.4–8.4 to 3.1–3.5 and an increase in (TTA) from 0.04–0.07 % to 0.35–0.4 % was observed after 24 h in CF. A slight decrease of β-carotene from pickle to relish in CF (4.0–6.6 %) and SF (4.4–6.6 %) in all the formulations. The highest β-carotene content of 155.3 mg/100g was observed in SV-SF pickle $p < 0.05$. A significant decrease ($p < 0.05$) in vitamin C for CF (88–90%) and SF (94–95 %) pickles a further reduction in vitamin C was in relish making. Fermentation substantially dec n phytate levels was also observed. Total phenols decreased by 26–29 % and 39–43 % in CF and SF, respectively. Fermentation significantly decreased chlorophyll from 52–57.8 g kg^{-1} to 31.9–48.3 g kg^{-1}. Also fermentation significantly reduced minerals, P (1080–1166 mg/100g to 264–439 mg/100 g), Ca (3113–3392 mg/100 g to 866–2445 mg/100g), Fe (148-185 mg/100 g to 61-82 mg/100 g) and Zn (5.5–8.0 mg/100 g 1.3–1.9 mg/100 g). Further reduction of minerals was observed in relish making. However, fermentation slightly increase nickel content from 0.26–0.29 to 0.3–0.86 mg/100 g. Fermentation can preserve ANS, with an increase in β-carotene and reduction in antinutrients. Both CF and SF can be recommended to small-scale farmers, small-scale processors, and households to improve their nutrition and livelihood. However, CF took a short time (3–4 days) to finish than SF (25–30days).

Keywords: Antinutrients, fermented pickle, minerals, vitamin C

Contact Address: Frank Sangija, Nelson Mandela African Institution of Science and Technology, Food and Nutritional Sciences, Tengeru, Arusha, Tanzania, e-mail: sangijaf@nm-aist.ac.tz

Physicochemical and Functional Properties of Anchote (*Coccinia abyssinica* (Lam.) Cogn.) Tuber Flour as Affected by Pre-Treatment and Drying Temperature

Adugna Mosissa Bikila[1], Yetenayet Bekele Tola[1], Tarekegn Berhanu Esho[2], Sirawdink Fikreyesus Forsido[1]

[1]*Jimma University, Post-harvest Management, Ethiopia*

[2]*Addis Ababa Science and Technology University, Industrial Chemistry,*

Transformation of agricultural outputs into semi-processed and processed food materials is one of the current scientific demands. There might be changes in original property and composition of the product during processing. As part of this concern, physicochemical and functional properties of raw and pre-treated (blanched and boiled) anchote (*Coccinia abyssinica* (Lam.) Cogn.) tuber flours prepared by drying at different temperatures (60, 80, 100 °C) were evaluated. Anchote is a potentially productive and nutritious starchy tuberous crop indigenous to Ethiopia. A factorial experiment in a completely randomised design was employed to run the experiment. Ranges of the results for pH, total soluble solids (TSS), water absorption capacity (WAC), oil absorption capacity (OAC), water absorption index (WAI), water solubility index (WSI), swelling power (SP), foaming capacity (FC), foam stability (FS), total polyphenol content, and total flavonoid content were 5.70–6.47, 5.37–10.8 °Brix, 2.42–4.21 g g^{-1}, 0.94–1.44 g g^{-1}, 3.40–5.42 g g^{-1}, 11.40–20.37 %, 4.56–7.20 g g^{-1}, 3.31–33.33 %, 1.89–20.00 %, 0.22–0.80 mg GAE/g, and 0.12–0.44 mg CE/g, respectively. The results showed that both pre-treatment and drying temperature significantly ($p < 0.05$) affected the pH, TSS, WAC, OAC, WAI, WSI, SP, TPC and TFC of the flours. The pH, TSS, WAC, WAI, WSI and SP were higher in the flour prepared from boiled anchote dried at lower temperature; while OAC, FC and FS were higher in the flour obtained from the raw tuber. The flour obtained from blanched and boiled anchote dried at lower temperature exhibited better functional property relative to the raw. Therefore, this value-added product could be used as a thickener and to improve texture compared to the flour made from the raw anchote in food formulations.

Keywords: Anchote flour, drying temperature, functional properties, physicochemical properties, pre-treatment

Contact Address: Adugna Mosissa Bikila, Jimma University, Post-harvest Management, Jimma Zone, P. O. Box 307 Jimma, Ethiopia, e-mail: a.mosissa@yahoo.com

Nutritional Profile of *Nomadacris semptemfasciata* and its Perpective Use to Fight against Malnutrition in Madagascar

Christian Tolojanahary Ratompoarison, Felamboahangy Rasoarahona, Jean Rasoarahona

University of Antananarivo, College of Agric. Scie., Dept. of Food Science and Technology, Madagascar

Biodiversity can contribute to the fight against chronic malnutrition in Madagascar. Edible insects are a resource potentially exploitable but mis-known by local consumers and stakeholders. The objective of this study is to explore the nutritional composition of *Nomadacris septemfasciata* and to discuss about its potential uses to enhance nutrition. N. semptemfasciata is a species available in large quantities during the hot and humid season in the highlands of Madagascar. Analysis shows that the protein content of N. septemfasciata is the highest among Orthoptera orders, 77.46 % of dry matter. It contains all essential amino acids, with an amino acid score in accordance with the FAO / WHO / UNU recommendation except for methionine. The high tryptophan content 6.17 g / 100 g of protein makes possible to use this insect as supplement to foods limited in this amino acid, such as rice and tubers, which are the staple food of Malagasy people. The lipid fraction represents 8.46 % of the dry matter with 14 fatty acids and the dominance of palmitic acid at 23.4 %, arachidic acid at 14.9 %, and 10.8 % α-linolenic acid. *N. semptemfasciata* powder contains a good omega 3 content 10.8 g / 100 g of fat. Similar in content to cod liver oil, fish oil and walnut oil. The content of iron 9.99 ± 1.00 mg / 100 g and zinc 21.16 ± 1.90 g / 100g makes it a potential source of mineral which can be used for food enrichment. Adding *N. semptemfasciata* powder to the daily ration can contribute to improve Malagasy diet quality, proteins and micronutrients intakes.

Keywords: Amino acids, edible insect, fatty acids micronutrients , nomadacris septemfasciata, protein

Contact Address: Christian Tolojanahary Ratompoarison, University of Antananarivo, College of Agric. Scie., Dept. of Food Science and Technology, Campus Universitaire Ankatso, 101 Antananarivo, Madagascar, e-mail: christianratompoarison@gmail.com

Nutritional Profile of Three Candidate Food Ingredients for Food-to-Food Fortification in Benin (West Africa)

Marius Affonfere, Flora Josiane Chadare, Finagnon Toyi Kevin Fassinou, Paulin Azokpota

National University of Agriculture, School of Food Science and Nutrition, Benin

Malnutrition especially micronutrient deficiencies (MNDs) among infants is an important public health problem. Food-to-food fortification using local food ingredients is a cost effective and sustainable approach to overcome this issue. This research aimed at characterising three selected food ingredients candidate for food-to-food-fortification to enhance micronutrients intake for children aged 6–59 months in Benin. The local food ingredients were selected based on their availability and use as traditional food fortificant. There were characterised for their dry matter, ash, iron (Fe), calcium (Ca), zinc (Zn), magnesium (Mg), phosphorus (P), copper (Cu), sodium (N), manganese (Mn), vitamin C and total phenolic compounds contents using standard methods. Pro-vitamin A, phytate and tannin contents of the selected food ingredients were collected from literature. *Adansonia digitata* fruit pulp, *Moringa oleifera* leaf powder and *Cochlospermum tinctorium* root powder were the selected food ingredients. Baobab fruit pulp mineral contents in mg / 100 dw were 9.9 $\pm$ 0.1 for iron, 0.9 $\pm$ 0.1 for zinc and 402.2 $\pm$ 3.4 for calcium. As *moringa* leaf powder and *Cochlospermum tinctorium* root powder are concerned, their iron, zinc and calcium contents in mg/100 g dw were 34.1 $\pm$ 2.2 and 26.8 $\pm$ 2.7; 9 $\pm$ 0.0 and 0.9 $\pm$ 0.0 and 2054.9$\pm$11.5 and 1061.3 $\pm$ 11.5 respectively. *Adansonia digitata* fruit pulp had 372.7$\pm$12.2 mg/100 g dw of vitamin C content, 2128.2 $\pm$ 44.5 mg eq AG/100 g dw of total phenol content and 287.5 $\pm$ 201.0 mg/100 g dw of phytic acid content. *Moringa oleifera* leaf powder and *Cochlospermum tinctorium* root powder had respectively 24.6$\pm$1.4 mg/100 g dw and 23.4$\pm$1.3 mg g^{-1} dw of vitamin C content, 2256.7$\pm$259.0 mg eq AG/100 g dw and 2694.6 $\pm$ 29.8 mg eq AG/100 g dw of total phenol content and 829.0 $\pm$ 23.0 mg /100 g dw and 500.0$\pm$200 mg / 100 g dw of phytic acid content. The present study demonstrated the nutritional potential of these local food ingredients for food-to-food fortification.

Keywords: *Adansonia digitata* fruit pulp, Cochlospermum tinctorium root powder, food-to-food fortification, minerals, *Moringa oleifera* leaf powder, vitamins

---

Contact Address: Marius Affonfere, National University of Agriculture, School of Food Science and Nutrition, BP. 177 Savalou, Abomey-Calavi, Benin, e-mail: mariusaffonfere@gmail.com

Proximate and Antioxidant Composition of Cocoyam Chips Flavored with Three Different Spices

Olaide R. Aderibigbe[1], Suad Mazidatul-Khayr Badmus[2], Oluseye O. Onabanjo[2]

[1]*National Horticultural Research Institute, Product Development Programme, Nigeria*

[2]*Federal University of Agriculture, Abeokuta, Nutrition and Dietetics, Nigeria*

Nigeria produces the largest amount of cocoyam (*Colocasia esculenta*) in the world; however, it is considered an underutilised food crop because of its limited use and most importantly, it is considered a food for the poor. Chips is one of the crunchy snacks with satisfying texture and taste; and is widely eaten by adults and children in Nigeria. The commonly consumed chips are developed from plantain and potatoes. The purpose of this study was to develop flavoured cocoyam chips, thereby increasing the utilisation of this underutilised crop. Seven spice solutions (ginger, garlic, turmeric, ginger+garlic, ginger+turmeric, garlic+turmeric, ginger+garlic+turmeric) were prepared by dissolving ginger, garlic and turmeric powders either singly or in equal combinations in warm water. Cocoyam slices were soaked in the spice solutions for 5 minutes, removed from the solution and drained before frying in vegetable oil. The control sample was prepared by frying salted cocoyam slices. All the samples were subjected to proximate, vitamin C and antioxidant analysis following standard procedures. The control sample showed a significantly higher ($p < 0.05$) protein, ash, fat and fibre content (2.8 %, 5 %, 32.5 % and 15.5 %, respectively) than the ginger-spiced (2.5 %, 3.5 %, 22.5 % and 6 % respectively), garlic-spiced (1.4 %, 3.7 %, 30.0 % and 5.1 % respectively), turmeric-spiced (2.7 %, 2.7 %, 30.7 % and 9.3 %, respectively) cocoyam chips or any of the other cocoyam chips flavored with combination of spices. However, all spice-flavored cocoyam chips had significantly higher phenol and flavonoid content compared to the control sample. The results showed that flavouring of cocoyam chips with these spices is beneficial in terms of reduced fat and increased antioxidant content thus indicating the benefits consumer may derive from consumption of cocoyam chips

Keywords: Underutilised, chips, cocoyam, spices

Contact Address: Olaide R. Aderibigbe, National Horticultural Research Institute, Product Development Programme, Jericho G.R.A., 200031 Ibadan, Nigeria, e-mail: connecttolaide@yahoo.com

Composition, Nutritional Value and Uses of *Ricinodendron heudelotii, Vitex doniana* and *Cleome gynandra* Seeds Oil

Founmilayo Nadjidath Adome[1], Flora Josiane Chadare[2], Fernande Honfo[1], Djidjoho Joseph Hounhouigan[1]

[1]*Faculty of Agricutural Sciences, University of Abomey-Calavi, School of Nutrition, Food Sciences and Technology, Benin*

[2]*National University of Agriculture, Laboratory of Food and Bio-Resources Science and Technology and Human Nutrition, Benin*

Ricinodendron heudelotii, Vitex doniana and *Cleome gynandra* are three indigenous plant species whose seeds contain oil that can provide interesting alternatives to conventional seed oil species. This review focused on the physicochemical characteristics of their seeds and the nutritional properties of the oils extracted from these seeds. For that purpose, scientific articles and reports were collected and needed information was extracted and synthesized. Results showed that, compared to *R. heudelotii* very little work has been done on V. doniana and *C. gynandra* seeds and their oil. The oil content of *R. heudelotii* seeds is about twice higher (51.83 mg/100 g) than that of *V. doniana* (28.55 mg/100 g) and *C. gynandra* (27.35 mg/100 g). Seed oils of all the three species are polyunsaturated oils. *R. heudelotii* seed oil is composed mainly of linoleic acid (28.3–51.1 %) and a polyunsaturated fatty acid found to be a-eleaostearic acid (49.3–51.1 %). *C. gynandra* seeds oil is composed mainly of linoleic acid (56.3 %–61.1 %) and oleic acid (19.6–23.9 %). No information is available on fatty acid composition of *V. doniana* seed oil. Several oil extraction methods have been tested for *R. heudelotii* seeds, including aqueous extraction and extraction by pressing while for *V. doniana* and *C. gynandra* seeds, the oils are extracted by soxhlet using petroleum ether (60–80°C). Some essays have been carried out on the possible applications of the oils. Thus, *R. heudelotii* seed oil is a potential renewable candidate for the preparation of fast drying binder and resins, suitable for industry and surface coating application. *V. doniana* seed oil can be used for the production of resin, paint and skin cream production. No application of *C. gynandra* seed oil has been stated.

Keywords: Applications, composition, nutritional value, oil, seeds

Contact Address: Flora Josiane Chadare, National University of Agriculture, Laboratory of Food and Bio-Resources Science and Technology and Human Nutrition, Sakété, Benin, e-mail: fchadare@gmail.com

The Influence of Regional Origin on Cocoa Butter from Colombia: Tumaco - Huila - Santander

Daniel J.E. Kalnin, Mathilde Bigot, Victorine Charpentier, Arthur Messager, Justine Freulard, Anaëlle Campaignolle

ISTOM; Ecole supérieur d'agro développement international, ADI-SUD, Innovation and Agro/Food-développement international, France

Theobroma cacao L., the cocoa tree is a tropical cauliflorous tree that produces pods containing cocoa beans. These beans undergo various stages of processing, resulting in a finished product: chocolate. Cocoa butter constitutes about 55 % of the dry weight of cocoa beans is composed of mainly triglycerides. The proportions of their fatty acids vary according to the variety of cocoa, due to climatic, geographical, and genetic factors. This leads to variations in physical properties, such as hardness or melting point, which in terms lead to different organoleptic properties and then health benefits.

Those physical properties influence the temperature that allows the crystallisation of cocoa butter to be processed, called tempering. This is one of the most important steps in the manufacture of chocolate, due to the diversity of fatty acids that make up cocoa butter blends. Indeed, these complex mixtures of triacylglycerols have different melting points. The fatty acid profile observed in cocoa butter is thus the main factor that can influence its organoleptic and rheological properties, as well as its processing qualities during the transformation stages. On the contrary, we investigate single-origin chocolate. We, therefore, studied chocolate from three different regions of Columbia. These premises allow the cocoa trees to be grown under similar climatic and soil conditions and agronomic practices. It is important to notice that the traceability of different kinds of cocoa butter remains difficult. In our case, Luker Chocolate provides chocolate. Studies show that cocoa butter extracted from hybrid cultivars of *Theobroma cacao* L. exhibit relatively constant melting characteristics. Based on genetic diversity but regional homogeneity, we investigated how this regional difference influences the organoleptic properties after the crystallisation of cocoa butter. Thereafter, we demonstrate a difference in phase behaviour. This will enable the local transformation of single-origin cocoa butter, hence favouring rural developent.

Keywords: Cocoa butter, consumer perception, local transformation

Contact Address: Daniel J.E. Kalnin, ISTOM; Ecole supérieur d'agro développement international, ADI-SUD, Innovation and Agro/Food-développement international, 4 rue Joseph Lakanal, 49100 Angers, France, e-mail: d.kalnin@istom.fr

Screening of Different Classes of Olive Oil to Provide a Potential Healthiest Alternative for Deep Frying

MAHDI FENDRI[1], CHAIMA BEJAOUI[2], MARIEM SUIDI[2], AJMI LARBI[1]

[1]*Olive Institute, Specialized Unit of Tunis, Tunisia*

[2]*Higher Institute of Biotechnology of Sidi Thabet, Biotech Pôlet, Tunisia*

Nowadays, deep frying absorbs a considerable amount of the edible oils produced in the world. The use of fats and oils rich in saturated fatty acids has decreased because of their negative effects on health, including their implication in cardiovascular diseases. Olive oil, which accounts for only about 3 % of the vegetable oils consumed in the world, is one of the few oils whose benefits on human health have been widely studied and documented. Nonetheless, the use of olive oil in deep frying is still very insignificant, not only because of its relatively high price, but also because of a generic idea which perceive it as an oil that should be consumed in raw, without referring to existent comparative studies with other plant species. Although there is a wide literature demonstrating its resistance during the frying process, only a few studies compare different commercial classes of olive oil and/or cultivars performance with regard to frying and/or resistance to temperature. From this two years' study, we aim at describing the impact of a prolonged intermittent heat application of 180°C on the behaviour of different commercial categories of three olive cultivars compared to other vegetable oils used as a reference. After a four day's treatment, the generated polar compounds and total phenolic fraction conserved in the oils demonstrated that all classes of virgin olive oils were more resistant to heat application, while free fatty acidity was lower in refined oils. Fatty acid composition changes were also observed. A better resistance of higher quality virgin olive oils has been clearly observed, indicating that the efficiency of olive oil under high temperature for a prolonged time also depends on the cultivar origin and intrinsic quality value. The results of this study can be considered as a first step towards the selection of the best virgin olive oils that could be suitably used as a healthiest alternative for frying or other cooking processes that involve high temperature. However, the interaction between the oil and each specific ingredient should be furtherly studied accordingly.

Keywords: Cultivars, healthy frying, olive oil classes, temperature resistance

Contact Address: Mahdi Fendri, Olive Institute, Specialized Unit of Tunis, BP208 Cité Mahrajène, 1082 Tunis, Tunisia, e-mail: fendrimahdi@yahoo.fr

Anchovy Processing in Ghana: Describing Quality and Food Safety Parameters in the Processing Chain

Laura Wessels[1], Theophilus Annan[2], Amy Atter[2], Marian Kjellevold[3], Felix Reich[1], Johannes Pucher[1]

[1]*German Federal Institute for Risk Assessment, Germany*

[2]*Council for Scientific and Industrial Research - Food Research Institute, Ghana*

[3]*Institute of Marine Research, Norway*

Sufficiently nutritious and safe food is not yet generally accessible for everyone in some parts of Africa. Both, nutritional quality of food and food safety are important aspects that play a role in food security. As a source of suitable nutrients for a balanced diet, small fish species like European anchovy (*Engraulis encrasicolus*) can contribute towards nutrition security. At the same time, fish is considered as highly perishable and different processing techniques are used to prevent spoilage and for preservation. In certain parts of Africa, infrastructure does not support freezing or canning of the fish to increase its´ shelf live. In such circumstances, low cost traditional methods such as smoking, drying, frying, fermentation or a combination of salting and drying are usually applied. Nonetheless, smoking and drying not only allow longer storage time of the fish, but also address consumers´ food preferences and traditional heritage.

Smoking in Chorkor or Ahotor ovens, sun drying on the floor or on elevated racks are the four main processing techniques used to preserve European anchovy in Ghana. Questionnaires were conducted with individual experienced processors to gain in-depth knowledge of the different steps of the processing procedure. The information collected covers the source of fresh fish, personnel, processing techniques and equipment, type of storage materials, storage period as well as how to assess the quality of a product. Process flow descriptions were generated with focus on the quality parameters applied.

The aim of this work was to identify critical points along the processing value chain and to develop improved processing practices for the production of higher quality, safer and nutritious anchovies.

Keywords: Ahotor oven, Chorkor smoker, food processing, small fish, sun-drying

---

Contact Address: Laura Wessels, German Federal Institute for Risk Assessment, Max-Dohrn-Straße 8-10, 10589 Berlin, Germany, e-mail: laura.wessels@bfr.bund.de

Microbial Contamination and Occurrence of Aflatoxins in Processed Baobab Products in Kenya

MARGARET JAMES, WILLIS OWINO, SAMUEL IMATHIU

Jomo Kenyatta University of Agriculture and Technology, Food Science Department, Kenya

Ready to eat snacks derived from baobab pulp are common products in a number of Sub-Saharan Countries. This study was conducted to investigate the microbial and aflatoxin contamination levels in ready-to-eat baobab products from selected formal and informal processors in specific counties of Kenya. Processed baobab products (pulp and candies) samples were randomly collected from formal and informal processors and analysed for the total aerobic count, Enterobacteriaceae, yeast and molds, ergosterol, aflatoxins, moisture, and water activity. The moisture and water activity of baobab pulp and candies from formal processors ranged between 7.73 % - 15.06 % and 0.532–0.740 compared to those from informal processors which ranged from 10.50 % - 23.47 % and 0.532–0.751 respectively. Baobab pulp from informal processors had significantly higher ($p < 0.0001$, $p < 0.0001$) Enterobacteriaceae, yeast and molds counts (3.1 ± 0.70 log10 CFU g^{-1} and 5.3 ± 0.11 log10 CFU g^{-1}) than those from formal processors (0.7 ± 0.29 log10 CFU g^{-1} and 3.1 ± 0.38 log10 CFU g^{-1}) respectively. Similarly, a significant difference ($p = 0.015$) was observed in terms of Enterobacteriaceae counts between candies from formal (0.0 ± 0.00 log10 CFU g^{-1}) and informal processors (1.8 ± 0.56 log10 CFU g^{-1}). The ergosterol content in these baobab product samples ranged between 0.46 to 1.92 mg/100 g while the aflatoxin content ranged between 3.93 to 11.09×10^3 $\mu g\,kg^{-1}$ respectively. Fungal and aflatoxin contamination was detected in 25 %, 5 %, and 5 % of the pulp from formal processors, informal processors, and candies from informal processors respectively. Fungal and aflatoxin contamination in baobab products may indicate poor storage and processing conditions. The food safety risks could be effectively mitigated by initiating training along the baobab value chain on; good hygiene practices, good manufacturing practices, hazard analysis critical control point as well as appropriate postharvest handling of baobab fruit and its pulp.

Keywords: Aflatoxin, baobab, ergosterol, microbial quality, microbiological limits

---

Contact Address: Kathrin Meinhold, Rhine-Waal University of Applied Sciences, Fac. of Life Sciences, Marie-Curie-Str. 1, 47533 Kleve, Germany, e-mail: kathrin.meinhold@hochschule-rhein-waal.de

Investigation of Phenolic Content, Antioxidant Capacity and Pomological Characterisation of Wild Sea Buckthorn (*Hippophae rhamnoides* L.) from the Walnut-Fruit Forest of Kyrgyzstan

Janyl Iskakova[1], Zhyldyz Oskonbaeva[2], Jamila Smanalieva[3]

[1]*Kyrgyz-Turkish Manas University, Environmental Engineering, Kyrgyzstan*

[2]*Rhein-Waal University of Applied Sciences, Germany*

[3]*Kyrgyz Technical University, Food Processing Technology, Kyrgyzstan*

Sea buckthorn (*Hippophae rhamnoides* L.) is widely used for soil, water, and wildlife conservation and anti-desertification purposes. The natural and climatic conditions of Kyrgyzstan make it possible to grow sea buckthorn in forest areas, particularly in sea and river coasts, in dry-sandy semi-desert sites, and high mountains. Walnut-fruit forests of Kyrgyzstan are unique for their great biological and genetic diversity, and constitute natural plantations of sea buckthorn. In Kyrgyz folk medicine, sea buckthorn is used for gastrointestinal system disorders and healing of burn wounds. Unfortunately, information on biologically active compounds, antioxidant activity of sea buckthorn berries from walnut-fruit forest of Kyrgyzstan in the scientific literature is very limited. This study aimed to investigate the chemical composition (dry matter, sugars, pH, titratable acidity, ash, fibre, vitamin C) and the radical scavenging activity, total polyphenols of sea buckthorn berries for documentation in the national food composition database and comparison with the all available recent studies on sea buckthorn.

The investigated samples of fresh sea buckthorn feature 66.03 g/100 g of moisture content, 1.00–1.13 g/100 g of invert sugars, 1.78–2.12 g/100 g of organic acids, 12.13 g/100 g of fibres, and 1.75 g/100 g of ash content. The moisture content of the studied samples was significantly lower than the moisture content in berries from Europe, but higher than in berries from Pakistan and India. Fresh sea buckthorn contained on average 181.9 mg/100 g of vitamin C and a large amount of total phenolic compounds of 386 mg/100 g fresh weight. The antioxidant activity of the sea buckthorn was high and found as 3.8 μg/ml, which corresponds to the literature data. The measured physical attributes and chemical composition of these berries are important in promoting the use of these products in the food and cosmetic industries as well as in medicine.

Keywords: Antioxidant activity, physical attributes, polyphenols, sea buckthorn, vitamin C

Contact Address: Zhyldyz Oskonbaeva, Rhein-Waal University of Applied Sciences, Thaerstasse 29, 47533 Kleve, Germany, e-mail: zhyldyz.oskonbaeva@gmail.com

Bacteria Contamination of Poultry Meat Processed by Smallholder Poultry Farmers

Timieteimo Douglas, Eyerin-Ebi H. Japan, Ruth Tariebi S. Ofongo
Niger Delta University, Dept. of Animal Science, Nigeria

Poultry birds are raised for the production of quality protein as meat or eggs for human consumption. Meat from table birds for human consumption can be contaminated with bacteria during slaughter and along the processing chain. *Salmonella, Escherichia coli, Campylobacter jejuni* are some bacteria which causes economic losses in the poultry industries and are also very dangerous to consumers health. In low medium income countries like Nigeria, poultry meat from smallholder farmers are processed manually in small scale. This method makes poultry meat readily available for low-income households that cannot afford, imported well packaged poultry meat. This study was designed to assess contamination of poultry meat processed by smallholder poultry farmers. Poultry meat samples such as head and thigh (drum stick and thigh) were purchased from vendors for smallholder poultry farms. Samples were chopped to fine particle using a sterile knife and 1gram weighed on an electrical weighing balance then transferred into 9 ml of normal saline solution to make 10 ml of stock solution. Serial dilution was performed and the third diluent was plated out on nutrient agar, incubated at 37°C for 24 hours. Colonies were picked at random according to their varying cultural characteristics, sub-cultured by streak method on fresh nutrient agar plates and incubated at 37°C for 24 hours to obtain pure isolates. Biochemical tests, gram staining technique and bacterial motility tests were carried out to identify bacterial isolates. All analysis were carried out in triplicates. Nine bacteria were identified on both meat samples which were obtained as pure culture. *Staphylococcus* (21 %), *Pseudomonas* (17 %), *Bacillus* (13 %), *Escherichia coli* (13 %), *Micrococcus* (12 %), *Salmonella* spp (8 %), *Protus* (8 %), *Shigella* (4 %), *Klebsiella* (4 %); respectively. Poultry meat processed by smallholder poultry farms are contaminated with a wide range of bacteria besides *Salmonella* spp and *E. coli*.

Keywords: smallholder poultry farm, bacteria contamination, poultry meat, processing

Contact Address: Ruth Tariebi S. Ofongo, Niger Delta University, Dept. of Animal Science, Wilberforce Island, P.M.B. 071, Yenagoa, Nigeria, e-mail: tariruth@live.de

Fortifying Fermenting Enset (*Ensete ventricosum*) with *Moringa* Leave and Garden Cress: Physicochemical and Microbial Profiling

Addisu Fekadu Andeta[1], Fantahun W. Misganaw[2], Gemechu Leta Debele[1], Ribka Ebrahim[3]

[1] *Arba Minch University, Microbial and Molecular Systems, Ethiopia*

[2] *Ethiopian Biotechnology Institute, Industrial Biotechnology Directorate, Ethiopia*

[3] *Arba Minch University, Inst. of Technology, Dept. of Water supply and Env't. Eng., Ethiopia*

Enset (*Ensete ventricosum*) is an important food security crop for over 20 million Ethiopian people after fermentation into kocho. The fermented food product is a good source of starch (65 %) and soluble sugars (9 %), but it is low in protein (< 4 %) and fat (< 1 %). The aim of this study was to assess the physicochemical and microbial dynamics during fortified-enset fermentation. To this end, six enset plants were processed using enset processing machines and fortified each with 3 % shed-dried *Moringa stenopetala* leaf powder (MS), *Moringa oleifera* (MO) leave powder, *Lepidium sativum* (LS) seed powder and the fourth treatments left as a Control (C). The fortified and control enset samples were allowed to ferment for two months in a Sauerkraut jars. Samples were taken on days 1, 7, 15, 30 and 60. The pH and moisture content were measured. Also, microbial dynamics were determined using the plate count method. The pH of the enset sample fortified with MS, MO, LS and C samples on day 0 were 5.99, 6.15, 5.83 and 6.63, respectively. On day 60 the pH reduced to 4.25, 4.00, 4.28 and 3.92 for MS, MO, LS and C samples, respectively. The total aerobic count attained a maximum of 9.23, 9.40,9.29 and 9.28 log cfu/g at day 0 for MS, MO, LS and C, respectively. Lactic acid bacteria count (LAB) were 8.10, 8.13, 8.10 and 8.10 log cfu/g on day 1. On day 60 the LAB count reduced to 7.34, 7.05, 7.73 and 6.70 for MS, MO, LS and C, respectively. Moreover, enset fortified with MS and MO leave powder inhibited the growth of undesirable microorganisms (*Clostridium* sp, Enterobacteriaceae and molds) from enset samples on day 15 and onwards. Addition of *moringa* leaves powder (MS and MO) to fermenting enset showed a significant improvement in the protein, calcium, iron, phosphorus and magnesium contents of kocho. In conclusion, addition of *moringa* leave powder to fermenting enset significantly decreased Clostridium, yeast and mold counts and Enterobacteriaceae) abundance, whereas it increased Lactobacillus abundance as compared with the untreated enset samples.

Keywords: Enset, fermentation, fortification, *Moringa*, lactic acid bacteria

Contact Address: Addisu Fekadu Andeta, Arba Minch University, Microbial and Molecular Systems, Secha Condominium, T4-3-11, 4400 Arba Minch, Ethiopia, e-mail: addisu.fekadu@amu.edu.et

Challenges and Opportunities toward Sustainable Consumption and Value Addition of Cashew Apples in Tanzania

NOEL DIMOSO, EDNA MAKULE, NEEMA KASSIM

Nelson Mandela African Institute of Science and Technology (NM-AIST), Food Biotechnology and Nutritional Sciences, Tanzania

Cashew apple is an important healthy fruit and yet is highly underutilised in developing countries. This study explored factors affecting utilisation of cashew apple among farmers in Lindi and Mtwara regions. Semi-structured questionnaire was used on 600 cashew farmers to collect information on cashew apple consumptions, processing and utilisation constraints. In addition, dried cashew apple product was developed, in which full matured, ripe and intact fruits were plucked from the cashew tree. Then they were washed, blanched, sliced and immersed in 70 % sucrose prior to drying on an oven or solar drier. As a result, majority of farmers reported to consume raw cashew apples. The frequency of consumption was more than five fruits a day (61.87 %) and almost every day (55.98 %) during the season. Traditional technologies for processing cashew apple porridge and alcohol were employed by about 43.7 % of farmers. Lack of knowledge on post-harvest handling (86.2 %) and processing technologies (82.7 %) were mostly claimed to hamper cashew apple utilisation. Both dried products showed no significant different ($p > 0.05$) on carotenoids (0.28–0.33g/100 g), vitamin C (0.73–0.85g/100 g) and tannins contents (266.59–267.95 mg/100 g). During storage at ambient temperature for 60 days: total phenolic, tannins and vitamin C were significantly reduced ($p < 0.05$) in both oven and solar dried products. Furthermore, both dried products showed similar ($p > 0.05$) overall sensory acceptability. The combination of blanching, osmotic dehydration and solar or oven drying provide economically feasible value added products that can be reproduced in both urban and rural settings to enhance reduction of postharvest losses of the fruit.

Keywords: Cashew apple, Hot air drying, Osmotic dehydration, post-harvest losses, Solar drying, Utilisation constraints, Value addition

Contact Address: Edna Makule, Nelson Mandela African Institute of Science and Technology (NM-AIST), Dept. of Food Biotechnology and Nutritional Sciences (FBNS), Arusha, Tanzania, e-mail: edna.makule@nm-aist.ac.tz

Exploring Date Palm, Desert Date, and Acacia as Promising Sources of Edible Oil in Moroccan Deserts Area

Said El Harkaoui[1], Zoubida Charrouf[2], Hanae El Monfalouti[2], Badr Kartah[2], Said Gharby[3], Bertrand Matthäus[1]

[1]*Max Rubner-institut (MRI), Dept. of Safety and Quality of Cereals, Working Group for Lipid Research, Germany*

[2]*Mohammed V University of Rabat, Lab. of Plant Chemistry and Organic and Bio-Organic Synthesis, Morocco*

[3]*University Ibn Zohr, Lab. of Biotechnology, Materials and Environment (LBME), Morocco*

Investigating new sources of food has been of growing concern because of the increase of the population, climatic changes, and diverse industrial applications. These factors led to the processing and studying of new sources depending on each country.
Morocco is a country that is known for the scarcity of rains and successive periods of drought. In the dry season, many rural people rely on plants growing in the deserts such as date palm (*Phoenix dactylifera*), deserts date (*Balanites aegyptiaca*), or Acacia (*Acacia Raddiana*). On one side these plants provide people with nutrients from different parts of the plant and on the other side, they can cope with the increasing problems of climate change with less and less rainfall resulting in erosion, desertification, deforested soils, and a decrease of biodiversity. In addition, they can help to overcome the nutritional situation, result in a better income and safe employment and improve the situation of women working in cooperatives. Therefore, it is crucial to explore each part of these plants seeking nutritional compounds such as edible oil.
The lipid fractions of date palm seeds, acacia seeds, and desert date kernels were extracted by solvent and analysed by gas chromatography (GC-FID or GC-MS) and high-performance liquid chromatography (HPLC). The composition of the oils in terms of triacylglycerols, fatty acids, phytosterols, and tocopherols was done, and the oil content was measured. The composition of the extracted oils was compared with well-known edible oils and the results were used to obtain an informative profile that will serve as the basis for further chemical investigations and nutritional evaluation.

Keywords: Acacia, date palm, desert date, fatty acids, sterols, tocopherols

Contact Address: Said El Harkaoui, Max Rubner-institut (MRI), Dept. of Safety and Quality of Cereals, Working Group for Lipid Research, 32756 Detmold, Germany, e-mail: elharkaouis@gmail.com

Presence of Listeria in Boursin and Chevrotin Goat Cheese Sampled in the Alfenas Region, Brazil

Felipe Damasceno Leandro[1], Andressa Santana Natel[2], Ariane Flávia Nascimento[1]

[1]*University José Do Rosário Vellano, Animal Science Department, Brazil*

[2]*University José do Rosário Vellano (UNIFENAS), Dept. of Agronomy, Brazil*

The production of goat's milk in Brazil shows an exponential and continuous increase, which drove the increase in the production of more elaborated dairy products, such as gourmet cheeses type Boursin and Chevrotin. However, there is no national legislation to characterise these products, adopting general standards. Another point is the microbiological quality of these cheeses, especially in relation to Listeria wich representes na important parameter of quality and risk to food health. The objective of this work was to evaluate the physical-chemical and microbiological characteristics of two types of Brazilian analog goat's cheeses type, Boursin and Chevrotin, in the Alfenas-MG region. Thereby, establish the physical-chemical characteristics of the local product, relating them to the legislation in force in the country, in order to meet the demand of improvements in the dairy sector, evidencing occasional divergences and their respective social and economic impact. For this purpose ten cheeses, 05 Boursin and 05 Chevrotin types, produced in Alfenas region, were purchased at the local market and the fat and humidity was determined using the Gerber method and oven drying, respectively, and analysed according to the current legislation. Also was analysed the presence of listaria spp., *Salmonella* spp and quantification of total coliforms. From the physico-chemical point of view, 57.2 % and 39.02 % of fat and 77.56 % and 47.36 % of moisture were observed for cheeses type Boursin and type Chevrotin, respectively, results appropriate to national legislation for high and medium humidity cheeses. As for the microbiological analyzes, it was determined that there was no *salmonella* in any of the products analyzed. Listeria monocytogenes was found in 40 % of the samples of both cheeses, and 60 % and 80 % of total coliforms in Boursin and Chevrotin cheeses. It is concluded that the cheeses type Boursin and type Chevrotin produced and packaged in the region of Alfenas / MG, although they are in agreement with the legislation in relation to the contents of fat and humidity, however they were positive for Listéria, being not indicated for consumption presenting health risk.

Keywords: Foodborne disease, listeria, microbiology

Contact Address: Felipe Damasceno Leandro, University José Do Rosário Vellano, Animal Science Department, Rua Padre José Grimminck 185, 37135018 Alfenas, Brazil, e-mail: felipe.leandro@aluno.unifenas.br

Comprehensive Study of Volatile Compounds of Tea

Filip Vojacek

Czech University of Life Sciences Prague, Department of Sustainable Technologies, Czech Republic

Tea is one of the major non-alcoholic beverages in the world, usually prepared by leaching the refined leaves of the plant *Camellia sinensis* in hot water. The tea leaves contain hundreds of chemical compounds. Most numerous in different types are volatile compounds. Many kinds of research about the chemical composition of tea leaves, predominantly for the food and pharmaceutical industry, were elaborated. But, because of the quantity and wide variability of volatile compounds, these researches often provide significantly different results. Therefore, until nowadays, the content and composition of volatile compounds and their changes in composition are not explored and explained enough. This study provides a complex analysis of these volatile compounds in refined tea leaves prepared for direct consumption of the final consumer. It monitors the volatiles composition according to the origin and refining method of the tea leaves, the storage method of the refined tea leaves, and the extraction method used for extraction of the volatile compounds. The analysed teas come from different Vietnam regions. Teas were stored at room temperature and packed in an anaerobic environment with the absence of light. Three extraction techniques, such as hydrodistillation, solvent extraction and headspace solidphase micro-extraction, were used for the extraction of the volatile compounds. In addition, this study examines if there are differences in volatile composition depending on the market price of refined tea. The extracted volatile compounds were analysed by GC - MS, and the results were further statistically analysed and in detail discussed.

Keywords: Gas chromatography, tea leaves, volatile compounds

Contact Address: Filip Vojacek, Czech University of Life Sciences Prague, Department of Sustainable Technologies, Strelecka 1063, 53701 Chrudim, Czech Republic, e-mail: vojacekf@ftz.czu.cz

Seed Storage Proteins in *Chenopodium quinoa* Germplasm

Lucie Dostalíková[1], Petra Hlasna Cepkova [2], Iva Viehmannová[1], Dagmar Janovská[2], Václav Dvořáček[2]

[1]*Czech University of Life Sciences Prague, Fac. of Tropical Agrisciences, Dept. of Crop Sciences and Agroforestry, Czech Republic*

[2]*Crop Research Institute, Gene Bank, Czech Republic*

Quinoa (*Chenopodium quinoa* Willd.) is a dicotyledonous pseudo-cereal from the Chenopodiaceae family that originated in the Andean Mountains of the South America region. In recent years, the production and consumption of quinoa seeds have increased. Cultivation has also spread to countries outside the original growing area, among others, to countries of temperate climate. However, the leading quinoa producers in the world are still Bolivia, Ecuador, and Peru. The main reason for the spread of quinoa is its remarkable adaptability to different climatic and soil conditions. Quinoa seeds are valued mainly for their relatively high content of gluten-free storage protein (14.6 %). In addition, the composition of amino acids is very balanced and approaches an ideal protein with a quality equivalent to the milk protein casein. Quinoa grain also contains the essential amino acids lysine, methionine, and threonine, which are considered limiting factors in other crops. With the aim of analysing the band spectra of seed storage proteins and detailed characterisation of their polymorphism and genotype relationship, 35 quinoa genotypes of different geographical origins were analysed by SDS-PAGE. A total of 20 strong allelic positions were detected in the molecular weight range from 10 to 175 kDa. The protein bands differed in intensity, and the main variability in band position among genotypes was found in the positions around 30 and 60 kDa. The evaluated samples showed some degree of heterogeneity at the level of overall seed protein polymorphism. Electrophoretic analysis of seed storage protein proved to be a helpful tool to discriminate quinoa genotypes as a first step in evaluating quinoa genetic resources, contributing general and specific valuable knowledge for breeding and selecting perspectives of quinoa genotypes.

Keywords: *Chenopodium quinoa*, genetic resources, total protein, variability

Contact Address: Lucie Dostalíková, Czech University of Life Sciences Prague, Fac. of Tropical Agrisciences, Dept. of Crop Sciences and Agroforestry, Kamýcká 129, 16500 Prague, Czech Republic, e-mail: dostalikova@ftz.czu.cz

The Nutritional Implications of Pericarp Color in Pigmented Rice through Modelling

Rhowell Jr Tiozon, Reuben James Buenafe, Nese Sreenivasulu
International Rice Research Institute, Consumer-driven Grain Quality and Nutrition Center, Philippines

Various health-promoting properties have been linked to the consumption of pigmented rice due to its polyphenolic and mineral content. As consumers become more aware of these benefits, they often shift from consuming white rice to pigmented rice. Pigmented rice is easily identified in the market through its pericarp colour and is classified on this basis. Because of this, consumers have always associated the colour of rice with its nutritional value. This study examined the relationship between rice pericarp colour and its polyphenolic content and antioxidant capacity through partial least squares regression (PLS) models. In addition, we proposed a new classification scheme for coloured rice based on its mineral and polyphenolic content, which could be predicted through its colour traits by employing artificial neural network (ANN) models. The results revealed that lightness (L*), redness (a*), yellowness (b*), and hue angle (h*) are important colour properties to consider in predicting the polyphenolic contents and antioxidant capacity. While the PLS model showed modest accuracies for predicting the total phenolics (TPC, $R^2 = 0.59$), total flavonoids (TFC, $R^2 = 0.60$) and total anthocyanin content (TAC, $R^2 = 0.63$), it showed low accuracy in predicting the antioxidant capacity ($R^2 = 0.19$). This shows that colour alone could be used to estimate the phenolic contents of rice but not its antioxidant capacity since there are other bioactive compounds found in rice that contribute to its DPPH scavenging activity but not necessarily a contributor to its pigment. Furthermore, we classified the pigmented rice into four general classes based on its polyphenolic and mineral content. Class A revealed that samples containing low TPC, TFC, and TAC have high values of Mn and Na. Conversely, class C revealed that samples with high polyphenolic content have low values of Mn and Na. The neural network predicted the classes using the previously identified important colour trait with an accuracy of 31.25 % and can be further improved by adding more hidden layers to the model.

Keywords: Anthocyanin, flavonoids, neural network, phenolics, pigmented rice

Contact Address: Rhowell Jr Tiozon, International Rice Research Institute, Consumer-driven Grain Quality and Nutrition Center, 258 B12 L5 Apartment 1 Lirio St. Lakeshore Subdivision Sto. Domingo, 4300 Laguna, Philippines, e-mail: r.tiozon@irri.org

Determination of Heavy Metals in White Rice of Tarom Cultivar in First and Second Cultivation in Mazandaran Province, Iran

Rahim Shaeri[1], Naghmeh Mobarghaee[2], Houman Liaghati[3]

[1]*Azad University Ayatolah Amoli of Amol, Agro technology, Science and Seed Technology, Iran*

[2]*Shahid Beheshti University, Environmental Planning, Iran*

[3]*Shahid Beheshti University, Environmental Sciences Research Institute, Iran*

Rice is one of the most important foods in the diet of people around the world. Statistics show that China, with the production of about 146 million tons of rice in the 2019–2020 crop year, has the highest production of this food in the world, and India and Indonesia with 118 and 34 million tons, respectively are located, in the second and third place in the world. In Iran, in 2020, 2.6 million tons of white rice were produced in the country at the area of about 800,000 hectares. Heavy metals are one of the most dangerous contaminants for health and enter the human body through food. Irrigation with sewage, pesticide and fertiliser application are the most important factors in increasing heavy metals in the soil and food.

Mazandaran province, as the centre of rice cultivation in Iran has the first rank in rice cultivation in Iran. recent years, due to the economic attractiveness and the provision of the necessary infrastructure, the second cultivation of rice, in one year, has become common among farmers. Due to differences in climate and harvest time of the first and second types, there are differences in the quantity and quality of rice produced. In this study, the difference between the amount of heavy metals in these two types of cultivation has been investigated.

Samples were collected from local farms in Amol and Mahmoudabad during harvested period. All the heavy metals were determined by wet digestion method using a Flame Atomic Absorption Spectrophotometer based on international standard method of AOAC. The results show the increase of heavy metal in second cultivation. Therefore, food safety monitoring system for rice should be performed considerably.

Keywords: Food safety, heavy metal, Iran, Mazandaran, rice, second cultivation

Contact Address: Naghmeh Mobarghaee, Shahid Beheshti University, Environmental Planning, Velenjak, Tehran, Iran, e-mail: n_mobarghei@yahoo.com

Effect of Drying Methods on the Nutritional and Antinutritional Content of African Nightshade (*Solanum* sp.)

Marynurce Kazosi, Frank Sangija, Haikael Martin, Athanasia Matemu

Nelson Mandela African Institution of Science and Technology, Food Biotechnology and Nutritional Sciences, Tanzania

African nightshade (ANS) is one of the luminary food plants, considered a cheap and potential dietary source for micronutrients and bioactive compounds. The study evaluated the effects of drying techniques on nutritional (minerals & vitamin C) and antinutritional (oxalates & phytate) contents of ANS (*Solanum scabrum* (SS) and *Solanum villosum* (SV)). The study employed three methods for drying; indirect solar drying (ISD), mixed solar drying (MSD), and open sun drying (OSD). Pre-treatment done was blanching (85 °C, 2 min) with and without 3 % NaCl; others were un-blanched for control purposes. From the results, vitamin C retention in OSD was 12.08 % in SS, and 12.28 % in SV; MSD (13.12 % SS and 17.79 % SV), and ISD was 14.76 % SS and 19.2 % in SV). A considerable amount of minerals was retained, specifically; calcium retention were as follows; OSD (84.76 % SS, 93.54 % SV); MSD (83.12 % SS, 91.29 % SV); ISD (92.60 % SS, 96.57 % SV). For iron the percentage retained were; OSD (59.31 % SS, 49.77 % SV), MSD (52.29 % SS, 50.94 % SV) and ISD (77.88 % SS, 71.56 % SV); while for Zn, the significance percent were; OSD (43.64 % SS, 74.53 % SV), MSD (59.23 % SS, 86.32 % SV) and ISD (86.94 % SS, 90.09 % SV). On the other hand, the drying methods significantly reduced the oxalate and phytate content. The results showed an ISD to be the best method for retaining vitamin C, minerals and reducing antinutrients than other methods. On treatment, the blanching led to less retention of micronutrients and more reduction of antinutrients in ANS. Therefore, ISD can be a good method in preserve ANS while retaining nutrients and reducing antinutrients.

Keywords: African nightshade, antinutrients, drying methods, minerals, *Solanum* spp., vitamin C

Contact Address: Marynurce Kazosi, Nelson Mandela African Institution of Science and Technology, Food Biotechnology and Nutritional Sciences, 16106 Arusha, Tanzania, e-mail: marykazosi.mf@gmail.com

Assessment of Traditional Processing Technologies of *Tamarindus indica* in Northern Benin

Rose Estelle Kanfon[1], Flora Josiane Chadare[1], Cokou Pascal Agbangnan[2]

[1]*National University of Agriculture, Laboratory of Food and Bio-Resources Science and Technology and Human Nutrition, Benin*

[2]*University of Abomey-Calavi, Applied Chemistry Studies and Research Laboratory, Benin*

Native plants and fruits have recently gained a lot of attention due to their nutritional and functional potential.This is the case with *Tamarindus indica* which is a nutritious fruit with a variety of uses. It has several appellations depending on the region. The pulp of *Tamarindus indica* remains the most used organ of the species but underutilised despite its interesting nutritional properties. Its high perishability makes it compulsory to process it into different derived products.The objective of this study is to assess traditional technologies for processing *Tamarindus indica* pulp in the municipalities of Bassila, Djougou and Material in northern Benin. Data were collected through a semi-structured survey and sixty-five tamarind processors were interviewed apartment for nine socio-linguistic groups. Descriptive statistics and correspondance analyse were used to link ethnofood knowledge to processes.

It arises from the investigations that, out of the three municipalities surveyed, the ethnic groups producing tamarind drinks are mostly Yoa-Lokpa (42.86 %), Ottamari and related (28.57 %) followed by the Dendi and Yoruba (10 %) of which fifty percent are women. Main derived products from tamarind were nectar, powder and sirup. Citation frequencies from informants showed that amarind nectar is mainly produced by Yoa -Lokpa ethnic group and related (58.79 %) followed by dendi ans related (16.28 %) while Tamarind powder is processed only by Ottamari and related ethnic groups.Tamarind sirup is a product specific to the Yoruba and related socio-lingustic groups of the municipality of Bassila .It is suggested that further characterisation of the identified product is performed for their better valorisation.

Keywords: Benin, Tamarind drinks, tamarind pulp, traditional technologies

Contact Address: Rose Estelle Kanfon, National University of Agriculture, Laboratory of Food and Bio-Resources Science and Technology and Human Nutrition, BP. 774 Sakété, Abomey-Calavi, Benin, e-mail: kanfonrose@gmail.com

Insects as food

Posters

What Role Did Edible Insects Play in the Traditional Practices and Nutrition in Sub-Saharan Africa? A Review on Insect Safety

Carolyne Kipkoech[1], John Mwibanda Wesonga[2], Julia Jaster Keller[1]
[1]*Federal Institute for Risk Assessment, Germany*
[2]*Jomo Kenyatta University of Agriculture and Technology, Kenya*

Insect rearing is gaining popularity worldwide and is likely to contribute to increased nutrient intake in the malnourished population, especially in times of climate change, the intensified use of insects as a source of food and feed is a sensible approach to mitigate the worsening crisis. There is a likelihood that interest in edible insect consumption for food and feed will increase due to its nutritional, nutraceutical, and medicinal potential. Wild harvested insects are traditionally consumed in many parts of the world, and cannot be sustained under the current population pressure. Food security and expensive animal protein sources are a major concern in developing countries and there is a need to search for alternative sources to curb malnutrition. Although nutrition is secured in developed countries, a major concern is food safety and the environmental sustainability of food production. This calls for new ways that can increase the availability of animal protein sources and increase other nutrients such as essential fatty acids and minerals while addressing climate change, food quality, and environmental sustainability. Insects can greatly contribute to providing high-quality animal protein in an ecologically sustainable way. Edible insect rearing is on the rise; however, policies to guide the edible insect industry are not yet fully developed, as many laws governing the use of edible insects are currently in the making. The results of the research already conducted will likely inform policy on the consumption and general adoption of edible insects as a novel food source. ContamInsect project with the support of the German federal ministry of food and agriculture did a review that focuses on the use of edible insects and the safety concerns raised in sub-Saharan Africa. Microbial contaminations in wild-harvested insects were closely linked to poor hygiene conditions, while chemical and physical contaminants were closely linked to the environment and instruments used for insect collection. With the widespread wild harvesting and few legislations that govern the edible insect industry, there is a need for further research to ensure acceptance of edible insects as safe food and feed in Kenya and other regions with a similar setup.

Keywords: Chemical contaminants, insect farming, wild harvesting

Contact Address: Carolyne Kipkoech, Federal Institute for Risk Assessment, Max-Dohrn-Str. 8-10, 10589 Berlin, Germany, e-mail: kipkoechcarolyne@gmail.com

Production of Yellow Meal Worm (*Tenebrio molitor*) (Coleoptera: Tenebrionidae) for Food and Feed in Myanmar

Aye Aye Myint[1], Tin Htut[2]

[1]*Yezin Agricultural University, Department of Animal Science, Myanmar*

[2]*Alliance for Agricultural Research, Rural Development and Advisory Service, Myanmar*

The yellow mealworm *Tenebrio molitor* is economically important edible insect and has been used as an alternative protein source for food and animal feed in Myanmar. Although the demand for edible insect market is increasing, most are seasonally available in the local market. The larvae of this species are often used as pet food, and they offer a promising alternative protein-rich animal feed. In this study, wheat bran (WB), wheat bran with chinese cabbage (WB+CC), rice bran with chinese cabbage (RB + CC) and rice bran (RB) were used to evaluate the growth performances of *Tenebrio molitor* larvae. Weight of larvae,length of larvae, number of dead, number of pupae, duration of pupal stage and pupal survival rate and development time of mealworms fed on different feeds were measured. Among the feeds, mealworm larvae fed on wheat bran with chinese cabbage and rice bran with chinese cabbage showed the heavier larval weight (37 %) and (26 %) than wheat bran (19 %) and rice bran (18 %) only. The highest number of pupae (53 %) was recorded from wheat bran with chinese cabbage and lowest number of pupae (5 %) was collected from rice bran only feed. The maximum survival rate (70 %) was observed from larvae feeding in wheat bran with chinese cabbage followed by wheat bran only (15 %) , rice bran with chinese cabbage(13 %) and lowest survival rate(3 %) was observed from rice bran feed. The shortest developmental time (79 Days) was found from larvae treated with wheat bran and chinese cabbage followed by (100 Days) from rice bran with chinese cabbage and wheat bran feeds and longest developmental time (111 Days) from rice bran only feed. According to this experiment, mealworm larvae fed on wheat bran supplemented with chinese cabbage had improved growth rate and increased the production efficiency. Therefore, our findings could be improved the diet formulation for mealworm mass production.

Keywords: Feed supplementation, growth performance, mass production, mealworm

Contact Address: Aye Aye Myint, Yezin Agricultural University, Department of Animal Science, Yezin, 15013 Zeyathiri Township, Myanmar, e-mail: ayeayemyint2006@gmail.com

Mini-Rearing of Edible *Gryllus bimaculatus* at Farmers' Level in Madagascar to Have Healthy and Sustainable Protein

Razafindrakotomamonjy Andrianantenaina, Ravaomanarivo Lala Harivelo

University of Antananarivo, Dept. of Entomology, Madagascar

In Madagascar, a third of households are in a situation of severe food insecurity and a large majority of the population has an insufficient diet in terms of quantity or quality. According to the National Office of Nutrition, the Amoron'i Mania Region is among the most affected regions with a food insecurity rate of 30 %. Insects can be used as a source of animal protein, the average content of which is 28 g / 100 g compared to meat of 19 g / 100 g of fresh material. Raising insects is much easier and cheaper than other types of farming. Crickets are the most successful example of insect farming for human consumption in the tropics. We chose the two-spotted cricket *Gryllus bimaculatus* which occurs naturally in the region. The objective of the study was to optimise the housing for insect production. Six rearing rooms with a dimension of 2 m × 2.50 m × 2.45 m each were built with specific structures to stabilise the temperature and relative humidity inside, to minimise the temperature variation between day and night, characteristic of a tropical climate. Various feeds were fed to the crickets raised such as pre-prepared foods for poultry and kitchen waste vegetables. The climatic conditions of the room during the experiment are 28 ± 2 °C for temperature, and 82 ± 10 % relative humidity. The results on feeds are: out of the 200 first instar larvae at the start, we obtained 130 to 132 adults with a development period of 30 to 43 days. The average adult fresh weight ranged from 1.06 to 1.50 grams. To sum up, this is a successful production, the room offers the optimal climatic parameters for rearing. The larval survival rate was 66 % which is questionable. Adult developmental period and fresh weight are comparable to previous laboratory results. The adults from this breeding were eaten directly after cooking or dried and processed into powder. The sustainability of this rearing will be ensured by the local availability of feed and the insulation performance of the room.

Keywords: Cricket, food security, insect consumption, production, rearing room

Contact Address: Razafindrakotomamonjy Andrianantenaina, University of Antananarivo, Doctoral School Life Sciences and Environment, National Center for Applied Research for Rural Development (FOFIFA), Antananarivo, Madagascar, e-mail: nantenaina73@gmail.com

Mainstreaming Insects for Food and Feed in Sub-Saharan Africa: Implication for Circular Economy and Green Growth

Zewdu Ayalew Abro[1], Menale Kassie[2], Chrysantus Mbi Tanga[2], Gebeyehu Manie Fetene [3]

[1]*International Center of Insect Physiology and Ecology (icipe), Social Sciences and Impact Assessment Unit, Ethiopia*

[2]*International Center of Insect Physiology and Ecology (icipe), Social Sciences and Impact Assessment Unit, Kenya*

[3]*Addis Ababa University, Dept. of Economics, Ethiopia*

Due to deteriorating natural resources and the threats of climate change, the supply of protein for food and feed is not on par with the growing demand for protein in sub-Saharan Africa. The region's fast-growing cities also produce tonnes of biowastes per annum but with little capability to manage these wastes. To address these problems, producing and processing insects for food and feed is emerging as a novel new economic activity. It can enhance environmental cleanup by recycling biowaste and reducing greenhouse gas emissions. It can increase the supply of proteins for animals and humans at cheaper societal costs and contribute to crop productivity by increasing the supply of quality biofertilisers. It can further create new jobs, increase food security and income, and contribute to sustainable economic growth and development while safeguarding the environment. Despite the benefits of insect production in strengthening circular economy and green growth, the sector remains in its infancy and receiving inadequate support from donors, policymakers, and researchers in sub-Saharan Africa. Our objective in this paper is to provide a unique perspective to stakeholders by reviewing experiences of sub-Saharan Africa in insect production, and the intent of producers and consumers to use edible insects. We will first critically evaluate the policy environment under which researchers and the private sector produce and process edible insects for food and feed. Second, we draw policy lessons by reviewing existing production technologies and practices to ensure food and feed safety and increase productivity and the environmental and socioeconomic benefits of the insect production sector. We aim to stimulate further research and debate on consumers' intent to use insects and their willingness to pay for insects for food and feed, environmental sustainability and greenhouse gas emissions, employment creation, nutrition security, and the value chain of insect products. We believe the study will contribute towards "shifting paradigms in agriculture for a healthy and sustainable future" in line with the Tropentag 2021 theme.

Keywords: Biofertilisers, circular economy, edible insects, food and feed protein, food security, green growth, insect production and processing, job creation

Contact Address: Zewdu Ayalew Abro, International Center of Insect Physiology and Ecology (icipe), Social Sciences and Impact Assessment Unit, Gurdsholla Campus Fogera Building, room number 48, Addis Ababa, Ethiopia, e-mail: zabro@icipe.org

Sensory Quality and Consumer Acceptation of Insect Incorporated Products in the Highlands of Madagascar

Randriamiarana Faniry, Christian Tolojanahary Ratompoarison, Felamboahangy Rasoarahona

University of Antananarivo, College of Agric. Scie., Dept. of Food Science and Technology, Madagascar

Food in Madagascar lacks diversity, especially for rural population. The carbohydrate rich diet is dominated by cereals and tubers constitutes. Protein source as meat and fish hold only a small part of the food ration. Edible insects are popular in rural area of Madagascar but their used is limited in traditional cooking methods like boiling and frying. The possibility of using edible insects as food ingredient was investigated in this study. The locust *Nomadacris septemfasciata* in a powder form was incorporated in two popular processed food products , a snack locally called "caca-pigeons" and French baguette. Locust powder was used in the preparation at rates of 0, 10 and 20 %. Proximate composition of end products was analysed. Sensory test was performed with a survey combinating food neophobia, food disgusted and assessment of the products acceptation. Participants willingness to buy insect-containing foods was surveyd using choice experiments method. Respondents was composed of 35 % from rural area and 65 % from urban area. As results protein and ash content was increased significantly. For 10 % insect incorporated French stick, protein content increased from 7.54 to 11.48 %, the ash from 1.40 to 1.52 %. The protein content of 10 % insect incorporated caca-pigeon increased from 12.81 to 17.46 % and the ash content from 1.66 to 1.85 %. Food neophobia and disgust influence negatively the participants' motivation to buy insect-containing food. Participants' willingness to pay for insect-containing product depends on the nutritional value of the presented product and the recommendation they had from official organisation. The sensory analysis shown that the product enriched with 10 % of locust powder was the most preferred. Food products with locust powder have potential to improve protein intake in Madagascar.

Keywords: Edible insect, food disgust, food neophobia, locust, Madagascar, protein, "caca-pigeon"

Contact Address: Christian Tolojanahary Ratompoarison, University of Antananarivo, College of Agric. Scie., Dept. of Food Science and Technology, Campus Universitaire Ankatso, 101 Antananarivo, Madagascar, e-mail: christianratompoarison@gmail.com

Socio-economics

Crises - poverty - perseverance

Oral Presentations

Posters

Health Shock, Household Consumption, and Over–Indebtedness: Does Having Access to Financial Market Matter?

LINH NGUYEN DUY, TUNG NGUYEN THANH, THANH NGUYEN TRUNG, ULRIKE GROTE

Leibniz University of Hannover, Inst. for Environmental Economics and World Trade, Germany

Dealing with health shocks is not only dealing with health risk but also with financial risk. In low and middle income countries, households may use costly coping strategies such as selling productive assets, lowering education expense, borrowing from moneylender in response to health shocks due to the non-existing or incomplete financial market. In turn, these coping strategies may have negative consequences on household economy in the future. Estimating the impact of health shocks is thus important for designing and promoting proper mitigating instruments. However, this is not trivial task. Health shocks can have very different implications depending on their length and severity. Additionally, the reverse causality between health situation and economic outcomes is also a challenge to address.

This paper examines the economic impact of severe ill health events which were reported as a major income shocks on rural households, and explores the mechanism that health shocks push households into over-indebtedness. Base on that, we investigates the role of credit and health insurance in mitigating the negative effects of health shocks. We employ the heteroscedasticity-based estimator to a four-wave balanced panel data in Vietnam to overcome the endogeneity issue. The results show that health shocks lead to a significant decline in off-farm income and a substantial increase in healthcare expenditure. These losses are then transmitted into consumption and make rural household unable to fully insure non-health consumption. In response to health shocks, households have to increase informal borrowings to smooth consumption, and thus resulting in over-indebtedness. Getting access to formal credit and health insurance helps households lower the decrease of non-health consumption. Moreover, health insurance can reduce the probability of being over-indebted.

As agricultural transformation is a growing process in developing countries, our study suggests that severe illnesses are an important income shock among rural households. Therefore, it is essential to eliminate barriers in access to formal credit to improve household self-insurance capacity. Meanwhile, promotion of healthcare and enhancement of healthcare services provision at local level should go together in order to bring more benefit to insured households and prevent them to fall into financial hardship

Keywords: Consumption insurance, credit access, health insurance, health shocks, heteroscedasticity-based identification, over-indebtedness

---

Contact Address: Linh Nguyen Duy, Leibniz University of Hannover, Inst. for Environmental Economics and World Trade, Königsworther Platz 1, 30167 Hannover, Germany, e-mail: linh.nguyen@iuw.uni-hannover.de

COVID-19 Pandemic: Changing Conditions for Local Food System Actors to Operate Towards Agroecology

Markus Frank[1], Mariano M. Amoroso[2], Brigitte Kaufmann[3]

[1]*German Institute for Tropical and Subtropical Agriculture (DITSL), Germany*
[2]*National University of Rio Negro, Institute for Natural Resources, Agroecology and Rural Development, Argentina*
[3]*German Institute for Tropical and Subtropical Agriculture (DITSL), Germany*

Measures implemented to control the COVID-19 pandemic are challenging global food systems, and weaknesses of prevalent food regimes are evidenced. Responses by local food system actors, such as alterative local marketing and farm-level diversification strategies, indicate that impacts of the pandemic provide new windows of opportunity for transforming local food systems, guided by the principles of agroecology. Agroecological transition studies stress the possible trigger effects of sudden change situations and necessary new conditions for shifts in production, marketing and consumption practices. Studying the diverse changing conditions that frame the individual room to manoeuvre of food actors for changing their actions and strategies is promising to understand what pushes actors to change from the usual. We encompass these emerging questions in our ongoing case study research on exploring transition pathways in mixed smallholder production systems in Northern Patagonia, Argentina. In a first step, in June 2020 we conducted a survey (n=30) to assess how local farmers and processors perceived the changing situation in the early stage of the pandemic (March – June 2020) and what strategies they proposed and implemented to adapt their production and marketing to the changing conditions. In line with other recent studies, our results indicate strong disturbances and impacts perceived by the actors. Immediate responses reveal adaptation strategies in agroecological niches that may strengthen local transition processes, such as developing civic food networks. Based on these results, in a second step, we are currently conducting in-depth interviews and group discussions with local marketing initiatives that were encouraged by the pandemic, to study i) which production and marketing conditions have changed in the course of the pandemic; ii) which principles and properties of the agroecological target system became more relevant for actors to operate; and iii) which capitals and capacities where employed by the actors to implement strategies. By presenting (preliminary) results, our contribution seeks to engage in the discussion about potentials of the current global crisis to facilitate shifting paradigms in localised food systems towards sustainability.

Keywords: Adaptation strategies, agroecology, civic food networks, pandemic, transition

Contact Address: Markus Frank, German Institute for Tropical and Subtropical Agriculture (DITSL), Steinstr. 19, 37213 Witzenhausen, Germany, e-mail: m.frank@ditsl.org

The Effect of COVID-19 on Welfare of Baobab Collectors in Rural Malawi

DENNIS OLUMEH, DAGMAR MITHÖFER

Humboldt Universität zu Berlin, Albrecht Daniel Thaer-Institut für Agrar- und Gartenbauwiss., Germany

Restriction measures of the COVID-19 Pandemic have disrupted food systems and detrimentally affected welfare of smallholder farmers in Sub-Saharan Africa (SSA). Underutilised Plant Species (UPS) are non-trivial in ensuring stability of food systems in SSA. In this context, this study assessed the impact of the Covid-19 Pandemic on business activities, income and food security of baobab collectors in Malawi, using a primary dataset from 864 baobab collectors collected in February-March 2021. Descriptive statistics and bivariate probit model were applied in the empirical analysis. We find many changes in activities of the baobab collectors due to COVID-19. Collectors reported changes in the quantity collected, transportation options and costs (including labour costs). Further, they highlighted changes in the way meetings were conducted with buyers, training officers, and avoiding movements within and outside their villages. We also find selling price differences in different selling outlets in the 2020 season compared to the average of the last two season (2018/19), with prices in the 2020 season being lower for both baobab whole fruit and pulp. With regards to income shock and food security, more than half of the respondents (54 %) experienced income shocks due to COVID-19 and at least 12 % reported worsened food security during the pandemic as compared to the situation pre-covid. Results from the bivariate probit regression show a strong correlation between collectors who suffered income shocks and food insecurity. Collectors who sold larger amounts of baobab were less likely to experience income shocks and worsened food security during COVID-19 pandemic. Male baobab collectors had a higher probability to experience food insecurity and income shocks. Interestingly, access to credit increased collectors' probability of suffering from income shock and food insecurity. To navigate these challenges most collectors adjusted their dietary patterns (59 %) and reduced non-food expenditure (63 %). These findings indicate that the on-going strategies for steering the recovery of food systems post covid-19 should go beyond the conventional food systems by making targeted polices for UPS chains as well.

Keywords: Bbivariate probit model, COVID-19, food security, income shocks

Contact Address: Dennis Olumeh, Humboldt Universität zu Berlin, Albrecht Daniel Thaer-Institut für Agrar- und Gartenbauwiss., Invaliden Str. 42, Berlin, Germany, e-mail: etemesideaty@gmail.com

Political Economy of Lockdowns in Food Insecure Countries

Nikola Blaschke, Regina Birner, Christine Bosch, Line Drescher
University of Hohenheim, Inst. of Agric. Sci. in the Tropics (Hans-Ruthenberg-Institute), Germany

Since January 2020 as a reaction to the COVID-19 pandemic, more and more countries implemented lockdowns to slow the spread of the disease. This happened most strikingly during March, when most countries in the world decided to restrict their population's freedom almost overnight to an extent never seen before in history. While lockdowns, when well designed, can be a meaningful tool to protect the population, some countries implemented lockdowns that had detrimental effects on the livelihoods of poor and food insecure populations.

To gain a better understanding of the reasons for the implementation of lockdown measures, several studies have statistically examined what determinants are able to explain the stringency of a lockdown that a country implemented in the initial phase of the pandemic, such as the level of GDP and democracy, and freedom of press.

A factor that has not yet been considered although it might have a high explanatory power due to its importance for the population is the level of food security in a country: In what relation does it stand with the stringency of implemented lockdowns? This might yield interesting insights especially in combination with the level of democracy in a country, due to the ability of democracies to better act in the interest of their population than authoritarian regimes do.

Therefore, we address this gap by using global data sets such as the Global Hunger Index and the World Governance Indicators, and discuss the relationship between food insecurity and lockdown stringency with the help of descriptive statistics and regression analysis. We then complement the statistical analysis with qualitative insights from selected developing countries' lockdown trajectories and point out the similarities and differences in countries' reaction to the pandemic. We end with discussing what role the level of food insecurity plays for political decisions.

Keywords: COVID-19, democracy, determinants, food security, lockdown stringency, policy response

Contact Address: Nikola Blaschke, University of Hohenheim, Inst. of Agric. Sci. in the Tropics (Hans-Ruthenberg-Institute), 70595 Stuttgart, Germany, e-mail: nikola_blaschke@yahoo.com

Disruptions and Resilience of Agri-Food Value Chains in Light of the COVID-19 Pandemic: Review of Evidence and Implications for Future Responses

Dietmar Stoian[1], Paswel Marenya[2], Jason Donovan[3]

[1]*CIFOR-ICRAF, Sustainable and Equitable Value Chains, Germany*

[2]*International Maize and Wheat Improvement Center (CIMMYT), Kenya*

[3]*International Maize and Wheat Improvement Center (CIMMYT), Mexico*

We reviewed 129 studies published by research and development organisations that examined the effects of the COVID-19 pandemic on agri-food value chains to better understand the state of knowledge and possible gaps in coverage. Relatively well covered were the effects of COVID-19 restrictions at production, retail and consumption levels. In contrast, we found limited evidence on the effects of COVID-19 restrictions in the midstream segments of value chains (processing, wholesale) or as regards the provision of inputs and services. Responses by public and private sector organisations were fairly well covered, as opposed to NGO interventions and their effects. Geographic coverage was strongest for Asia and Africa but much weaker for Latin America and the Caribbean. Value chains of perishable products, such as fruits and vegetables, dairy and other animal products, were more affected than those of more durable products, such as cereals, roots and tubers. We did not find a clear-cut picture as regards the resilience of domestic vis-à-vis global value chains. Gaps in evidence were largely due to insufficient thematic or geographic coverage, not necessarily to limited effects of the pandemic as such. In few cases, baseline data were available against which observed disruptions could be assessed. Pre-existing conditions which may compromise value chain performance in addition to the effects of the pandemic and lockdown measures often went unnoticed. Looking forward, results highlight the need for addressing relevant coverage gaps and knock-on effects, such as yield reduction due to limited use of fertilisers and agrochemicals, demand-supply shifts, business closures, restructuring (reshoring, digitalisation), availability of finance and credit, and impacts on human nutrition and health. We present a conceptual framework for addressing such gaps and effects and conclude with suggestions how governments, businesses and NGOs can prepare better for future crises by increasing the resilience of value chains, or segments thereof, which proved particularly vulnerable to the effects of the pandemic.

Keywords: Public and private policy, research gaps, resilience

Contact Address: Dietmar Stoian, CIFOR-ICRAF, Sustainable and Equitable Value Chains, c/o Global Crop Diversity Trust, Platz der Vereinten Nationen 7, 53113 Bonn, Germany, e-mail: d.stoian@cgiar.org

A Multi-Attribute Model of Smallholder Resilience: COVID-19 Impacts on Cocoa Farmers in Ecuador and Uganda

Moritz Egger, Michael Curran, Lina Tennhardt, Christian Schader

Research Institute of Organic Agriculture (FiBL), Dept. of Socioeconomics, Switzerland

The concept of resilience has gained substantial relevance in a world rocked by the COVID-19 pandemic and governments' heterogeneous responses. Nevertheless, there exists a lack of consensus on how resilience is defined and measured in a comprehensive and meaningful way. We propose a new indicator-based, actor-orientated evaluation model to operationalize the concept at the farm and community level according to the three resilience capacities of Absorption, Adaption, and Transformation. The approach consists of a multi-attribute utility model using 32 resilience indicators organised into a hierarchical, qualitative decision tree, implemented in the DEXi Multi-Attribute Decision Making model. The utility function weights of qualitative attribute combinations were developed and verified by a group of experts from research and practice. We illustrate the model using a selection of relevant indicators from our extensive sustainability data set collected with the SMART Farm Tool from 395 smallholder cocoa farms in Ecuador and Uganda. The data was collected just prior to the COVID-19 outbreak and used to develop a baseline evaluation of resilience capacities for each farm. The evaluation model is calibrated using literature data on the likely impacts of the pandemic combined with sensitivity testing and scenario evaluation. A planned second step of empirical validation is described to compare the predicted resilience capacities of the model with information on actual impacts of the pandemic for each farm. This will provide a unique opportunity to verify and validate the model in a real world setting, which is very rare in the field of applied resilience research. We hypothesise that cocoa farms with high performance in the resilience capacity evaluation were able to cope better with the effects of the COVID-19 pandemic than cocoa farms with low resilience capacity performance.

Keywords: Cocao, COVID-19, DEXi, multi-attribute model, resilience, resilience capacities

Contact Address: Moritz Egger, Research Institute of Organic Agriculture (FiBL), Dept. of Socioeconomics, Bachstrasse 14, 5072 Oeschgen, Switzerland, e-mail: moritz.egger@fibl.org

Direct Marketing Activities during COVID-19 Lockdown and Potential for Rural-Urban Linkages in Bengaluru, India

NEDA YOUSEFIAN[1], M. SOUBADRA DEVY[2], K. GEETHA[3], CHRISTOPH DITTRICH[1]

[1]*Georg-August Universität Göttingen, Inst. of Geography: Human Geography, Germany*

[2]*Ashoka Trust for Research in Ecology and the Environment (ATREE), India*

[3]*University of Agricultural Sciences (GKWK), India*

Rural-urban linkages are vital elements in a sustainable food system. As the COVID-19 pandemic unfolded, supply chains were disrupted and fear of infection impacted food shopping decisions, pushing consumers to seek local and safer options for procuring fresh produce. Direct marketing arose as a promising alternative for both consumers and producers. We undertook a study in Bengaluru, India, in order to understand how direct marketing activities have unfolded with the COVID-19 pandemic. Media reports highlighted the plight of farmers struggling to market their harvest during lockdown as well as the farm to fork initiatives and lockdown farmers markets that were created as a response. This study conducted online and telephone surveys with both consumers and producers in Bengaluru to explore the elements of supply and demand that have fostered and hindered direct marketing schemes. We targeted urban, middle-class consumers for their role in driving consumption trends and their targetability through online data collection methods. We surveyed Resident Welfare Associations to understand their role in fostering direct marketing between urban households and peri-urban farmers. We also surveyed Farmer Producer Organisations to better understand how they pivoted and adapted their marketing activities to the changes caused by the pandemic and lockdown. We found that consumers are interested in sourcing fruits and vegetables directly from farmers, but communication and logistics between consumers and producers are major hindrances. Although producers are diversifying their marketing strategies, operating these different channels at economically viable scales is essential to ensure long-term success. We find that the role of technology, specifically messaging apps, can streamline direct marketing activities and remove the barriers that currently hamper rural-urban linkages. Finally, existing community and farmer organisations have the size and scale to make direct marketing schemes a worthy endeavor for both consumers and producers and bolster the overall sustainability of the local food system.

Keywords: City region food system, producer organisation, supply chains

---

Contact Address: Neda Yousefian, Georg-August Universität Göttingen, Inst. of Geography: Human Geography, Goldschmidtstr 5, 37077 Göttingen, Germany, e-mail: neda.yousefian@uni-goettingen.de

What Makes Me Want You Here? Hosts' Opinion Towards and Contact with Refugees for Integration – An African Settlement Setting

Brigitte Ruesink, Steven Gronau

Leibniz University Hannover, Institute for Environmental Economics and World Trade, Germany

Many of the world's refugees remain in Africa, where they stay long-term mainly in neighbouring countries. Present directions point to integration, in which host society and political surrounding play a key role. The paper aims to investigate how public opinion towards and contact with refugees support integration processes. This is not only an important analysis to manage the refugee crisis but also to use the crisis as an opportunity. With the refugees, additional human resources arrive to support rural development in generally poor hosting areas. We apply the research to a settlement setting in rural Zambia, a recent dataset of 275 households from 2018 and an econometric analysis. This is the first study dealing with a comprehensive set of factors that affect hosts opinion towards and contact with refugees in an African camp context, also with respect to the Comprehensive Refugee Response Framework by the United Nations. Our results show particularly religiosity, trust, life satisfaction, food insecurity, agricultural ownership and natural resource uses of the host society as main factors that need policy consideration for promoting refugee integration. Literature suggests refugees affect their host communities in multiple ways. They can support rural development by increasing human capital in the area but also their effect on the host community's food security is not negligible. Either directly through trade on food and good markets or indirectly by their effect on the economy, refugees can influence the host's nutrition. In our analysis we find no significant effects between contact with refugees and relevant variables like host's land size, livestock possession, savings or food security. However, these variables exhibit a significant effect on the hosts' opinion about refugees, indicating a fear to interact with them. To counter this fear, policies have to be implemented to enable hosts and refugees to achieve unrealised economic opportunities for rural development and increase food security.

Keywords: Africa, contact, host society, integration, opinion, refugees, rural development

---

Contact Address: Brigitte Ruesink, Leibniz University Hannover, Institute for Environmental Economics and World Trade, Königsworther Platz 1, 30167 Hannover, Germany, e-mail: ruesink@iuw.uni-hannover.de

Upgrading Village Chicken Value Chains in Ghana: An Application of Spatial Group Model Building

Dolapo Enahoro[1], Charles Mensah[1], Gregory Cooper[2], Karl Matthew Rich[1]

[1]*International Livestock Research Institute (ILRI), Policies, Institutions and Livelihoods, Ghana*

[2]*University of London, Centre for Develop., Environ. and Policy, United Kingdom*

While village chicken production holds much promise for improved incomes, reduced poverty, and better household nutrition among poor households across Africa, the value chain associated with this poultry sub-sector is hampered by low productivity, the lack of market opportunities, and weak buyer-supplier linkages, among others. These constraints limit the potential for smallholder chicken producers to seize opportunities presented by the growing consumer demand for poultry products in rural and urban areas. Improving the sector and its competitiveness will require approaches that increase production efficiency, reduce retail prices, and maintain good quality of products. Our study investigated the role of aggregation systems for improving efficiency of local chicken value chains and local nutrition, focusing on high chicken-producing areas in Northern Ghana. By consolidating the collection of farm produce at farm-gate, aggregation systems could significantly reduce the marketing costs that farmers would otherwise face in selling products to local markets. Our study applied a systems-thinking approach to investigate whether system characteristics of the village chicken sector are amenable to innovative and nutrition-sensitive marketing programs, identifying potential bottlenecks to uptake. Our method of qualitative system mapping employed innovative adaptations of participatory processes known as group model building to account not only for spatial attributes of the system under investigation, but for challenges associated with in-person focus groups due to Covid-19. A novel, hybrid offline and online approach facilitated interactions with stakeholders in both asynchronous fashion and through targeted, short virtual sessions. A key result emerging from the process was that strong interconnectedness and feedbacks within the system ensure that it may not be possible to improve production quantities and qualities alone (e.g., via the introduction of productivity-enhancing technology), without considering the various constraints imposed by local livelihoods, diseases dynamics (such as Newcastle disease), and marketing processes. As such, policymakers should not expect to be able to simultaneously commercialise the local chicken value chain and improve the availability of produce within local markets without intervening at multiple points within the food system. Our study also gave interesting highlights on facilitating group model building processes in low-resource agricultural systems without the benefit of in-person interactions.

Keywords: Ghana, poultry, spatial group model building, system mapping, value chains

---

Contact Address: Dolapo Enahoro, International Livestock Research Institute (ILRI), Policies, Institutions and Livelihoods, Accra, Ghana, e-mail: d.enahoro@cgiar.org

The Socioeconomic Determinants of Poverty in Saudi Arabia

Miriam Al Lily, Hermann Waibel

Leibniz University Hannover, Development and Agricultural Economics, Germany

The discovery of oil in the 1930s transformed Saudi Arabia from one of the poorest countries in the world to one of the richest. However, the economic prosperity did not improve the lives of all nationals equally. An estimated 20 % of the Saudi population is thought to be excluded from the oil wealth and living in poor conditions. The existing literature has recorded descriptive statistics of poor households without making a comparison to non-poor households. The present research paper seeks to address this limitation by comparing both poor and non-poor households. It applies a binary logistic regression model to analyse several dimensions of poverty; namely demographic, human capital, economic, health and social dimensions. Poverty is conceptualised as relative poverty based on the country's national poverty line of $6 per person per day. The empirical basis of the study is a socioeconomic household survey administered among 496 households in Dammam in 2019. The results show that education and unemployment are crucial determinants of poverty outcomes. In addition, large family sizes combined with the tradition of having a single breadwinner pushes households into poverty. Female-headed households are particularly vulnerable due to their traditional role in Islam. Furthermore, social capital positively impacts households' welfare, whereas being of African descent has a negative influence. On the contrary, health, personal attitudes and being of Bedouin origins are not significant variables in the model. The social welfare system is able to mitigate some of the disadvantages, but not all of them. In the research sample, social welfare payments lift one third of the poor households out of poverty. Of the remaining poor households, some receive no social welfare payments and some do not receive sufficient payments to escape poverty. Based on these findings, a couple of policy implications can be proposed: first, increasing investments in educational support initiatives, especially in poor communities; second, creating more low-skilled jobs for nationals; third, strengthening the role of women in society; fourth, providing family planning awareness campaigns; and fifth, increasing the coverage and amount of social welfare payments.

Keywords: Middle East, poverty determinants, Saudi Arabia, social exclusion

Contact Address: Miriam Al Lily, Leibniz University Hannover, Development and Agricultural Economics, Königsworther Platz 1, 30167 Hannover, Germany, e-mail: miriam.allily@gmail.com

Impacts of Agricultural Upgrading Strategies on Smallholder's Vulnerability to Poverty and Food Insecurity: Panel Evidence from Rural Tanzania

Claudio Paul Ngassa[1], Luitfred Kissoly[2], Anja Fasse[1]

[1] *Weihenstephan-Triesdorf University of Applied Sciences, Environmental and Development Economics, Germany*

[2] *Ardhi University, Economics and Social Studies, Tanzania*

Poverty and food insecurity continue to characterise a considerable number of households in Sub-Saharan Africa. Empirical evidence shows that rural agricultural households constitute majority of the poor and food-insecure. Owing to this, multitudes of agricultural interventions have been implemented to enhance food security and reducing poverty. However, past adoption studies shows huge focus on static welfare analyses with scare evidence on dynamic impacts of technology adoption along the traditional value chain. This study evaluates impacts of adopting upgrading strategies along the traditional value chain on households' vulnerability to poverty and food insecurity. We use a balanced panel data (n=457) obtained from households' survey of 187 adopters and 270 controls collected in 2014, 2016 and 2018 in sub-humid (Morogoro) and semi-arid (Dodoma) regions in Tanzania. We establish households' dynamic poverty and food insecurity status based on the vulnerability as uninsured exposure to risk (VER) approach. We analyse adoption impacts by modelling probabilities of the households' ordered poverty status and food insecurity status using the conditional-mixed-process (CMP) framework with instrumental variables ordered probit model to account for possible self-selection bias. Our results show that adoption of upgrading strategies is associated with decreased probability of households to stay poor or vulnerable and positively influence households' probability to remain non-poor. Further, adoption of upgrading strategies increased the probabilities of extreme food insecure to remain in their category, and increased the probability of the transient food insecure to exit the food vulnerable category. Future research will deal with the question which innovations could improve food security status among the extreme food insecure. These findings imply that adoption of upgrading strategies along the traditional agricultural value chains has substantial benefits on reducing vulnerability among rural households. Policy efforts should therefore focus on continued implementation of agricultural upgrading strategies to improve rural households' welfare.

Keywords: Adoption, food security, households

Contact Address: Claudio Paul Ngassa, Weihenstephan-Triesdorf University of Applied Sciences, Environmental and Development Economics, Am Essigberg 3, 94315 Straubing, Germany, e-mail: claudio.ngassa@hswt.de

Shock Experience, Risk Aversion, and Farm Production: Evidence from Rice Farmers in Thailand

Manh Hung Do, Trung Thanh Nguyen, Ulrike Grote

Leibniz Universität Hannover, Institute for Environmental Economics and World Trade, Germany

Rice is one of the most important crops for food security and rural livelihoods in developing countries. However, the current rice farming practices heavily rely on synthetic fertilisers and pesticides to achieve a higher rice productivity. Although rural households in developing countries are living in a highly vulnerable environment with a wide range of adverse shocks such as weather shocks and pest/disease shocks, there are few papers taking these aspects into account when simultaneously estimating their impacts on risk attitude and crop production. Farmers' behaviour under risks might explain low agricultural productivity and vicious cycles of poverty. Indeed, uncertainties caused by adverse shocks can affect rural households' risk attitude that leads to improper applications of inputs and, therefore, reduce their farming technical efficiency. This paper aims to (i) identify determinants of farmers' risk preferences, and (ii) investigate the influence of risk preferences on input use and farming technical efficiency in the context of weather and pest/disease shocks. We use a panel dataset of 1200 identical rice households collected in 2013 and 2017 from the Thailand – Vietnam Socio-Economic Panel (TVSEP) to address these research problems. The results from fixed-effects estimations with instrumental variables show that fertilisers and pesticides are risk-decreasing inputs in Thailand. In other words, rice farmers, who are more unwilling to take risks, tend to apply more fertilisers and pesticides. Besides, in the context of shocks, farmers appear to apply more inputs, however, these shocks are additional factors that negatively affect farming efficiency of rice production. More importantly, the estimation results from the four-quartile groups of asset values show that the correlation of farmers' risk attitude and technical efficiency are only significant in the two poorer groups. This further implies that more risk-adverse farmers in less wealthy groups overuse inputs and this improper application leads to lower farming efficiency. We suggest that the stimulation of policies on providing production insurance mechanisms and enhancing farmers' awareness of proper application is critical to mitigate adverse impacts of shocks and reduce overuse of chemical inputs.

Keywords: Farming efficiency, input application, rice, risk preferences

Contact Address: Manh Hung Do, Leibniz Universität Hannover, Institute for Environmental Economics and World Trade, Königsworther Platz 1, 30167 Hannover, Germany, e-mail: hung@iuw.uni-hannover.de

Making Smart Choices – Behavioural Traits and Resilience to Environmental Shocks among Farming Households in Thailand

Rasadhika Sharma, Robyn Blake-Rath, Ulrike Grote
Leibniz Universität Hannover, Institute for Environmental Economics and World Trade, Germany

Climate change, in the form of intensified environmental shocks, adds significantly to the existing challenges of small-scale farming households in emerging economies. Households can adopt various response strategies, based on their resilience capacity, to mitigate the impact of shocks on their overall welfare. These can be categorised as absorptive, adaptive or transformative, depending on the intensity of change that they entail. Furthermore, the chosen strategies cannot always be considered as positive as they could negatively impact the household's welfare outcomes in the future. Therefore, understanding the decisions pertaining the household's choice of response strategies demands more attention. Literature identifies household financial capital as an important determinant of its resilience capacity. However, evidence on the role of human capital, especially behavioural traits, is scarce. Additionally, most findings on behavioural traits and resilience are based on data from developed countries and may not hold in the context of emerging economies. Therefore, the aim of this paper is to investigate the role of behavioural traits in the household's choice of response strategies to environmental shocks in rural Thailand. In particular, we examine how behavioural traits influence the decision of households to adopt (i) absorptive, adaptive or transformative and (ii) positive or negative response strategies.
We use primary household level data on around 2000 households from the Thailand Vietnam Socio Economic Panel from 2017 and 2019, in combination with spatial data on rainfall to obtain causal effects. A set of response strategies such as diversification of crop patterns, use of child labour, migration, and selling of productive assets is considered to capture the diverse nature of household responses. After categorising these strategies, behavioural traits measures of openness, conscientiousness, extraversion, agreeableness, and neuroticism (Big Five Model) as well as risk preference and patience are included in the analysis. Seemingly Unrelated Probit regressions are used to estimate both research questions. We expect our results to show a greater role for behavioural traits, especially openness, risk and patience. A better understanding of this decision-making process can aid in designing policies and developing programmes that encourage the choice of smart response strategies among households and promote agricultural resilience.

Keywords: Behavioural traits, environmental shocks, household decision-making, resilience

Contact Address: Robyn Blake-Rath, Leibniz Universität Hannover, Institute for Environmental Economics and World Trade, Königsworther Platz 1, 30167 Hannover, Germany, e-mail: blake-rath@iuw.uni-hannover.de

Risk-Management to Reduce Multidimensional Poverty? Comparative Evidence on the Effects of Crop Diversification on Poverty in Southeast Asia

Eva Seewald, Oliver Schulte, Ulrike Grote

Leibniz Universität Hannover, Institute for Environmental Economics and World Trade, Germany

Vietnam and Thailand are both heavily affected by climate change. According to the Global Climate Risk index, which measures the exposure to extreme weather events, Vietnam and Thailand rank 6th and 9th respectively for the period from 1999–2018. Higher temperatures, greater variability in rainfall patterns and altered growing seasons negatively affect agricultural production and, thus, farmers' income. Crop diversification can help to mitigate income loss from altered weather conditions. The effect of crop diversification on farmers' income and poverty has already been studied by using income-based measures that rely on an assessment of what amount of income would normally be sufficient to meet minimum needs. This study focuses on a multidimensional approach to measure poverty. Multidimensional poverty indices provide a more detailed picture about patterns of poverty. The empirical analyses use a uniquely large and long-term data set of 4,400 rural Vietnamese and Thai households from a socio-economic panel (TVSEP) ranging from 2007 to 2017. This data is used to calculate the multidimensional poverty index as well as the Simpson Index of Diversification to measure crop diversification at the household level. The multidimensional poverty index contains information on health, education, standards of living, and income. For the analysis, a panel data regression method will be implemented using the multi-dimensional poverty index as dependent and the Simpson Index for Diversification as well as self-assessed weather shocks as control variables next to other socio-demographic controls. In order to deal with the endogeneity issue between self-assessed weather shocks and crop diversification as well as the stochastic nature of income, precipitation data is used to construct an index indicating extreme heat or wet seasons. We find a positive impact of crop diversification on poverty, meaning that crop diversification helps not only to reduce monetary poverty but also positively impacts other dimensions of poverty such as health, education, and standards of living. Thus, policy makers should consider an extension of services promoting crop diversification.

Keywords: Income poverty, panel data, precipitation, shocks

Contact Address: Eva Seewald, Leibniz Universität Hannover, Institute for Environmental Economics and World Trade, Königsworther Platz 1, 30167 Hannover, Germany, e-mail: seewald@iuw.uni-hannover.de

The Effect of Risk Perception on the Demand for Index Insurance in Mongolia

Lukas Mogge, Kati Krähnert
Potsdam Institute for Climate Impact Research, Research Department 2: Climate Resilience, Germany

Climate change is increasing the intensity and frequency of extreme weather events. Such extreme weather events impede development, increase the risk of poverty, and widen existing within-country inequalities. Smallholder farm households in low- and middle-income countries that depend on natural resources for their living are particularly affected. Not only are these households more geographically exposed to extreme weather events, they are also less resilient when hit by such shocks.
For rural farm households, access to formal insurance is an important means to adapt to increasing weather risks. Yet, in many low- and middle-income countries, conventional, indemnity-based agricultural insurance failed and insurance markets remain underdeveloped. A potential solution that is discussed with much optimism among policy stakeholders and the academic community alike is index-based weather insurance. Despite high hopes among policymakers, most index insurance programs struggle with low take-up rates. This paper provides novel panel data evidence on how households' risk perception shapes demand for index-based weather insurance. The focus is on Mongolia, where index insurance is offered to pastoralists threatened by extreme weather events that cause high livestock mortality. Using a household fixed effects approach, we show that households are significantly more likely to purchase index insurance when they live in an area exposed to adverse weather conditions in the months preceding the sales period. Similarly, more pessimistic expectations on future weather are associated with higher insurance take-up. As insurance payouts did not play a major role during the study period, we argue that the observed relationship is driven by households updating their risk perception in response to recent weather risks

Keywords: Extreme weather events, index insurance, livestock, risk perception

Contact Address: Lukas Mogge, Potsdam Institute for Climate Impact Research, Research Department 2: Climate Resilience, P.O. Box 60 12 03, 14412 Potsdam, Germany, e-mail: mogge@pik-potsdam.de

Relationship between Farmers' Climate Risk Vulnerability and Food Security Status

Mustapha Yakubu Madaki, Miroslava Bavorova

Czech University of Life Sciences Prague, Fac. of Tropical AgriSciences, Dept. of Economics and Development, Czech Republic

Climate changes are expected to alter pest and disease outbreaks, increase the frequency and severity of droughts and floods, and increase the likelihood of poor yields and crop failure, and livestock mortality. Smallholder farmers in developing countries are one of the most vulnerable social groups to climate change as they lack the resources to adapt. This study analysed the relationship between farmers' vulnerability to climate risk and their food security status based on the survey in Nigeria conducted between October 2020 and February 2021. A multi-stage sampling method was used to select the respondents. First, six states were selected from six Agroecological Zones (AEZs) purposively. Second, two Local Governments Areas (LGAs) were selected from each state chosen. Third, two wards were drawn from each selected LGA and finally, 45 farming households were randomly selected from each selected ward which forms 1,080 households for the study. A structured questionnaire was used for collecting data on farmers' exposure, sensitivity, and impact of climate risk (Climate Risk Vulnerability), Fifteen questions were used to capture the climate risk vulnerability in the last 10 years and quantity of consumed food in the last 7 days (Food Consumption Score) through a face-to-face interview Chi-square test was used to analyse the association between climate risk vulnerability and food security status. Results revealed that more than 65 % of farming households experienced drought, flood, and extreme temperature, shortage of livestock feed, crop pest and disease outbreak, crop yield reduction, and complete crop failure in the last 10 years. A significant statistical association was found between climate risk vulnerability and farmers' food security as 55.7 % of farmers with poor food security status experienced high climate risk while only 13.7 % of the farmers with acceptable food security status experienced high climate vulnerability. Farmers with high climate risk vulnerability should be supported with climate risk adaptation to make their agriculture resilient to the climate risk and in turn, improve their food security.

Keywords: And food security, climate risk, exposure, impact, sensitivity

---

Contact Address: Mustapha Yakubu Madaki, Czech University of Life Sciences Prague, Fac. of Tropical AgriSciences, Dept. of Economics and Development, Kamycká 129, 165 00 Prague 6 - Suchdol, Czech Republic, e-mail: yakubu_madaki@ftz.czu.cz

Strategies to Close the Living Income Gap of Different Coffee-Grower Archetypes in Kaffa Region, Ethiopia

WYCLIFFE ODERA MAGOWA, INGRID FROMM

Bern University of Applied Sciences (BFH), School of Agricultural, Forest and Food Sciences (HAFL), Switzerland

In Ethiopia, small-holder coffee farmers are confronted with numerous challenges as a result of low income from farming. In order, to intervene, it is important to find out the what the farmers earn, their sources of income and challenges facing their livelihoods as far as living income is concerned. This study in the southern part of Ethiopia helps to calculate an estimate of actual income of coffee farmers and compare it to the living income reference value for Ethiopia to establish the living income gap. Expert interviews, focus group discussions and individual farmer interviews were conducted to collect the income data and marketing strategies of the farmers. These farmers were classified into different producer archetypes. The farmer archetypes were defined by the coffee production type of a farmer. The results showed that the small-holder coffee farmers have an income which is below the reference value. The living income gap depends on the archetype a farmer belongs to. There is overdependency on coffee farming among the farmers which increases the risk of huge loss in case of the crop failure or sudden price drop. Results from the discussions bring out various challenges which the farmers are facing in regard to production and marketing of their products. The results from the study proved the need for intervention by various actors to help combat poverty in the region. Due to these challenges associated with coffee farming, it is necessary to promote crop diversification among these farmers. It is also important for the stakeholders in the sector to help the farmers break the long market chain and intervene in providing education on better production systems.

Keywords: Coffee, Ethiopia, income gap, living income, poverty

Contact Address: Wycliffe Odera Magowa, Bern University of Applied Sciences (BFH), School of Agricultural, Forest and Food Sciences (HAFL), Bernstrasse 111, 3052 Bern, Switzerland, e-mail: wycliffeodera.magowa@students.bfh.ch

A Discourse Analysis on COVID-19 Impacts on Food and Nutrition Security in Kenya and Ghana

Jemima Magak, Regina Birner, Christine Bosch
University of Hohenheim, Inst. of Agric. Sci. in the Tropics (Hans-Ruthenberg-Institute), Germany

For most developing countries, food insecurity is a major problem resulting from weak food systems. Majority of their populations' livelihoods were disrupted by COVID-19 through lock-downs and restriction measures put in place by the country governments to control the spread of the pandemic. The impact of COVID-19 was similar to Ebola and the Food Price crisis, and it is not clear what we learnt from these past crises regarding the impacts on food security and nutrition. This led to find out how the organisations reacted similarly or differently to the pandemic and if they learnt from Ebola. Looking at three major agricultural international organisations, the Food and Agricultural Organisation of the United Nations (FAO), the World Food Programme (WFP) and the International Food Policy Research Institute (IFPRI), the study aimed to: (i) identify the role played by governments and international organisations to combat food and nutrition insecurity during the pandemic, (ii) assess the policies adopted to reduce the spread of the Covid-19 and (iii) outline the impacts of COVID-19 on food and nutrition security. This was evaluated by conducting a discourse analysis of the press releases, official statements and newspaper articles from these organisations and the two study countries, Kenya and Ghana. A comparison between presence and absence of other crises, COVID-19 cases and the Stringency Indexes were used as criteria for the country selection. The results show that the pandemic slowed down economic activities which affected peoples' livelihoods through loss of employment and reduced incomes. Movement restrictions affected trade which in turn affected producer incomes and further production, thereby reducing food availability. FAO and IFPRI talked more on the impacts on food availability as compared to WFP which emphasised more on food access. Referring to past crises, these organisations majorly recommended countries to secure trade flows, strengthen social protection mechanisms and improve resilience in the medium and long term. The study concludes that the restriction policies implemented by the countries affected food and nutrition security of the most vulnerable populations. The given organisation recommendations were feasible but rather came after the countries had already implemented the restriction policies.

Keywords: COVID-19, food access, food and nutrition security, food availability, international organisations, restriction policies

Contact Address: Jemima Magak, University of Hohenheim, Elfenstrasse 40, 70567 Stuttgart, Germany, e-mail: jemimamagak@gmail.com

Rice and Bean Consumption in Brazil During COVID-19 Pandemic

Alcido Elenor Wander[1], Cleyzer Adrian Cunha[2]

[1]*Brazilian Agricultural Research Corporation (EMBRAPA), Brazil*
[2]*Federal University of Goiás (UFG), School of Economics, Brazil*

Rice and beans are important staples for food security and nutrition in Brazil. With the Covid-19 pandemic and the restrictions to economic activities, food prices increased and income of socially vulnerable groups decreased. Thus, the objective of this study was to follow the behaviour of the Brazilian consumer of rice and beans during the Covid-19 pandemic. Therefore, an online survey was carried out with 328 consumers from different Brazilian states between November 2020 and January 2021. Information was raised regarding the consumption of rice and beans during the pandemic. The data were submitted to frequency analysis, descriptive statistics, and mean tests. The main results were (a) that during the pandemic in 2020 there were no major difficulties in finding rice and beans to purchase; (b) that the amount of rice and beans consumed in households is equivalent to the pre-pandemic period; (c) that the main reasons for the increase in the prices of rice and beans are related to high demand, increased exports and inelastic supply in the short term; (d) that the most consumed products form polished or white rice, carioca beans, and black beans; (e) that price and brand, in this order, are the main criteria for purchasing rice and beans; (f) that the emergency aid paid by the Federal Government in 2020 did not change the consumption habits of rice and beans; and (g) that the majority of consumers interviewed intend to maintain rice and beans consumption habits in the post-pandemic. Overall, only minor changes in consumption pattern of rice and bean during the pandemic were observed.

Keywords: Consumption habits, demand, household consumption, public policy, supply

Contact Address: Alcido Elenor Wander, Brazilian Agricultural Research Corporation (EMBRAPA), Rodovia GO-462, km 12, 75375-000 Santo Antonio de Goias, Brazil, e-mail: alcido.wander@embrapa.br

Gender relations and power in agricultural production

Power Relations and Cooperation in Polygynous Households in Rural Burkina Faso

VIVIANE YAMEOGO, THOMAS DAUM, REGINA BIRNER

University of Hohenheim, Inst. of Agric. Sci. in the Tropics (Hans-Ruthenberg-Institute), Germany

How best can interventions be implemented to ensure intrahousehold cooperation for improved productivity and food security? Actors within the gender and rural development space continue to grapple with this question in an attempt to ensure that their interventions do not deepen inequalities but foster cooperation that yields optimal and equitable benefits for all household members. While literature has focused on the outcomes and processes of intrahousehold cooperation, insights into why and under which conditions household members cooperate are rare, especially for polygamous households for which the prevailing orthodoxy is that such households are conflict-ridden. To explain why cooperation occurs in some households, and not in others, this paper examines the intrahousehold power dynamics, the nature of the collective dilemmas, and the institutional arrangements that shape intrahousehold interactions. The Institutional Analysis and Development framework was applied as an analytical tool to study collective action among the Fulani and the Mossi polygynous farming households in Burkina Faso. Data were collected through ethnographic instruments, including participant observation, supplemented with focus group discussions, net-map exercises, and in-depth interviews. The study found that intrahousehold cooperation is contingent upon the nature of the problems to resolve, the transaction costs, and trade-offs involved in performing joint endeavours. The results demonstrated that norms and rules, through rewards and sanctions, have the power to articulate intrahousehold collective action. Furthermore, the relations of production, subsumed in the conjugal and inter-generational contracts,define the rights and obligations of each household member, with implications for cooperation and division of labour. Projects that aim at enhancing productivity and ensuring the adoption of improved technologies should be wary of the intrahousehold dynamics and their potential effects on the likelihood of households embracing innovative tools for increased productivity and food security. The results also point to the importance of equity concerns in tailoring policy interventions in rural areas.

Keywords: Cooperation, Fulani, IAD, institutional arrangements, intrahousehold, Mossi, polygyny

---

Contact Address: Viviane Yameogo, University of Hohenheim, Inst. of Agric. Sci. in the Tropics (Hans-Ruthenberg-Institute), 70593 Stuttgart, Germany, e-mail: guesy2003@yahoo.fr

Gender Dynamics and Food Security in the Kenyan African Indigenous Vegetables Supply Chain

Luzia Deissler[1], Henning Krause[2], Ulrike Grote[1]

[1]*Leibniz University Hannover, Institute for Environmental Economics and World Trade, Germany*

[2]*Lower Saxony Chamber of Agriculture, Horticulture Business Devision, Germany*

In Sub Saharan Africa, producing and selling food crops, such as African Indigenous Vegetables, has traditionally been under the control of women. However, in the last decade food crop production has become more commercialised. The process of commercialisation generally results in an increasing engagement of male farmers. At the same time, the bargaining power of women within households may change with commercialisation. This paper aims at analysing the distribution of work among gender in the African Indigenous Vegetables value chain and the intra-household decision-making and bargaining dynamics of small-scale farming, rural households. The analyses are based on 570 small-scale producers of African Indigenous Vegetables in rural and peri-urban Kenya. We investigate factors that are influencing the female intra-household bargaining power and evaluate its impact on household welfare and food security, including numerous indicators. Our results show that most of the work in the African Indigenous Vegetables value chain is still done by women, irrespective of the degree of commercialisation. The multidimensional logit regression reveals that female bargaining power is negatively influenced by an increasing farm size, a higher asset score of the household and female off-farm work. A positive impact was observed by tertiary female education, female landownership, male off-farm work, a high share of female income in the overall household income and the location of a household in Nakuru. With Propensity Score Matching, we find that increased female intra-household bargaining power has no significant influence on household expenditure, but some inconsistent influence on the food security of a household.

Keywords: Decision-making, food security, gender inequality, intra-household dynamics, welfare

Contact Address: Luzia Deißler, Leibniz University Hannover, Institute for Environmental Economics and World Trade, Königsworther Platz 1, 30167 Hannover, Germany, e-mail: deissler@iuw.uni-hannover.de

Gender Differences in Access to Information and Adoption of Climate-smart Agriculture Practices in Uganda

Rashid Khan[1], Christine Bosch[1], Edward Kato[2], Elizabeth Bryan[3], Regina Birner[1]

[1]*University of Hohenheim, Inst. of Agric. Sci. in the Tropics (Hans-Ruthenberg-Institute), Germany*

[2]*International Food Policy Research Institute (IFPRI), Uganda*

[3]*International Food Policy Research Institute (IFPRI), United States*

Agriculture in developing countries is facing the twin challenges of climate change while there is still the need to increasing yields and thereby contributing to farmers' incomes. Climate-smart agriculture (CSA) practices offer farmers the means to increase productivity and profitability, adapt to the negative effects of climate change, and mitigate greenhouse gasses. One enabling factor for the adoption of CSA practices that has been mentioned is access to agricultural and climate information. Adoption of CSA practices is still low. Women often face greater constraints in accessing information, which has led to lower adoption of climate-smart practices, relative to men. Therefore, women often have less access to information than men on a range of productivity-enhancing technologies and practices, which obviously limits their participation in household decision-making on agricultural production. Due to less access to information, and different roles in agriculture, women are less aware of climate-smart agriculture practices. This paper discusses the types of agricultural and climate information and extension services that Ugandan farmers access, and then examines the gendered differences in the level of awareness and adoption of climate-smart agricultural practices using intrahousehold survey data from 720 households collected in 2020. It examines the determinants of awareness and adoption, focusing on the extent to which indicators of women's empowerment influence access to information, awareness, and adoption of climate-smart agricultural practices. Due to selection bias, we will run a Heckman model results on awareness and adoption. Sub-indicators and aggregate measures from the abbreviated Women's Empowerment in Agriculture Index (A-WEAI) are used to measure empowerment. The expected results could be useful for increasing smallholder women farmers' access to information and adoption of climate-smart practices in the Ugandan context.

Keywords: Adoption, climate-smart agriculture, empowerment, information

Contact Address: Rashid Khan, University of Hohenheim, Inst. of Agric. Sci. in the Tropics (Hans-Ruthenberg-Institute), 70593 Stuttgart, Germany, e-mail: rkhan.p@gmail.com

A Gendered Analysis of Small-Scale Cocoa Production in Uganda

MICHAELA KUHN[1], LINA TENNHARDT[2]

[1]*University of Göttingen, Dept. of Agricultural Economics and Rural Development, Germany*

[2]*Research Institute of Organic Agriculture (FiBL), Dept. of Socioeconomics, Switzerland*

Agriculture is an important accelerator for economic growth, food security, and poverty alleviation in many developing countries. In the specific case of the cocoa sector, which underwent a rapid transformation in recent years due to the steadily increasing demand for cocoa beans, the majority of smallholder cocoa farmers live below the international poverty line. The sector does not exploit its full potential because, amongst others, female farmers, who make up a large proportion of farm managers, provide a notable amount of agricultural labour and contribute to the rural economy, face multifaceted constraints that reduce their productivity. Therefore, female farmers can be identified as the group that is largely missing out on the positive development and empowerment due to cultural, social and, institutional gender-based disparities. This research uses primary cross-sectional survey data of smallholder cocoa farmers in Uganda to investigate a potential gender gap based on a holistic statistical approach. It first draws on empirical evidence to what degree women participate in agriculture. Subsequently, the data analysis reveals that the sample's female farmers are disadvantaged in various key aspects of farming, such as access to land, credit, training, and other productive resources. In addition, there are differences in the role distribution and decision-making between male and female managed farms, where women are generally excluded from input decisions and female farm managers are dependent on a female workforce. Furthermore, regression models confirm a gender gap in cocoa revenue generation for the sample group. A formal bank account, a greater workforce and a larger cocoa area can be identified as the key determinants that significantly influence revenue. These properties represent areas where the female farmers of the sample are at a comparative detriment to their male counterparts.

Keywords: Cocoa production, female empowerment, gender, revenue generation, smallholder agriculture, Uganda

Contact Address: Michaela Kuhn, University of Göttingen, Dept. of Agricultural Economics and Rural Development, Platz der Göttinger Sieben 5, Göttingen, Germany, e-mail: info@michaela-kuhn.com

Suicide and Suicide Attempts from Pesticides: A Major Problem for Women in the Amhara Region, Ethiopia

BIRTUKAN ASMARE[1], BERNHARD FREYER[2], JIM BINGEN[3]

[1]*University of Natural Resources and Life Sciences, Vienna (BOKU), Social and Economic Sciences, Austria*

[2]*University of Natural Resources and Life Sciences, Vienna (BOKU), Div. of Organic Farming, Austria*

[3]*Michigan State University, Dept. of Community Sustainability, United States of America*

Suicide and suicide attempts by pesticides remain alarming in the Amhara region of Ethiopia. This study was based up on qualitative data that combines in-depth interviews and focus group discussion with women smallholder farmers. The easy availability of pesticides through illegal parallel markets, unlabeled smaller containers and storage inside home induced suicidal behavior that was triggered by a wide range of corona-virus pandemic behaviors that creates isolation, rape, unwanted pregnancies, and the disruption of livelihoods. The interview reveals that the major pesticides responsible for many of the deaths and attempted suicides were rodenticides (Zinc Phosphide), fumigant (Aluminum Phosphide) and insecticides (Endosulfan, Dimethoate, Chlorpyrifos), while some survivors were not able to recall the name of pesticides used for committing suicide. Many of the pesticides used stored inside each home/fields. Drinking soap, cow-dung and yogurt were used to remove the ingested pesticides from the stomach. For cultural and religious reasons, many cases of suicide and attempts were hidden. Survivors are often excluded from society that also denies them marriage and other social relations. In order to reduce the incidence of suicide and suicide attempts, this study suggests i) bio-pesticides that are less hazardous, ii) local pesticide market control, iii) public education on pesticide hazards, iv) well-equipped health services and facilities that diagnose poisoning symptoms.

Keywords: Ethiopia, pesticides, rural women, suicide, suicide attempts

Contact Address: Birtukan Asmare, University of Natural Resources and Life Sciences, Vienna (BOKU), Social and Economic Sciences, Vienna, Austria, e-mail: birtukan.asmare@students.boku.ac.at

Gender Inequalities in Cocoa Farming and Farmland Ownership in Ghana

Kwabena Buabeng, Katharina Löhr, Stefan Sieber
Leibniz Centre for Agric. Landscape Res. (ZALF), Sustainable Land Use in Developing Countries" (SusLAND), Germany

Women produce more than fifty percent of the food grown worldwide. In Ghana, like in many other cocoa-producing countries, to be recognised as a cocoa farmer is primarily associated with land ownership. As land titles mostly belong to men, the role of women in agricultural production remains unrecognised. Traditional patriarchal roles have favoured men and promoted structural gender inequalities in terms of decision-making, access to and control over land for agricultural activities, and the ability to engage in more productive activities. This study assesses the role of women in the case of cocoa production and their access to cocoa farmlands. Accurate data on gender roles in cocoa production is scarce and thus makes it challenging to assess the effectiveness of gender strategies and policies on women empowerment. Drawing from the Capabilities and Vulnerabilities Analysis framework, the study aims to assess the differences between male and female participation in cocoa farming activities, focusing on the types of farm activities women participate in, why they participate, and how they significantly differ in terms of the challenges they face on the farmlands. Following the Capabilities and Analysis framework theory, a semi-structured questionnaire will be administered to farmers in the deciduous rainforest zones in the Ashanti, Western and Eastern Regions of Ghana and analysed by descriptive statistics and nonparametric statistical tests. The findings to be presented will bridge the research and policy gap of gender-sensitive research designs with appropriate sampling, which accounts for the heterogeneity among women involved in cocoa farming and, most importantly, includes hard-to-reach and especially disempowered women. The approach seeks to generate more reliable data based on the entire network of labour and kinship relations surrounding a cocoa farm and not just the household to gain a deeper understanding of the power relations revolving around cocoa production. This will allow identifying ways to effectively engage with all women involved in cocoa farming while assessing possible women-specific risks and vulnerabilities.

Keywords: Gender equality, Land ownership, Participatory resource management

Contact Address: Kwabena Buabeng, Leibniz Centre for Agric. Landscape Res. (ZALF), Sustainable Land Use in Developing Countries" (SusLAND), Regatttastrasse 62, 12527 Berlin, Germany, e-mail: buabengk@gmail.com

Climate Change' Beliefs and Risk Perception: A Gender Perspective from Iran

Ameneh Savari Mombeni[1], Masoud Yazdanpanah[1], Moslem Savari[1], Tahereh Zobeidi[2], Stefan Sieber[3]

[1]*Agricultural Sciences and Natural Resources University of Khuzestan, Iran*

[2]*University of Zanjan, Dept. of Agricultural Extension, Communication and Rural Development, Iran*

[3]*Leibniz Centre for Agric. Landscape Res. (ZALF), Sustainable Land Use in Developing Countries (SusLAND), Germany*

Climate change' belief and risk perception are the preconditions for farmers' adaptation response. Therefore, identify the perceptual and cognitive processes of farmers is very important to encourage adaptation to climate change. There is an assumption that men and women have different perceptions and consequently adaptive behaviours. However, there is so far little literature that addresses the gender differences in beliefs and risk perceptions of climate change in rural areas particularly south west of Asia. This study uses a qualitative approach to compare belief in climate change and perceived risk among men and women farmers. The research population consisted of farmers in Baghmalek county of Khuzestan in southwestern Iran. Semi-structured interviews were made with 18 men and 15 women farmers who were purposefully selected. The results showed farmers all acknowledged that climate change had occurred in their area. Both men and women believed reduced rainfall (20 respondents) and increased temperature (18 respondents) as the main signs of climate change. Men cite increased pests and plant diseases, increased drought, and drying up river water as other signs of climate change. While women consider the loss of vegetation in the region and the drying up of groundwater (wells, aqueducts, springs) as signs of climate change. In addition, climate change risk perception has varied by gender. Women perceived social problems such as loss of agricultural jobs (8 respondents), migration (6 respondents), and health (4 respondents), however, men mainly refer to livelihood and economic risks such as reduced access to healthy food (10 respondents), reduced agricultural yields (7 respondents) and the increase of diseases (6 respondents). It seems, most men and women perception the occurrence and risks of climate change. Although there are similarities in the perceptions of the two groups, female farmers were mostly afraid of unemployment and the loss of agriculture, while men were concerned about the probability of declining incomes. These results can be used as a basis for developing appropriate interventions to adapt to climate change in the agriculture sector.

Keywords: Adaptation, belief, climate change, qualitative method, risk perception, rural women

Contact Address: Masoud Yazdanpanah, Agricultural Sciences and Natural Resources University of Khuzestan, Mollasani, 744581 Ahvaz, Iran, e-mail: masoudyazdan@gmail.com

The Role of Women in Beekeeping Activities and the Contribution of Bee-Wax and Honey Production for Livelihood Improvement

Tariku Olana Jawo

Czech University of Life Sciences Prague, Fac. of Tropical AgriSciences, Dept. of Crop Sciences and Agroforestry, Czech Republic

The study was conducted in Arsi Negelle District of Oromia Regional State, Ethiopia, to assess the role of women in beekeeping activities and the contribution of honey production for household livelihood improvement. Using a purposive sampling technique, 90 households were included in the survey. A combination of RRA tools (key informant interview, in-depth semi-structured interview, group discussion and observation) were employed to collect primary data from beekeepers and peasant associations. The result revealed that the main purpose of keeping honey bees was for both income generation and household consumption. The number of beehives/colony owned by the bee-keepers in traditional, transitional and modern beehives was 81.7 %, 12.3 % and 6.01 %, respectively. This indicated that the majority of the farmers in the study area depends on traditional methods of honey production systems. The average honey yield per year/colony was 6.09±0.35, 12.7 ±0.62 and 19.7±0.67 kg for traditional, transitional and modern beehives, respectively. Even though women participation in beekeeping activities was low (34.6 %) as compared to men (65.4 %), it was promising. Absconding, pesticides and herbicides application, honey-bee pests (ants, wax moth (*Galleria mellonena*), lizard), high cost of modern beehives, shortage of improved bee forage, lack of beekeeping equipment, dependence on a traditional production system and lack of credit access were the main constraints in beekeeping development in the area. Among the beekeeping constraints and/or threats in the study area absconding, pesticides and herbicides application and shortage of bee forage during the dry seasons were the most pertinent factors accounting for 19 %, 11 % and 9 % of the sample respondents, respectively.

Keywords: Beehives, extension, gender, honey-bee, livelihood

Contact Address: Tariku Olana Jawo, Czech University of Life Sciences Prague, Fac. of Tropical AgriSciences, Dept. of Crop Sciences and Agroforestry, Kamycka 129, 165 00 Prague, Czech Republic, e-mail: jawo@ftz.czu.cz

Women Empowerment and Intra-Household Nutritious Food Distribution and Consumption in Crop-Livestock Production Systems: Empirical Evidence from Bangladesh

Fatema Sarker, Thomas Daum, Regina Birner

University of Hohenheim, Inst. of Agric. Sci. in the Tropics (Hans-Ruthenberg-Institute), Germany

Despite world leaders' commitment to end hunger by 2030, malnourishment remains rather high in many developing countries. While in South Asia, there is progress in food security, health services, and some other development indicators, but malnourishment is still high, which researchers called as 'Asian enigma of malnutrition.' Women's disempowered situation is the root cause of these contradictory findings. Focusing on this fact, some studies posit that empowering rural women through livestock interventions can set them on the path to better achieve nutritional outcomes within their households. However, it is unclear how 'the triple linkage' of livestock-empowerment-nutrition unfolds in reality and how it shapes the intra-household nutritious food consumption where discriminations against girls in food allocation are largely set in literature. This study explores 'the triple linkage' within villages in rural Bangladesh that have adopted livestock rearing as a means to their empowerment, adopting a mixed-method study approach. Quantitative data was collected from 287 farmers along four dimensions: livestock production decision-making, marketing of livestock products, access and control over income from livestock, and decision-making over food choices. We found that the children and partners of empowered women by the livestock intervention had better protein intake with a reduction in the women's own protein intake and the protein food intake of the girls and boys from the household with higher empowerment level of women is much equal than the others. The results from the 23 gender-disaggregated focus group discussions revealed that livestock farming has contributed to the milk intake of every household member considerably. The socio-cultural norms, strong patriarchal influence, poor economic conditions, a large number of family members, poor participation in training or social groups are the reasons for unequal food distribution. While livestock interventions may not be necessary upset gender norms, it imposes new labour demands on women, with negative implications for their nutrition. Development agencies need to implement safeguards to mitigate this negative spillover.

Keywords: Livestock, nutrition, women's empowerment

Contact Address: Fatema Sarker, University of Hohenheim, Agricultural Science in the Tropics and Subtropics, Stuttgart, Germany, e-mail: fatema.sau@gmail.com

Impact of Income on Food and Nutrition Security: A Case of Youth Employment Promotion Program in Sierra Leone

Lucy Apiyo Adundo[1], Martin Petrick[2], Eleonore Heil[3]

[1]*Justus-Liebig University, Center for International Development and Environmental Research, Germany*

[2]*Justus-Liebig University, Inst. for Agricultural Policy and Market Research, Germany*

[3]*Justus-Liebig University, Dept. of Consumer Research, Communication and Food Sociology, Germany*

The agricultural sector plays an important role in food security and the economic well-being of most developing countries. The sector contributes 15 percent of the total Gross Domestic Product (GDP) in Sub-Saharan Africa. Agriculture is also the primary source of livelihood for 10 to 25 percent of urban households. Youth employment in connection with agricultural transformation in developing countries in Africa has gained attention in recent years as one way to promote youth involvement towards sustainable improvement of their livelihoods. Income is likely to influence dietary diversity if a household share of food expenditure is significant or if the income is used to source factors of production to diversify their production. Subsequently, Sierra Leone is a country with a young population, and facilitating youth participation in agriculture has the potential to drive widespread poverty reduction among youths and adults. This study evaluated the influence of the youth employment promotion (Business Loop) programme on food and nutrition security in three districts of Sierra Leone. Minimum Dietary Diversity Score (MDD-S) 24-Hour recall indicator was chosen for this study because it is designed for settings where other dietary diversity assessment methods are unfeasible in developing countries. Random quantitative and qualitative cross-sectional data were collected from 134 youth agricultural entrepreneurs where the questionnaire considered fourteen compulsory and three optional food groups adapted to local food items during the survey. Ordinary Least Square (OLS) regression showed that monthly income (sales revenue) had a significant net effect on MDDS. This means that the entrepreneurs spend a significant proportion of their income on food as confirmed during the Focus Group Discussions. Income increased with the probability that a respondent was female although 10 percent of the female population achieved an MDD-S between 1 – 4 showing a lower micronutrient intake. Even though an increase in income is crucial to improving dietary diversity in developing countries, outcomes can be more visible if programs focus on women inclusion and enhancing food and nutrition knowledge.

Keywords: Dietary diversity, income, youth employment program

Contact Address: Lucy Apiyo Adundo, Justus-Liebig University, Center for International Development and Environmental Research, Senkenbergstrasse 3, 35390 Gießen, Germany, e-mail: lucy.apiyo@outlook.de

The Gendered Yield Gap and Women's Empowerment: Evidence from Smallholder Farmers in Uganda's Central Region

Lukas Welk[1], Christine Bosch[1], Regina Birner[1], Elizabeth Bryan[2], Edward Kato[3]

[1]*University of Hohenheim, Inst. of Agric. Sci. in the Tropics (Hans-Ruthenberg-Institute), Germany*
[2]*International Food Policy Research Institute (IFPRI), United States*
[3]*International Food Policy Research Institute (IFPRI), Uganda*

Despite a substantial increase in agricultural productivity in developing countries during the past decades, there is evidence of a significant gap between men's and women's agricultural productivity, estimated at roughly 25 %. There are still knowledge gaps on the determinants of this gap and how it can be closed. There is particularly little information on how women's empowerment influences agricultural productivity and yields. This study tries to contribute to filling this knowledge gap by using quantitative survey data collected for the project 'Reaching Smallholder Women with Information Services and Resilience Strategies to Respond to Climate Change' led by the International Food Policy Research Institute (IFPRI). The objective of the proposed study is to understand the determinants of gendered differences in agricultural productivity. By exploring potential linkages between women's empowerment and gendered yield gaps, the study aims to identify which indicators of women's empowerment have a significant potential to narrowing the agricultural productivity gender gaps. The evidence will be drawn from an empirical analysis of a recent intrahousehold survey conducted in Uganda. To measure empowerment the Abbreviated Women's Empowerment in Agriculture Index (A-WEAI) developed by IFPRI is used. It consists of 5 domains of empowerment, including control over use of income and workload. Additionally, descriptive statistics will be used to show the yield and productivity differences between men and women cultivated plots as well as other potential relevant explanatory variables. To measure the impact of women's empowerment and potential influencing variables a Kitagawa-Oaxaca-Blinder decomposition will be conducted. In this model, the gendered yield gap is decomposed into differences in the mean values of the endowments of the two groups and the group differences in the returns to these endowments. Results are expected to differ from similar and previous approaches, as gender differences are considered in greater detail, utilising several measures from the A-WEAI. This will provide in-depth understanding of the influence of women's agency on yield and productivity gaps. The results may not only help to identify reasons for the productive and yield gaps between men and women but also help in finding solutions to reducing the gap and increasing the productivity.

Keywords: Gender yield gap, Kitagawa-Oaxaca-Blinder-decomposition, women's empowerment in agriculture

Contact Address: Lukas Welk, University of Hohenheim, Salbeiweg 16, Stuttgart, Germany, e-mail: lukas.welk@uni-hohenheim.de

Participatory *ex-ante* Impact Assessment of Nutrition-Sensitive Interventions in Madagascar: Differences by Gender and Location

Sarah Tojo Mandaharisoa[1], Jonathan Steinke[2], Christoph Kubitza[2], Narilala Randrianarison[1], Arielle Sandrine Rafanomezantsoa[1], Denis Randriamampionona[1], Harilala Andriamaniraka[1], Stefan Sieber[3,2]

[1]*University of Antananarivo, Tropical Agriculture and Sustainable Development Dept., Madagascar*

[2]*Humboldt-Universität zu Berlin, Thaer-Institute of Agricultural and Horticultural Sciences, Germany*

[3]*Leibniz Centre for Agric. Landscape Res. (ZALF), Sustainable Land Use in Developing Countries (SusLAND), Germany*

Participatory ex-ante impact assessment of intended interventions can support the successful implementation of development projects. It can help to design and target interventions according to beneficiaries' needs, especially where high diversity both in wealth and social norms are observed. In the South East of Madagascar, 11 gender-disaggregated workshops were organised at the end of the second lean period, each with six to seven future beneficiaries of a nutrition-sensitive intervention project. The 80 participants first ranked eight predefined impact criteria by their perceived importance. Second, the expected impacts of 14 proposed interventions on each food security criterion were rated. Third, participants selected their preferred interventions from among the proposed interventions. The results show that men and women prioritise the same impact criteria, especially, increased income and increased food self-sufficiency. In the same way, for both, most positive assessments are observed with kitchen gardens and poultry. However, both gender and location influence the perception of project impacts. Women expect more positive impacts with voucher distribution, whereas men prioritise cash crop and poultry activities. Participants from the coastal region positively assess VSLA and farmer's organisation promotion, unlike those from the inland, who support interventions more related to production such as kitchen garden, farmer field school, and food storage. Thus, overall, men and women have different preferences regarding the specific interventions to achieve food security. Compared to men, women are more interested in the promotion of diet diversification and hygiene sensitisation. Men mainly choose technical interventions such as farmer field school, and storage and transformation training. Nevertheless, support to poultry production and farmer field schools are among the most attractive interventions for both genders. Our results, revealing the perception of beneficiaries, are useful to inform the design and targeting of nutrition-sensitive interventions to different target groups.

Keywords: Beneficiaries' perception, criteria ranking, food security interventions, impact rating, Madagascar, participatory assessment

Contact Address: Sarah Tojo Mandaharisoa, University of Antananarivo, Tropical Agriculture and Sustainable Development Dept., Antananarivo, Madagascar, e-mail: tojmands@yahoo.fr

System Approaches for Understanding Rural Women's Agri-Food Entrepreneurship in Oyo State Nigeria

Ugochi Geraldine Akalonu[1,3], Simon Oyegbile[2,3], Oluwakemi Omowaye[3], Margareta Lelea[1,3], Oliver Hensel[1], Brigitte Kaufmann[4,3]

[1]*University of Kassel, Fac. of Organic Agriculture, Germany*
[2]*Forestry Research Institute of Nigeria (FRIN), Nigeria*
[3]*German Institute for Tropical and Subtropical Agriculture (DITSL), Germany*
[4]*University of Hohenheim, Inst. of Agric. Sci. in the Tropics (Hans-Ruthenberg-Institute), Germany*

Women in rural communities are at the forefront of harsh economic impacts. Women take care of the home and are entrusted with household food and nutrition requirements. If equipped with the right resources such as sufficient income, women work to make sure their families are food secure. To this end, rural women engage in entrepreneurial activities such as food processing and trade as a means of earning additional income. Policies and intervention programs designed to enhance the entrepreneurial capacities of these rural women mostly take a reductionist approach and use standard recommendations and thereby fail to understand the full context and motivations behind these entrepreneurial decisions. This also leads to a lack of understanding of how the business functions under the prevailing and often adverse conditions. This research was carried out with the Osanetu women's group in Oyo State, Nigeria. The purpose of this research is to employ a systems approach for the analysis of rural small scale agri-food entrepreneurship including Participatory Monitoring and Evaluation (PME) to involve the women in analysis. The essence is to deeply understand the motivations behind their entrepreneurial engagement and how disturbances affecting their processing are managed. We used a qualitative research design incorporating participatory methods and tools. We employed adapted second-order cybernetics as a system approach to understand fully why these entrepreneurs do what they do.
Findings revealed that these rural women were extrinsically motivated, opportunity-, and socially-driven entrepreneurs. Through constant observation of their production system, these women entrepreneurs regulate/control their group business production processes, observed disturbances, and proffered context-specific solutions to counter these disturbances. PME proved valuable as a method that helped the women to systematically observe their production processes through monitoring and hence to gain better knowledge about them. This research will help policymakers and change agents to better tailor their strategies to enhance the entrepreneurial capacities of rural women entrepreneurs. Using a systems approach to understand rural entrepreneurship can lead to the creation of more profitable and sustainable businesses in rural communities.

Keywords: Entrepreneurship, monitoring & evaluation, motivation, participatory, second-order cybernetics, systems approach

Contact Address: Ugochi Geraldine Akalonu, University of Kassel, Fac. of Organic Agriculture, Witzenhausen, Germany, e-mail: ugochi.akalonu@yahoo.com

Gendered Perceptions in Maize Supply Chains: Evidence from Uganda

Anusha De[1], Bjorn Van Campenhout[2]

[1] *KU Leuven, LICOS Centre for Institutions and Economic Performance (FEB), Belgium*
[2] *International Food Policy Research Institute, Development Strategy and Governance Division, Uganda*

Faced with imperfect information, economic actors use judgment and perceptions in decision-making. Inaccurate perceptions or false beliefs may result in inefficient value chains and systematic bias in perceptions may affect inclusiveness. In this paper, we study perceptions in Ugandan maize supply chains. A random sample of maize farmers was asked to rate other value chain actors – agro-input dealers, assembly traders, and maize millers – on a set of important attributes such as service quality, price competitiveness, ease of access, and overall reputation. These other value chain actors are tracked and asked to assess themselves (provide self-ratings) on the same attributes. These ratings are used as a proxy measure for perceptions and beliefs. Interestingly, price competitiveness is scored the lowest and we find that input dealers, traders, and millers assess themselves more favourably than farmers do. Traditionally, women have not been playing a significant role in the input dealing, processing, and trading activities for maize in Uganda. Systemic gender bias in perceptions can demotivate the entry of women into the food supply chains, creating better opportunities for men to flourish in these industries. Issues like discrimination against women, barriers to market access for women, mistrust, lower credit availabilities, etc. are some of the consequences of such gender bias in perceptions. Thus, we also zoom in on heterogeneity in perceptions related to gender and find that women rate higher than men. The sex of the actor being rated does not affect the rating and gender-based homophily is not present in the perceptions of the farmers.

Keywords: Gender, maize supply chain, perceptions, ratings

Contact Address: Anusha De, KU Leuven, LICOS Centre for Institutions and Economic Performance (FEB), Waaistraat 6 Bus 3511, 3000 Leuven, Belgium, e-mail: anusha.de@student.kuleuven.be

Intergenerationality, Gender and Youth Aspirations in Zambia: Shifting Agricultural Paradigms for a Sustainable Future

Oluwafemi Ogunjimi, Thomas Daum, Juliet Kariuki, Regina Birner

University of Hohenheim, Inst. of Agric. Sci. in the Tropics (Hans-Ruthenberg-Institute), Germany

Many researchers and policymakers argue that agriculture can help to generate the much-needed jobs for Africa's growing population. Supporting this dominant narrative, several studies have explored youth aspirations towards farming. However, little is known about the factors and actors shaping youth aspirations, in particular the role of parents, and the role of gender differences in this regard. Shifting this paradigm, this paper follows a unique "whole family" approach, which builds on mixed-methods data collection from 348 parents (household heads and spouses) and corresponding adolescents (boys and girls) in rural Zambia to explore these hitherto neglected aspects. The study finds that parents strongly shape children's aspirations – they are much more influential than siblings, peers, church, or media. While male youth are more likely to envision a career in farming (full or part-time), female youth are more likely to envision a non-farming-based livelihood. This reflects their parent's aspirations and is reinforced by the patriarchal system of land and asset transfers through the generations. The multinomial logistic regression shows that farm satisfaction and the desired place of residence (rural or urban) are significantly associated with the likelihood of adolescent boys and girls envisioning a career in farming. Household characteristics such as the household method of farmland cultivation (animal traction), fathers' aspiration, and mothers' livelihood aspirations in agriculture for sons and daughters are statistically significant determinants of male and female youth's livelihood aspirations in farming. The study recommends a "whole-family" approach, which takes into account the influential role of parents, for policies, programs, and projects focusing on the rural youth, and a stronger focus on gender aspects.

Keywords: Aspirations, gender, intergenerationality, intra-household dynamics, youth

Contact Address: Oluwafemi Ogunjimi, University of Hohenheim, Inst. of Agricultural Science in the Tropics (Hans-Ruthenberg-Institute), Wollgrasweg 43, 70599 Stuttgart, Germany, e-mail: fmogunjimi@yahoo.com

From whom do Mothers Receive their Nutrition Knowledge? Participatory Stakeholders' Analysis in Northern Benin

Chérif Issifou[1], Waliou Amoussa Hounpkatin[1], Irène Médémè Mitchodigni[1], Brigitte Kaufmann[2]

[1]*University of Abomey-Calavi, Faculty of Agricultural Sciences, Benin*

[2]*German Institute for Tropical and Subtropical Agriculture (DITSL), Germany*

Child Feeding (CF) is mostly a task done by mothers, who conduct it based on their knowledge and capacities. However, little is known about how this knowledge builds up and from whom they receive information and advice. Learning about these actors and their roles will be useful to facilitate future nutritional education interventions. This study aims to identify stakeholders and their importance in CF knowledge acquisition by mothers.

In northern Benin, in the framework of a transdisciplinary project, stakeholder analysis was conducted by developing a series of Venn diagrams with various mothers, fathers and managers of government structures/NGOs involved in nutrition in peri-urban and rural areas of Banikoara and Nikki Districts. This was complemented by individual interviews and focus group discussions. The recordings of the interviews were transcribed and then analyzed using content analysis.

Actors identified to be important in CF knowledge acquisition by mothers can be categorized into three main groups: family members, health officials and other services. Four different types of relationships between the stakeholders and the mothers were identified: i) acquisition of information (advice, knowledge) in CF; ii) provisioning (financial and food resources); iii) childcare (e.g. washing, dressing); and iv) medical treatment of children. In terms of acquiring knowledge about CF, mothers of children under five years refer first to their mother-in-law, then to the oldest woman in the household, the father-in-law, the brother-in-law, the husband, the Community Health Volunteer (CHV), the Social Promotion Center (SPC), the doctor, the traditional practitioner, the friend, and finally the NGO. Only the importance of the CHV and the SPC varies according to the environment (rural or peri-urban).

As the previous generation has a lot of influence on the knowledge acquisition by mothers in child feeding, nutritional education interventions should involve mothers of mothers to enhance knowledge uptake.

Keywords: Benin, child feeding, mothers, participatory, stakeholders

Contact Address: Chérif Issifou, University of Abomey-Calavi, Faculty of Agricultural Sciences, Abomey-Calavi, Benin, e-mail: ic.issifou@yahoo.com

Learning - teaching - co-operating

Agroecology Peasant Schools as a Social Learning Approach for Post-Conflict Reconstruction on the Colombian Andes

Giovanna Chavez Miguel[1], Michelle Bonatti[2], Alvaro Acevedo Osorio[3], Katharina Löhr[2], Stefan Sieber[2]

[1]*Leibniz Centre for Agricultural Landscape Research (ZALF), Land Use and Governance, Germany*

[2]*Leibniz Centre for Agric. Landscape Res. (ZALF), Sustainable Land Use in Developing Countries (SusLAND), Germany*

[3]*Universidad Nacional de Colombia, Rural and Agro-food Development, Colombia*

In Colombia, nearly six decades of war have devastated the countryside and induced land-related issues that perpetuate marginalisation of rural populations. Although the Peace Agreement and the Integral Rural Reform enacted in 2016 offer new prospects for resolving legacies of conflict, the reconstruction of the countryside requires augmented participation of farmer communities. Social movements advance agroecology in their propositional work as a framework for guiding intentional collective processes of socio-ecological transformation and contributing to the needed reconstruction of rural areas. Agroecology proposes an integrated approach for sustainable territorial management that applies contextualized ecological farming practices and recognises the potential of traditional agricultural forms to confront current post-accord and socio-environmental challenges. Agroecology peasant schools (ECAs) are farmer-led formative initiatives that emerge as a response to the hazards induced through conflict and operate as systems of solidarity and community support conformed by groups of farmers organised around productive, market and political processes of agroecological transformation. This study investigates the potentialities of ECAs as a grassroot strategy to confronting current post-conflict challenges through the development of new knowledge and learning systems. Through a systematisation of experiences, this qualitative study analyses the contexts in which ECAs emerge, their contributions to rural reconstruction and their pedagogical proposals. Results evidence that, although their approaches are locally-developed and thus context-specific, ECAs provide vital assistance to rural communities as they advance community-led territorial development projects and natural resource management schemes, thereby providing pertinent farmer education alternatives in rural areas. By visibilising social learning initiatives at a grassroot level, we argue that the pedagogical proposals of ECAs go hand in hand with the post-accord peacebuilding goals and advance agroecology education as a pertinent approach for pursuing the needed reconstruction of rural territories.

Keywords: Extension, family farming, popular education, rural reconstruction

---

Contact Address: Giovanna Chavez Miguel, Leibniz Centre for Agricultural Landscape Research (ZALF), Land Use and Governance, Oderberger Str. 15, 10435 Berlin, Germany, e-mail: giovanna.chavez.m@gmail.com

Experiential Learning: Groundwater Games and Collective Action in Ethiopia

Hagar Eldidi[1], Wei Zhang[1], Fekadu Gelaw[2], Dawit Mekonnen[3], Seid Yimam[3], Caterina De Petris[4], Natnael Sosa[5]

[1] *International Food Policy Research Institute (IFPRI), Environment and Production Technology Division, United States*
[2] *Haramaya University, School of Agricultural Economics and Agribusiness, College of Agriculture and Environmental Sciences, Ethiopia*
[3] *International Food Policy Research Institute (IFPRI), Environment and Production Technology Division, Ethiopia*
[4] *Humboldt-Universität zu Berlin, Germany*
[5] *Independent Consultant, Ethiopia*

Groundwater management is highly complex, with declines not directly visible and many users sharing the same resource and not realising their interconnectedness. Behavioural games that simulate real-life common-pool resource use have shown promise as a social learning tool for improving resource governance. This study adapts and implements a groundwater governance game, originally developed for India (Meinzen-Dick et al. 2018), in Ethiopia, to assess the potential of using game intervention to help raise awareness of groundwater depletion and improve understanding of the importance of collective governance. In each game round, players make individual choices on cultivating a water intensive or water saving crop and see the effect of their collective decisions on the water table. In 15 treatment villages surrounding the Meki River catchment, the game was played with groups of five men and five women separately. Participant surveys were conducted before and after the game to capture individual mental models regarding groundwater use and management, as well as any immediate learning effects. A community-wide debriefing discussion was held in each village after the game to reflect on the process and lessons learned, and stimulate discussions about groundwater governance. The findings indicate a clear learning effect for participants who played the games in terms of understanding of groundwater dynamics, the joint effect of diverse water uses and users, and the importance of collective resource governance. Data from the games also shows group-level resource management evolves between the no communication, communication, and rule-making rounds of the game, with gendered differences in decision-making. We discuss the creative ways in which players communicate to manage groundwater in the game, including applying crop rotation,

Contact Address: Hagar Eldidi, International Food Policy Research Institute (IFPRI), Environment and Production Technology Division, Washington, United States, e-mail: h.eldidi@cgiar.org

quota rules, monitoring and sanctioning. Unlike other places where competition over groundwater and over-extraction has reached critical levels, small-scale groundwater irrigation in Ethiopia is still at a young stage of uptake and is simultaneously being promoted as a means of livelihood security for farmers. Game interventions in the context of Ethiopia thus presents a unique opportunity to 'plant the seeds', i.e., influence learning and collective action that can help prevent groundwater depletion in the future before reaching a water crisis situation.

Keywords: Behavioural games, commons, Ethiopia, governance, groundwater

Passivity and Activeness of Cooperative Members: A Case of Rice Farmers in Western Zambia

Samuel Mwanza, Jiří Hejkrlík

Czech University of Life Sciences Prague, Fac. of Tropical AgriSciences, Czech Republic

Measures to achieve agricultural development among smallholders elevates cooperatives as an important channel to improving their incomes and market access due to the postulated inherent advantages expressed theoretically in the form of reduced transaction cost, raised social capital and economies of scale. However, the advantages of cooperative membership in Zambia still remains a theory with number of members having only formal membership and passive role in the group. Using a theoretically grounded framework on benefits and challenges of cooperatives, this study assesses and links key theoretical aspects that determine active and passive involvement of farmers in rice producer groups. Simple descriptive statistics and the framework deductive content analysis were employed to analyse the quantitative and qualitative data respectively, from a total of 215 passive and active respondents of rice cooperatives in Limulunga and Mongu districts of western Zambia. Results on quantitative analysis reveals longer formal education for active members and a greater number of males being active. In addition, assets and total land holdings are found to be more among the active members. Active members have also higher share capital value investments, higher selling price, less sales through middlemen, better access to agricultural extension services, and share higher perceived cooperative benefits. Qualitative results from content analysis reveals links to the theories of reduced transaction costs, build-up of social capital and better economies of scale as theories explaining the benefits of cooperative membership, hence, influencing member activeness. However, theories related to governance and decision-making problems, investment related problems, and cooperative asset related problems were found to be critical in explaining passivity of members in Zambian cooperatives. In addition, general dormancy of cooperatives, low production quantities of rice, different levels of commitment and failure to benefit from subsidised inputs contributes all to passivity of members. Capitalising on trainings and awareness creation targeting change of members' approach towards their cooperatives can positively contribute to cooperative development in western province of Zambia.

Keywords: Commitment, content analysis, farmer groups, smallholder farmers

Contact Address: Jiří Hejkrlík, Czech University of Life Sciences Prague, Dept. of Economics and Development, Prague, Czech Republic, e-mail: hejkrlik@ftz.czu.cz

Social Network Analysis with the Net-map Tool - A Systematic Review

Athena Birkenberg[1], Regina Birner[1], Santiago Morales[1], Christine Bosch[1], Lilli Scheiterle[2]

[1]*University of Hohenheim, Inst. of Agric. Sci. in the Tropics (Hans-Ruthenberg-Institute), Germany*

[2]*International Fund for Agricultural Development (IFAD), Research and Impact Assessment Division (RIA), Italy*

As participatory and visual social network mapping tool, Net-Map combines participatory action research, stakeholder and social network analysis with a sociological analysis of power relations. During group or individual interviews, Net-Maps help in identifying and visualising what actors are involved in a given network, how they are linked, how influential they are with regards to a specific outcome, and what their goals are. Since its development in 2006, there has been an increasing number of applications of Net-Map as research method and as tool to support project management and organisational development. Remaining challenges in the application of the tool include group dynamics which can hinder systematic data collection, and the aggregation and comparability of maps during data analysis. The proposed paper has the objectives to (i) describe the origins of Net-Map, (ii) review its current applications, and (iii) derive implications for its future use. We carry out a systematic review of 86 peer-reviewed articles using Net-Map.

We give a detailed overview on the topics for which the tool has been used, how authors integrated the tool with other methods, how they analysed and used the insights gained from the net-map sessions, which limitations are mentioned and which quality assurance techniques are suggested. Preliminary results show that peer-reviewed published articles using Net-Map show applications in a wide variety contexts of social action, ranging from natural resource management, to agricultural and rural service provision, and innovation, research and policy processes. Objectives range from collecting network data to eliciting governance challenges. Most authors used the method within a qualitative case study approach and in combination with other qualitative methods. Few quality assurance techniques are applied and the application of the tool is often not well described and discussed.

Based on these results we discuss the suitability of Net-Map for analysing these diverse contexts and identify benefits and limitations of Net-Map. To support the future application of Net-Map, we provide a framework to facilitate its systematic use and clarify basic principles of quality control.

Keywords: Governance, mixed methods, social networks

Contact Address: Christine Bosch, University of Hohenheim, Inst. of Agric. Sci. in the Tropics (Hans-Ruthenberg-Institute), Wollgrasweg 43, 70599 Stuttgart, Germany, e-mail: christine.bosch@uni-hohenheim.de

Incentivized Payments in Experimental Games Can Lead to Behavioural Change

LARA BARTELS[1], THOMAS FALK[2], BJÖRN VOLLAN[3]

[1]*ZEW – Leibniz Centre for European Economic Research, Germany*
[2]*International Crops Research Institute for the Semi-Arid Tropics (ICRISAT), Innovation Systems for the Drylands, India*
[3]*University of Marburg, Sustainable Use of Natural Resources, Germany*

Unstainable water usage is a widespread environmental problem in developing countries. As the gap between water supply and demand is widening, local to global conflicts increase in number and intensity. Addressing unsustainable use of water requires an understanding of users' behaviour and how the same can be influenced. Our research paper investigates environmental behaviour in the context of agricultural water management challenges.
Especially economic experiments are a powerful tool to study behaviour and causal mechanisms in human interactions. Lately, more and more experimenters observe that these experiments also support learning and behavioural change. Consequently over the last years, games have been increasingly applied as intervention tools in natural resource governance with a strong capacity development purpose promoting behavioural change. While there appears to be a growing understanding on how monetary incentives influence decision-making within experiments, less research investigates how monetary incentives matter or interfere with collective action outside the experiment.
In a framed field experiment about water management in rural India, we compare individual incentivized payments with non-incentivized payments. Our results show little evidence for different behaviour in the game but we find some evidence that incentivized payments increase the probability to observe real-life behavioural change after the game compared to a control group without game intervention. This effect most likely arises as participants take the exercise more serious and deliberate with a higher degree of commitment. Our work is relevant concerning methodological advances in the field of experimental methods to study environmental behaviour but more importantly, has implications on the design of more effective and efficient behavioural change intervention tools.

Keywords: Collective action, games for sustainability, impact assessment, India, monetary incentives, social learning, water management

Contact Address: Lara Bartels, ZEW – Leibniz Centre for European Economic Research, Mannheim, Germany, e-mail: lara.bartels@zew.de

Growing Transformational Future Livestock Sector Professionals - Emerging Impacts of the ILRI CapDev Grand Challenge

Wellington Ekaya

International Livestock Research Institute, Capacity Development Unit, Kenya

In 2019, the International Livestock Research Institute (ILRI) launched a new initiative, the 'CapDev Grand Challenge', an approach to grow cohorts of transformational future livestock sector professionals. The target group is graduate students and researchers from national organisations in LMICs where ILRI works. Annually ILRI hosts about 100 PhD and MSc students embedded within its research programs and co-supervised by professors from degree-awarding universities where the students are registered. Designed to complement graduate research and technical/scientific training, the Challenge process aims at boosting soft skills and systems thinking capability.

The CapDev Grand Challenge covers a 16-to-18-month period, starting with a 3-minute research pitching contest. This is followed by soft skills training involving 10 courses in 10 months, and a deepening phase when participants apply their acquired skills in the workplace, while virtually mentored and supported to attend international conferences. The final activity is a comprehensive impact tracking survey.

This paper presents results from an impact tracking survey conducted in April 2021 on the inaugural CapDev Grand Challenge process involving 71 participants from 24 countries in Eastern, Southern and West Africa and South, East and Southeast Asia.

Over 80 % of participants strongly agreed that the CapDev Grand Challenge process broadened their thinking and;

- Boosted their confidence and effectiveness in communicating science to non-technical audiences,
- Strengthened their confidence and ability to engage with intention to influence decision makers,
- Increased their confidence and effectiveness in engaging and working with researchers outside their own discipline,
- Boosted their motivation and confidence to take up leadership roles,
- Made them more knowledgeable, confident and effective in designing and discussing project impact pathways,
- Increased their confidence and effectiveness in mapping out actors in a project, and explaining their roles.

Contact Address: Wellington Ekaya, International Livestock Research Institute, Capacity Development Unit, P.O. Box 30709-00100 Nairobi Naivasha Road Uthiru., N/A Nairobi, Kenya, e-mail: w.ekaya@cgiar.org

Results demonstrate a strong potential of the Challenge to create systems thinkers, and effective communicators able to contribute, not just new research evidence, but also to development outcomes by deploying scientific evidence to influence decision makers. The process catalyses strategic outreach for impact, thus fostering synergies and sustainability in complex livestock food systems.

Keywords: Capacity development, Livestock sector, next-generation professionals, soft skills, systems thinking, transformational

Perspectives on the Performance of Farmer Cooperatives in Northern Ghana

Jang-Hee Im, Irene Egyir, Daniel Sarpong, Akwasi Mensah-Bonsu
University of Ghana, Dept. of Agricultural Economics & Agribusiness, Ghana

Ghana has a history of cooperative development since 1928, but the capacity of farmer cooperatives in Ghana has been found low. It was necessary, therefore, to find out measures to address the current problem facing the farmer cooperatives. The research attempted to answer three questions: 1) How should success of farmer cooperatives be measured in the Ghanaian context? 2) What determines the success of the farmer cooperatives? and 3) What should be done to strengthen the capacity of the farmer cooperatives? Panel data were collected from 281 farmer cooperatives in 21 districts of Northern, North East, Savannah, Upper East and Upper West Regions for the periods of 2016 and 2019/2020. Additionally, 99 officers and 24 private agribusiness entities were interviewed to inform the analysis of cooperative policy in Ghana. A profile analysis was conducted to test whether the eight success indicator candidates (revenue, net surplus, asset, dues (periodic membership fee), member satisfaction, membership size, community service, longevity (years of existence)) are distinct with significance over three clusters of farmer cooperative namely Success, Neutral and Failure. The analysis resulted in the Wilks lamda of 0.4613, approximately 32.594 of F statistic at a highly significant level. Therefore, the eight success indicators were found to explain differences in performance of the three clusters. An ordered probit model was established to identify determinants of success of farmer cooperatives in northern Ghana. Each of the eight success indicators were estimated as dependent variables over nine independent variables namely number of meeting, number of training, paid staff, joint sale or buying, joint commodity, contribution to community, number of extension service visit, project beneficiary and gender composition. All the success indicators except longevity were explained with significance by the independent variables. Government agencies and development partners are recommended to use the seven success indicators as a diagnostic tool to assess performance of farmer cooperatives and subsequently inform policy decisions.

Keywords: Performance, policy, succes determinants

Contact Address: Jang-Hee Im, University of Ghana, Dept. of Agricultural Economics & Agribusiness
current address: 2nd Floor Allianz Plaza 96 Riverside Drive, Nairobi, Kenya, e-mail: joynjoyim@koica.go.kr

Producer Organisations Key Instrument for Strengthening the Agricultural Sector in Colombia

Adriana Marcela Santacruz Castro

Corporation of Colombian for Agricultural Research - Agrosavia, Marketing and Technology Transfer, Colombia

Advancing towards the end of sustainable agriculture is no longer an optional task, since urgent actions are required not only in development, but also in the adoption of practices conducive to the conservation of ecosystems and their natural resources. Currently there are multiple alternatives in favour of the sustainable use of agrobiodiversity and other practices to that end, however, only some producers manage to implement them, among other things due to availability, complexity in use or cost.

Producer organisations in Colombia, where different production systems, natural resources, cultures, economies, and different processes of territorial governance coexist and are potential structures that can favour the implementation of sustainable agriculture. However, these social structures are usually conceived by the institutional framework as recipients of knowledge and assistance, mainly in technical-productive issues, and not as an actor involved, interested and with multiple capacities to guarantee the rational use of natural resources for agricultural production.

This presentation seeks to show how interventions for more than two years with several organisations, that initially sought to obtain a higher agricultural productivity, changed their perspective during the implementation. The interventions were accompanied by the strengthening of skills related to planning, communication and leadership. Besides, the visualisation of new alternatives for participation in new markets and development of a marketing strategy; have showed pathways for the improvement of social, productive, environmental and economic management and finally the empowerment of producer organisations to take better and informed decisions for their communities and the competitiveness of agricultural sector in Colombia.

Keywords: Governance, institutions, producer organisations, sustainability

Contact Address: Adriana Marcela Santacruz Castro, Corporation of Colombian for Agricultural Research - Agrosavia, Marketing and Technology Transfer, Km 14 Way Mosquera, 111931 Mosquera, Cundinamarca, Colombia, e-mail: amsantacruz@agrosavia.co

Agricultural Support for Coca Substitution in La Montañita and Puerto Rico, Caquetá, Colombia 2018–2019

Erika V. Wagner-Medina, Claudia Patricia Rendon Ocampo, Laura Cristina Romero Rubio

Corporación Colombiana de Investigación Agropecuaria – Agrosavia, Colombia

Since 1974, Colombia have been subjected to several strategies to eradicate illicit use crops, those involve: aerial spraying of glyphosate, manual eradication and conditional crop substitution, measures that have not yet solved the problem. After the peace agreement in 2016 arose The Comprehensive National Program for the Substitution of crops for illicit use (PNIS in Spanish) with the same approach as previous initiatives. In 2019, an analysis of this programme took place in the municipalities of La Montañita and Puerto Rico in Caquetá, with the technical teams of the programme through surveys (50) and semi-structured interviews (6) that were described and coded with the QDA Miner lite® free software. During PNIS implementation it was identified that the technical territorial capacity allows its strengthening due to the employment of local technical labour that facilitated the bonds of trust with beneficiaries. Some already known sustainable agricultural practices that were not implemented before due to lack of financial resources were in place and awareness on better management of natural resources towards a sustainable approach of the agroecosystems was raised. Furthermore, the programme served people located in distant places who have never received technical accompaniment or financial support. However, the programme did not synchronise with the local capacities and resources, territorial economy and among its own activities. There were delays in the delivery of the financial resources, and the contracting of technical operators to work continuously. For the program, the total eradication of illicit use crops was mandatory, ignoring the complexity of the transition from an illicit economy to a legal one (that has not establish yet), causing, among other things, that some producers were left adrift, subjected to displacement, hunger and uncertainty. The complexities of its implementation reveal the particularities of the technical agricultural teams, the conditions of the Immediate Attention Plan, the agricultural development and value chain gave elements to rethink on implementation approaches and the transformative potential of technical accompaniment considering the National Agricultural Innovation System that was sanctioned by law in 2017 to create the agricultural extension service to promote productivity, competitiveness, and sustainability with innovation as its axis.

Acknowledgments: Interviewed, Fondo Colombia en Paz and Ministerio de Agricultura y Desarrollo Rural de Colombia-MADR

Keywords: Agricultural development, alternative development, drug policy, peacebuilding.

Contact Address: Erika V. Wagner-Medina, Corporación Colombiana de Investigación Agropecuaria – Agrosavia, Centro de Investigación Tibaitatá. Km. 14, vía Mosquera, Bogotá, Colombia, e-mail: erika.wagnerm@gmail.com

Farmers Producer Organisation (FPOs) Are Instrumental for the Upliftment of Socio- Economically Vulnerable Tribal Farmers

Smita Joshi, Tanay Joshi, Amritbir Riar

Research Institute of Organic Agriculture (FiBL), International Cooperation, Switzerland

Scheduled Tribes (ST) are among the most vulnerable and disadvantaged socio-economic groups in India. The Government of India has made numerous efforts to empower these tribes through schemes for the development and progressive legislation. Farmers Producer Organisations (FPOs) in collaboration with non-government and government organisation, has a vital role to play in the upliftment of agriculture for livelihood improvement. However, this role needs to be re-assessed and finetuned to work at the interface of biophysical and socio-economic barriers. For this, Five FPOs were studied in the state of Madhya Pradesh in central India. Madhya Pradesh is the home for 21 % of the Schedule Tribal population in India, and Jhabua district has 87 % of the ST population in the state. Most of the rural population in the district depend on agriculture for their livelihood. Different FPOs were selected based on different criteria like commodities/crops preferences, activity and services portfolio and operating mechanisms for promoting institutions etc. Out of these 5 FPOs, ten farmers each were selected by adopting a random sampling technique to know the socio-economic impact of FPOs on their member farmers. The data was collected from the respondents on a well-designed and adopted survey questionnaires and from the organisations through telephonic interviews. Reflection on data was analysed in the light of secondary sources consultation and literature review to understand the socio-economic performance & constraints of FPOs. Interviews with the farmer members suggest that the FPO has been successful on many fronts as an institution for collective action. The success in bargaining for lower input supply prices, the ability to pool produce to get a higher price for outputs, and the innovative methods in training and information dissemination have resulted in significant benefits to the members in enhancing their incomes. However, the main challenge appears to be the inability to access capital, which, to some extent, is undermining the advantages of collectivisation. The results of the study highlighted the significant contribution of farmer organisations towards developing the socio-economic conditions of farmers, thus making them self-sufficient and self-reliant.

Keywords: FPO, livelihood, Madhya Pradesh, schedule tribes

Contact Address: Amritbir Riar, Research Institute of Organic Agriculture (FiBL), Dept. of International Cooperation, Frick, Switzerland, e-mail: amritbir.riar@fibl.org

Impacts of Social Cohesion on Farmers' Soil Conservation Behaviour: Analyzing the Case of Bushehr, Iran

Masoud Yazdanpanah[1], Maryam Tajeri Moghadam[2], Tahereh Zobeidi[3], Stefan Sieber[4], Katharina Löhr[4]

[1]*Agricultural Sciences and Natural Resources University of Khuzestan, Iran*

[2]*University of Tabriz, Dept. of Extension and Rural Development, Iran*

[3]*University of Zanjan, Dept. of Agricultural Extension, Communication and Rural Development, Iran*

[4]*Leibniz Centre for Agric. Landscape Res. (ZALF), Sustainable Land Use in Developing Countries (SusLAND), Germany*

Land degradation in the form of soil erosion is a serious threat to food security, sustainable agricultural production, and environmental sustainability. Soil conservation is considered as a precondition for achieving food security and adopting environmental policies. Iran is a vulnerable country in terms of land degradation and soil erosion. In Iran, many activities have been conducted for soil conservation but did not meet the intended effects. Social cohesion, in terms of levels of trust, cooperation and socio-economic inclusion, is considered a key factor and catalysts of adoption practices such as soil conservation. However, limited knowledge exists on how social cohesion impacts on the acceptance of soil conservation practices at the community level. Therefore, this study investigates the impact of social cohesion on the soil conservation behaviour of farmers in Bushehr County, Iran. Based on random sampling, a face-to-face survey with a total of 180 farmers was conducted. In this study, we examine seven different components of social cohesion (1) public trust, (2) trust in institutions, (3) trust in communities, (4) norms, (5) formal networks, (6) informal networks, and (7) network size. The results show that (1) public trust, (3) trust in communities, as well as trust in (5) formal and (6) informal networks are the key to adopting soil conservation. These components explain 30 % of changes in the acceptance of soil conservation behaviours. (7) Network size, (4) norms, and (2) trust in institutions were not significant in accepting soil conservation behaviour. Accordingly, trust seems to have higher importance for the adoption of new practices than the network communication dimension. Trust is the catalyst that converts information into usable knowledge. Networks provide the context of information exchange. A network with a high level of trust is able to do more because networks are built on trust and confidence, thereby reducing risk by providing a safety net for people. The results of this study help to understand farmers' complex decisions about accepting soil conservation and help design strategies and policies to further accept soil conservation.

Keywords: Behaviour, network, social cohesion, soil conservation, trust

Contact Address: Masoud Yazdanpanah, Agricultural Sciences and Natural Resources University of Khuzestan, Mollasani, 744581 Ahvaz, Iran, e-mail: masoudyazdan@gmail.com

The Agricultural Extension as a Mechanism for Community Growth

Jose Carlos Montes Vergara, Claudia Patricia Rendon Ocampo

Corporation of Colombian Agricultural Research - Agrosavia, Colombia

Strengthening the social capital of the dairy sector in Colombia is one of the most important challenges that must be solved in order to increase its profitability and thus improve the living conditions of rural inhabitants. The New Zealand model, as one of the world's leading dairy producers, has been based on learning by doing and, farmer-to-farmer learning techniques to achieve the results it has available to date. With the aim of validating and implementing New Zealand methodologies in Colombia, during the intervention of the Proyecto Cadena de Valor Láctea ColombiaNueva Zealand (Dairy Value Chain Project), ten milk-producing families were selected in the municipality of Cumbal, Nariño so that , during the period between 2016 and 2019, they will work with an agricultural extension scheme that was based on the principles of "the farm as a business" and "learning by doing". During the evaluation period of the project, indicators were determined that would allow measuring the productive and social change. The results obtained allowed us to infer that the producers appropriated the technologies and methodologies validated by the project on their farms, generating business income four times higher and obtaining the family purposes that were reflected in the farm development plan for each year of intervention. The results show that the extension strategies made it possible to strengthen the farm management skills of the producers, favouring the wellbeing of each family and the association of which they were part. The improvements in the implemented practices allow to have greater agency capacity and make decisions based on data.

Keywords: Families, learning, pilot farms, profitability, technology

Contact Address: Claudia Patricia Rendon Ocampo, Corporation of Colombian Agricultural Research - Agrosavia, Avenue 84 #33-120, 50031 Medellín, Colombia, e-mail: crendon@agrosavia.co

School Gardens for Nutrition and Food Security: Case Study from Switzerland, Ethiopia, Peru and Myanmar

Dumas Margot, Giuliani Alessandra, Zbinden Karin
Bern University of Applied Sciences (BFH), School of Agricultural, Forest and Food Sciences (HAFL), Switzerland

More and more schools all over the world are using gardens to teach and foster different skills. However, literature on the impacts of school gardens is still scant and scattered. This study investigates the impact of school garden programs can have on children looking at impact domain of Nutrition and food security, civic participation, and Environmental Consciousness. Eleven selected key-experts, who work as teachers or coordinators of school gardens in Switzerland, Ethiopia, Peru and Myanmar, and who facilitated their respective projects have been interviewed. Students' drawings have been analysed to consider the view of the school children involved in the programs. The Food and Insecurity Experience Scale (FIES) and the Prevalence of Undernourishment (PoU) were used to evaluate nutrition and food security. The level of knowledge and practice in the four countries has been compared to the specific situation of the gardens involved in the research. Observations and questionnaires in Ethiopia and Peru have proven to be useful tools to assess the positive impact the school gardens have on the children's diet. Children enrolled in such programs are less likely to drop out of school. Inexpensive and easy to handle tools as the FIES which can be conducted or completed by any member of a household in a written or oral form are very interesting and require little resources, but delivers key information on a household or even more differentiated results between inhabitants of the same household. The results show that school gardens have a positive influence overall on all the impact domains, but they often lack tangible proof to promote their concept and objectives. In order to face this issue, it would be necessary to promote connections between the school gardens worldwide and to promote and define some common impact measurement tools, which could be implemented regardless of the location. The projects often face the same problems and are individually creating tools and methods that cater to the specific context of their school garden, rather than developing universal school garden assessment methods together.

Keywords: FIES, impact assessment, nutrition, PoU, school garden

Contact Address: Dumas Margot, Bern University of Applied Sciences (BFH), School of Agricultural, Forest and Food Sciences (HAFL), Ch. Du Bois des Rittes 32, 1723 Marly, Switzerland, e-mail: margot.dumas.md@gmail.com

Designing a Cooperation Model for Groundwater Governance in Coastal Odisha, India

Surajit Haldar, Ernst-August Nuppenau, Joachim Aurbacher
Justus-Liebig University Giessen, Inst. of Agric. Policy and Market Res., Germany

Failure of a theoretically efficient and empirically proven market mechanism in groundwater management has led to a search for alternative arrangements. Cooperation showed mixed results, while government intervention in terms of subsidy scheme inflated the successes. In the coastal aquifers of Odisha state in India, a government subsidy scheme in medium deep tube well installation addressed the economic scarcity of groundwater procurement while its management through cooperation among the member farmers by forming a water user association (WUA) seemed to solve the long-standing distribution issue. Sooner or later, improper management of the resource system and the physical structure collapsed the cooperation in water sharing and thereby a defunct WUA. In this study, we attempted to introduce an incentive scheme to coordinate the individual farmers' action to save water and its distribution through community action in a principal-agent model. Empirical evidence indicates that the community has the higher bargaining power to determine the benefit share in terms of the redistribution of profit after distributing the fixed initial incentive and the variable part of the incentive, that is individually earned by member farmers by adopting a water-saving cropping patter. Further simulations through water price increase by two to three-fold from the initial level do not increase the gross water-saving, implying a dominant type of cropping pattern in the farming system. We further investigated the model behaviour by farm types of highly diversified irrigated agriculture for direct market supply (HDIAM) and least diversified irrigated agriculture for contract sale (LDICS). HDIAM farms are mostly sensitive to water price hike and are flexible to adopt a maximum water saving cropping pattern when we assume a reasonable reservation price in favour of their crop enterprise decision. On contrary, LDICS farms showed indifferent water use behaviour with increasing water price, because of their higher acreage under perennial water-intensive crop enterprise, such as sugarcane. Hitherto, model results infer that in the second round of incentives, there should be a lower redistribution of water due to its higher marginal value that also accounts scarcity value of water.

Keywords: Cropping pattern, groundwater, incentives, principal-agent

Contact Address: Surajit Haldar, Justus-Liebig University Giessen, Inst. of Agric. Policy and Market Res., Senckenbergstrase 3, 35390 Giessen, Germany, e-mail: surajit.haldar@agrar.uni-giessen.de

The Water Licenses Issue in Goiás, Brazil: Inequalities, Disputes and Brasilian Productive Model Criticisms

Dedierre Gonçalves Silva[1], Lucas Soares Fraga[1], Valquíria Duarte Vieira Rodrigues[2], Hayanne Rodrigues Carniel Cavalcante[1], Jaqueline da Costa Paixão[1], Luciana Ramos Jordão[2], Thiago Henrique Costa Silva[2]

[1]*University Center Alves Faria, Law College, Brazil*

[2]*Federal University of Goiás, Law College, Brazil*

Agribusiness is a model of Brazilian rural development that is constantly defended, especially by the political classes, as the base and vocational future of the State of Goiás, located in brazilian mid-western. In a model of capital accumulation through the predatory usage of nature, agribusiness can be associated with intensive livestock production, technological packages (chemical and genetic), and has been interpreted as synonymous with development and strength. However, this system excludes other productive models and ways of life that do not fit productivist practices, expanding the concentration of income and land and converting land and water into mere resources, as if they could be dissociated from nature, and did not matter to keep life on the planet. Because the state of Goiás has a significant agricultural contribution to the country's economy, since it is one of the largest producers of soy, corn and sugarcane plantations, which depend on a large volume of irrigation, it was selected as a standard case to understand the reproduction of agribusiness in the Brazilian rural environment. Thus, this research discusses whether the productive model of excessive water use is correlated with conflicts over water, through the analysis of licenses for irrigation and disputes over water in the state of Goiás. An investigation into irrigation concessions was carried out, noting that, in 2020, there were 1,853 catchment points active in Goiás, which could be associated with the expansion of water conflicts. According to the Pastoral Land Commission and the National Water Agency, between 2016 and 2020 there were 16 conflicts in Goiás that reached 1,752 families. By data comparison it is possible to say that the conflicts are not related to water scarcity, but to their poor distribution and usage, and in regions with great number of pivots grants they are accentuated. It concludes that it is not possible to understand water without considering the right to life, since it is necessary to establish a purposeful agenda between the various sectors of society and the public administration for the best use of water resources, so abundant, but, due to economic power, so uneven in accessibility.

Keywords: Central pivot grants, development, human right to water, water availability, water commodification

Contact Address: Thiago Henrique Costa Silva, Federal University of Goiás (UFG), Law College, Avendida Virgilio Joaquim Ferreira, 74860615 Goias, Brazil, e-mail: thiagocostasilva.jur@gmail.com

Games for Triggering Collective Changes in Natural Resource Management: Four Cases from India

Thomas Falk[1], Wei Zhang[2], Ruth Meinzen-Dick[2], Lara Bartels[3]

[1]*International Crops Research Institute for the Semi-Arid Tropics (ICRISAT), Innovation Systems for the Drylands, India*
[2]*International Food Policy Research Institute (IFPRI), Environment and Production Technology Division, United States*
[3]*ZEW – Leibniz Centre for European Economic Research, Germany*

As resource users interact and impose externalities onto each other, institutions are needed to coordinate resource use, create trust, and provide incentives for sustainable management. Coordinated collective action can play a key role in enabling communities to manage natural resources more sustainably. But, when such collective action is not present, what can be done to foster it?

There is growing awareness that the governance of natural resources has to be adapted to the specific context. Interventions are often implemented at small scale, and the potential to scale up facilitation intensive approaches is limited. Moreover, sustainable resource management frequently fails to emerge or breaks down after the project ends.

To date, researchers have typically used behavioural games to study cooperation patterns of communities. Recently, games have been adapted as learning and stakeholder engagement tools to improve management of the commons, strengthen self-regulation of resource use, and enhance constructive interactions among resource users. Combining games with other interventions and tools and facilitated discussions has been proposed as a promising approach to improve collective action institutions through experiential learning — a classic approach in education.

This paper reviews existing literature and synthesizes lessons learned from a series of studies testing the use of behavioural games for institutional capacity development in India. We conclude that, while games alone will not be the solution to all natural resource management challenges games can provide a structured and therefore replicable approach for influencing behaviour. They can also improve system understanding, raise awareness, influence norms, facilitate dialogue, train for crisis response, and increase legitimacy of decisions.

Keywords: Behavioural change, facilitation tools, forest, india, sustainable natural resource management, water

---

Contact Address: Thomas Falk, International Crops Research Institute for the Semi-Arid Tropics (ICRISAT), Innovation Systems for the Drylands, Hyderabad, India, e-mail: t.falk@cgiar.org

Understanding acceptance and implementation of agricultural innovations

Comparing Farmers' Willingness to Pay and Multiplication Costs for Clean Sweetpotato Seed: Evidence from Kenya

Christine Mwangi[1], Josiah Ateka[1], Robert Mbeche[1], Elijah Ateka[2], Luke Oyugi[1]

[1]*Jomo Kenyatta University of Agriculture and Technology, Dept. of Agricultural and Resource Economics, Kenya*

[2]*Jomo Kenyatta University of Agriculture and Technology, Department of Horticulture and Food Security, Kenya*

The production of orphaned crops such as sweet potato (Ipomea batatas) has the potential to promote ecosystem resilience and address food insecurity and malnutrition in the context of climate change. However, sustainable production of the crop is hampered by poor access to quality seed. Production of sweet potato is currently dominated by use of recycled planting materials often sourced from local social networks. Efforts to improve access to quality seed in Kenya have mostly focused on varietal improvement through quality seed selection, breeding and scaling up of clean seed multiplication. However, success and sustainability of the efforts will largely depend on the farmers' willingness to pay for clean seed. In this article, we assess the level and determinants of farmers willingness to pay (WTP) for clean seed among smallholder sweet potato farmers in Kenya. We then compare the estimated WTP with the cost of seed multiplication – to determine whether the clean seed business would be economically viable in the context of smallholder production systems. Data for the paper were collected among 383 sweet potato farmers and 30 sweet potato seed multipliers in Kenya. WTP was estimated using the double bounded contingent valuation method while the determinants of WTP were analysed using an ordered probit regression model. Results show that the mean WTP for clean seed was KES 647.74 (about US$ 6.35) for a 90 kg bag with variations in WTP observed across geographical regions, gender and wealth categories. Results further show that WTP increases with prior use of clean seed and experience in sweet potato production. Conversely, WTP declined with age and household size. The cost of clean seed multiplication was KES. 594.78 (US$ 5.83), which is lower than the estimated WTP. This implies that what the farmers would be willing to pay or can at least meet the cost of seed multiplication. These results suggest that seed multiplication business may be economically viable but this will depend on other transaction costs and establishment of efficient distribution systems.

Keywords: Climate change, cost of clean seed multiplication, double bounded contingent valuation, economic viability, ordered probit, willingness to pay

---

Contact Address: Christine Mwangi, Jomo Kenyatta University of Agriculture and Technology, Dept. of Agricultural and Resource Economics, 62000, 10200 Nairobi, Kenya, e-mail: christinemwangi.cm@gmail.com

Profitability of Small-scale Cocoa Production in Ecuador

Michael Worku Ketema[1], Lina Tennhardt[2]

[1]*Georg-August University of Göttingen, Dept. of Agricultural Economics, Extension and Rural Development, Germany*
[2]*Research Institute of Organic Agriculture (FiBL), Dept. of Socioeconomics, Switzerland*

The cocoa industry in Ecuador is an essential source of livelihood, generating income for more than 100,000 families. Cocoa is Ecuador's fourth-largest export crop and the country is the world's leading exporter of premium cocoa, accounting for 54 % of the world market share. While cocoa in Ecuador is produced predominantly by small-scale farmers, they receive a small share of the final chocolate price. However, to reduce the risks associated with market fluctuations, to boost the economic condition of small-scale cocoa producers, and later to ensure the future supply of (fine or flavor) cocoa from Ecuador, understanding and improving smallholder cocoa profitability is essential. This research focusses on assessing the profitability of small-scale cocoa production and identifying influencing factors in Ecuador. The analysis was based on a data set with farm-level economic data from 172 cocoa producers in northwest Ecuador. Descriptive statistics, Net Present Value (NPV), Benefit-Cost Ratio (BCR), and Multiple Regression Analysis were the major analytical tools applied for this research. Results indicate that cocoa production in northwestern Ecuador is not profitable (average NPV of -248 USD per ha, mean BCR 0.73). The duration of membership within the corporate sustainability program, the share of cocoa revenue, the number of family laborers per hectare, the total cocoa area, the produced cocoa variety, and use of fertiliser were identified as factors significantly influencing the profitability of cocoa production in the study area. The results indicate that small-scale cocoa producers in northwestern Ecuador are in a penury situation, and chocolate companies and other stakeholders involved in the supply chain could consider improving their effort to lift cocoa farmers out of poverty.

Keywords: Benefit-cost ratio, cocoa, ecuador, net present value, profitability, smallholder

Contact Address: Michael Worku Ketema, Georg-August University of Göttingen, Dept. of Agricultural Economics, Extension and Rural Development, Platz der Göttinger Sieben 5, 37073 Göttingen, Germany, e-mail: m.ketema@stud.uni-goettingen.de

Effects of Communication Channels on Adoption of Orange Fleshed Sweetpotatoes in Uganda

Brenda Namulondo[1], Philip Nyaga[2], Kimpei Munei[3]
[1]*National Agricultural Research Organization, Uganda National Genebank, Uganda*
[2]*University of Nairobi, Veterinary Pathology, Microbiology and Parasitology, Kenya*
[3]*University of Nairobi, Dept. of Agricultural Economics, Kenya*

The need to shift paradigms to a positive relationship between humans, environment, ecology and nature in agriculture is of global concern. Efforts have been made by several organisations to produce only products in line with the need and disseminated through various pathways; however, most of the technologies hardly reach farmers. The study examined the effects of communication channels in adoption of Orange Fleshed Sweet potatoes (OFSP) in Gulu district, Uganda. Household survey research design was adopted as the main investigative design, using Semi structured questionnaires, FGDs and Key informant interviews, administered to 218 respondents, based on purposive sampling method. The results indicated that the most common communication channels used in order of importance by respondents were interpersonal – farm demonstrations (88 %), followed by mass media – radio (10 %). The adoption rates of OFSP were found to raise from 2 % in 2009, 13 % in 2010 and 85 % in 2011. In relation to the most informative source, 98 % strongly agreed that it was extension agent whereas 2 % simply agreed; 75 % strongly agreed that seminar was accessible while 25 % strongly disagreed; 80 % strongly agreed that radio had better coverage capacity while 2.6 % disagreed; 100 % strongly agreed that other channels (community trainers and print media) were frequently used. This implied that mass media was more effective in OFSP dissemination but auxiliary details were communicated through interpersonal channels. All the farmers who used the channels were adults of which 84 % were married. Of these, 65 % were female, 36 % had some source of income, 30 % had some formal education while 56 % belonged to a social group. In conclusion, radio and field demonstrations were the main sources of information. Therefore, the study recommended that multiple channels; specifically, radios, print media (Mass media) and farm demonstrations, focus groups (Interpersonal) should be considered as strategies for agricultural information dissemination and communication, respectively. The study also recommended that farmers' socioeconomic characteristics should be considered in technology adoption to effectively save the planet and enhance a healthy and sustainable future.

Keywords: Information dissemination, interpersonal communication, mass media, technology adoption

---

Contact Address: Brenda Namulondo, National Agricultural Research Organization, Uganda National Genebank, P.O Box 40, Berkeley street , Entebbe, Uganda, e-mail: belithalee@gmail.com

Farmers' Preferences for Sustainable Intensification Attributes in Sorghum-Based Cropping System: Evidence from Mali

Felix Badolo[1], Bekele Hundie Kotu[2], Oyakhilomen Oyinbo[3], Karamoko Sanogo[1], Birhanu Zemadim[1]

[1]*International Crops Research Institute for the Semi-Arid Tropics (ICRISAT), Mali*
[2]*International Institute of Tropical Agriculture (IITA), Ghana*
[3]*Ahmadu Bello University, Zaria, Department of Agricultural Economics, Nigeria*

Adoption of agricultural intensification technologies is on average low, which results in a vicious cycle of low agricultural growth, persistent rural poverty and food insecurity in many rural areas of sub-Saharan Africa, including Mali. Amongst other constraints, the limited adoption of intensification technologies could partly be because technology deve- lopment does not properly consider the technological traits that farmers value most. On the other hand, there are growing concerns that promoting widespread adoption of inten- sification technologies may be associated with unintended adverse effects on the cropping systems and livelihoods of smallholder farmers, as experienced in the Asian Green Re- volution. This gave rise to the growing interest for cropping systems to be sustainably intensified. Using a discrete choice experiment, this study assessed farmers' preferences for technology attributes in sorghum-based cropping systems from the lens of sustainable intensification. The choice experiment was implemented among 567 smallholder sorghum-producing farmers in southern Mali. We considered six technology attributes corresponding to different domains of sustainable intensification namely: grain yield, yield instability, soil fertility, nutrition, labour requirement, and fodder yield. We used mixed logit model to analyse the data. The findings reveal that farmers have strong preferences to change their current cropping practices in favour of a more sustainable sorghum production system, indicating the importance of new R&D initiatives to respond to farmer's needs. More specifically, farmers prefer technologies which increase the grain and fodder yields, have positive or a neutral effect on soil fertility and household nutrition. However, the farmers are not homogenous in their preferences but reveal substantial heterogeneities, including the pooled sample and sub-samples by agroecology and social networks of farmers. This suggests that development actions should be adapted to local contexts to address farmers' preferences. More importantly, the farmers placed more value on nutrition outcome associated with uptake of sustainably intensified sorghum-based cropping systems, which calls for more R&D interventions along the nut- rition domain, such as biofortification of sorghum and legume-based intercropping. Thus, development partners should strongly invest in interventions around nutrition outcomes, as they are likely to gain widespread adoption among farmers.

Keywords: Agricultural intensification, choice experiments, farmer preferences, Mali, sorghum

Contact Address: Felix Badolo, International Crops Research Institute for the Semi-Arid Tropics (ICRISAT), ICRISAT - Mali, 320 Bamako, Mali, e-mail: F.Badolo@cgiar.org

Protect Culture to Maintain Traditional Crops: A Case Study among Tribal Farmer Communities in India

Marijn Voorhaar[1], Vincent Garin[1], Krithika Anbazhagan[1], Prashanth Ramini[2], Vittal Rao Kumra[2], Jana Kholová[1], Erwin Bulte[3], Robert Lensink[3]

[1]*ICRISAT, Crop Physiology, India*
[2]*Centre for Collective Development (CCD), India*
[3]*Wageningen University, Social Sciences, The Netherlands*

Similar to other traditional crops, sorghum cultivation is declining in several regions of the world. In this study, we investigated the main socio-cultural drivers that explain the persistence of sorghum cultivation by tribal farmers in the district of Adilabad (Telangana, India). For these indigenous communities, sorghum still plays a prominent role in their tradition and culture. Our study was conducted in Utnoor, a small region (350 km^2) with a high level of agro- and cultural biodiversity. Consisting of both hills and plain areas, it covers the main crops cultivated in India (rice, maize, wheat) with a substantial share for sorghum and chickpea. With a tribal population of almost 60 %, the Utnoor region is an agri-cultural hotspot where conclusions about big tendencies characterising Indian agriculture can be drawn. We interviewed 110 farmers (male: N=89 / female: N=21) living in 11 different villages representative of the three main agricultural systems present in the region (sorghum, chickpea, and rice/maize). We collected data about demography, agronomy, environmental pressure and access to technology. The most salient feature of our study is the adaptation of the Hofstede scale to measure cultural values in these farming communities. To the best of our knowledge, it is a first attempt to measure cultural differences in the (Indian) agricultural context with this well-established tool. Our findings show that sorghum is mostly grown for household consumption and fodder purposes, whereas chickpea, rice and maize are cultivated for the market. The sorghum market is mostly informal, making it less profitable for farmers. Sorghum farmers tend to live further away from a water source, compared to the other farmers. A strong correlation was found between culture and crop choice. Compared to chickpea, and rice/maize farmers, sorghum communities showed less hierarchy, a more collective attitude and attachment to the group, and were less likely to change cultural behaviours. Therefore, cultural aspects should be taken into consideration to design interventions aiming to preserve traditional crops.

Keywords: Culture, Hofstede, socioeconomics, sorghum, traditional crops, tribal communities

Contact Address: Marijn Voorhaar, ICRISAT, Crop Physiology, ICRISAT Campus, building 501B, Pigeonpea 121, RC Puram, 502324 Patancheru, India, e-mail: marijn.voorhaar@wur.nl

Indigenous Knowledge and Inclusive Innovations: a Conceptual Framework

Branwen Peddi[1], David Ludwig[2], Joost Dessein[3]

[1]*Ghent University / Research Foundation Flanders (FWO), Agricultural Economics, Belgium*
[2]*Wageningen University & Research, Knowledge, Technology and Innovation Group (KTI), The Netherlands*
[3]*Ghent University, Dept. of Agricultural Economics, Belgium*

Inclusive innovations in agriculture have become widely embraced as a new way forward whilst including local stakeholders and their indigenous and local knowledge (ILK) traditions. Especially in the Global South, given a history of counterproductive attempts at so-called modernisation, the inclusion of ILK in innovation processes and policies has been promoted as a way of creating more balanced, sustainable and just human-environmental relations. Nevertheless, the inclusion of actors does not equate the inclusion of their needs or knowledge in this process, nor does it ensure an equal treatment of different knowledge traditions. The issue of power is one that cannot be overlooked. Furthermore, the concepts of inclusive innovations and ILK are often ambiguously employed, so there is a need to further clarify different perspectives on these concepts and how they interrelate.
Thanks to our review of the literature on ILK and agricultural inclusive innovations, we are able to present a conceptual framework that allows to analytically distinguish and reflect on different modes of how ILK is treated in inclusive innovation processes. These modes encompass the views that innovations are either an inherent part of this body of knowledge (internal to ILK), externally induced (external to ILK), or a combination of both (hybrid). Combined with the established narratives of ILK being either a body of knowledge fixed in time and space ("static") or one that is continuously changing and adapting ("dynamic"), we discern five main frames or mental models when considering ILK and innovations. These five frames allow for a productive analysis of the development of inclusive innovations, looking at whose needs and which actors are included – and by whom –, and discusses possible pitfalls for each frame. The framework is also intended as a reflexive tool for researchers, practitioners and policymakers that are interested in ILK and inclusive innovations. Therefore, we aim to take a meta-perspective on the conference topic, and provide important reflections to consider while moving forward towards a healthy and sustainable future.

Keywords: Epistemic justice, inclusive innovation, indigenous knowledge, power

Contact Address: Branwen Peddi, Ghent University / Research Foundation Flanders (FWO), Agricultural Economics, Rue des Comtes de Salm 66, 6690 Vielsalm, Belgium, e-mail: branwen.peddi@ugent.be

Profitability Analysis and Characterisation of Non-Carbonized Baobab Fruit Shell Briquettes in Malawi

Dalitso Kafumbula[1], William Dumenu[2], Kathrin Meinhold[2], Victor Kasulo[1], Matthias Kleinke[2], Dietrich Darr[2], Lusayo Mwabumba[1]

[1]*Mzuzu University, Dept. of Forestry and Environmental Management, Malawi*
[2]*Rhine-Waal University of Applied Sciences, Fac. of Life Sciences, Germany*

Baobab (*Adansonia digitata* L.) tree has multi-purpose uses and different parts of the tree are reported to be useful. However, baobab fruit shells are regarded as wastes which have no economic value and as such they are wantonly discarded around factory and harvesting sites in Malawi. Briquette production is sought to be an ideal venture to add economic value to the shells.
This study investigates profitability and characterisation of non-carbonized briquettes from baobab fruit shells and mixture with other biomass materials in Malawi. The study is mixed methods and conducted in three major cities and two other districts. Nine types of briquettes, comprising varying ratios of baobab fruit shells, rice husks, groundnut shells and dried tree leaves were produced. Total production cost was found to be MK4, 568,124.8 per tonne, with fixed and variable cost amounting to MK4, 426,000.40 and MK142, 124.40 per tonne respectively. The various briquette types were characterised in terms of calorific value, shatter index, ash content, density and moisture content. For example, briquettes from baobab fruit shells only had calorific value of 18.753 KJ/Kg and moisture content of 5.5 %. The energy value of 18.753 KJ/Kg refers to the fact that baobab fruit shell briquettes are competitive to firewood and fossil fuels by their calorific value. A market survey which involved 246 individuals was carried out to investigate market potential of briquettes. Random and convenience sampling techniques were used in selecting individuals. Gross margin and break-even point analysis were performed to estimate profitability. SPSS was employed to analyse briquette demand. Pearson Correlation and Chi-square statistical tools were used to establish the relationships and associations respectively regarding briquette usage. The 5 point-likert scale was used to assess respondents' attitudes and perceptions on briquettes. One-way ANOVA was used to determine significant differences between types of briquettes.

Keywords: Baobab, biomass, briquette, characterisation, fixed cost, variable cost

Contact Address: Dietrich Darr, Rhine-Waal University of Applied Sciences, Faculty of Life Sciences; Sustainable Food Systems Research Centre, Marie-Curie-Str. 1, 47533 Kleve, Germany, e-mail: dietrich.darr@hochschule-rhein-waal.de

The Business Model for a Digitized Extension Service and Mechanisation Technology in the Rice Sector

Rico Amoussouhoui[1], Aminou Arouna[2], Jan Banout[1], Miroslava Bavorova[1], Haritini Tsangari[3]

[1]*Czech University of Life Sciences Prague, Fac. of Tropical AgriSciences, Dept. of Sustainable Technologies, Czech Republic*

[2]*AfricaRice Center, Ivory Coast (Cote d'Ivoire)*

[3]*University of Nicosia, School of Business, Cyprus*

Farmers in developing countries often struggle with the adoption of new technologies. This study aims to design an innovative and flexible business model for the sustainable scaling of new technologies. A stakeholder and profit-oriented Canvas business model was used by adding two additional segments (i) profitability and (ii) sensitivity to the original Canvas. The model was designed based on an upstream preliminary analysis of the business environment and its competitiveness. We use a case study based on technologies (Personalized extension application called RiceAdvice, and a threshing machine called ASI thresher from the initial of the partners who contributed: AfricaRice; the Senegal River Valley National Development Agency and the Senegalese Institute of Agricultural Research) developed by AfricaRice. The results show that the designed improved Canvas business model is reliable and profitable when both technologies are used separately. However, higher profitability is observed when both technologies are combined in one business model. In this case, the business has a Net Present Value of about $17,381.84 and an Internal Rate of Return (IRR) of 33 %. The study shows that the business model is very sensitive to the price of the service. Therefore, we recommend an evaluation of the business model to find out the appropriate price and payment method for both the service recipient and the service provider. This study contributes to the business model literature by providing a path for the sustainable adoption of new technologies through a sustainable business model approach.

Practical implication: The outcome of the study is a flexible and operational model for the sustainable adoption of new technologies. The model can be directly used and implemented by actors interested in agribusiness.

Theoretical implication: The theoretical approach of the business model was conceptualised to propose a practical and flexible business model.

Originality: Based on the theoretical approach, the study proposes a practical framework to provide an accurate tool for a sustainable adoption process of new technologies.

Keywords: Adoption, business model, digitized extension service, new technology

Contact Address: Rico Amoussouhoui, Czech University of Life Sciences Prague, Fac. of Tropical AgriSciences, Dept. of Sustainable Technologies, Kamýcká 129, 16500 Prague, Czech Republic, e-mail: amoussouhoui@ftz.czu.cz

Financial Analysis of Small Scale Artisanal Chocolate Processing in Ghana

Bernard Kwamena Cobbina Essel, Miroslava Bavorova

Czech University of Life Sciences Prague, Fac. of Tropical AgriSciences - Dept. of Economics and Development, Czech Republic

Ghana produces up to 950,000 tonnes of cocoa beans per year, accounting for around a quarter of global production, but produces very little chocolate (Huellen & Abubakar, 2021). The government has long sought to close this gap, mainly through Ghana's National Export Development Strategy, which aims to export at least 40 % of processed Cocoa in the next five years(2021–2025) (Mabe et al.,2020). With this policy, the government also seeks to address unemployment among cocoa-growing communities in Ghana. As a result, the government has encouraged young people to start businesses, especially in the cocoa sector, by providing financial support or training (Ichikowitz Family Foundation, 2020). However, the country's investment in cocoa processing is still low. This can be due to the high initial capital costs of establishing such businesses (Hütz-Adams et al., 2016). This study, therefore, seeks to analyse the costs and returns of artisanal small-scale chocolate processing in Ghana. The study will be conducted in the Ga East District of the Greater Accra Region of Ghana using a case study approach. Data will be obtained from administering structured questionnaires through face-to-face interviews in May 2021. Information to be captured includes capital expenditure to establish an artisanal small-scale chocolate processing venture, annual operating costs and the returns from sales of the final product. Discounting measures of investment worth, including Net Present Value (NPV), the Internal Rate of Return (IRR), and the Benefit-Cost Ratio (BCR), will be used to analyse the economic viability of establishing Artisanal Small-Scale Chocolate processing in Ghana. The findings will have implications for the cocoa policy as it aims to understand the need for interventions to support the development of small-scale chocolate processing.

Keywords: Artisanal small-scale chocolate, cocoa, entrepreneurship, Ghana, value addition

Contact Address: Bernard Kwamena Cobbina Essel, Czech University of Life Sciences Prague, Fac. of Tropical AgriSciences - Dept. of Economics and Development, 16500 Kamýcká 129 , Czech Republic, e-mail: esselb@ftz.czu.cz

Is Sustainable Tomato Production on Floating Gardens on the Inle Lake (Myanmar) Possible?

Wolfram Spreer[1], Frank Giesel[2], Vicha Sardsud[3], Win Pa Pa Soe[2], Alexander Kunze[4], Johannes Max[4]

[1]*Maejo University, International College, Thailand*

[2]*GIZ, Ethiopia*

[3]*Freelance Consultant, Thailand*

[4]*Hochschule Weihenstephan-Triesdorf, Department of Plant Nutrition, Germany*

For over 100 years tomato production on floating gardens has a tradition on the Inle Lake in Southern Shan State, Myanmar, making up for a unique scenery with a touristic potential. Over the past decades, tomato production on the lake contributed to the national supply, especially during the dry season. During the phase of opening the country under a democratic government, agricultural production in Myanmar developed fast and due to the improved protected cultivation, more fresh vegetables became available on the domestic market. At the same time, concerns about the excessive use of mineral fertiliser and agrochemicals on the floating gardens were raised, which caused pollution of the lake and limited the touristic potential.

After the German Development Cooperation (GIZ) ran an analysis of the potentials to turn tomato production on the lake more sustainable, the establishment of a "Tomato Knowledge Center (TKC)" together with local farmers was initiated. The aim was to secure the livelihoods of local people belonging to the ethnic minority of Intha and at the same time promote environmental protection and touristic development. For the intervention four main areas were identified: a) Production of healthy seedlings, b) Apply improved cropping techniques, including placed fertilisation and IPM, c) promote varietal diversification and crop rotation, and d) Provide training on post-harvest techniques and management.

The TKC was established on a site on the lake and during one cropping cycle, all improved techniques were applied. Farmers substantially increased their yields and produced a level of fruit quality that allowed them to obtain tenfold price due to direct marketing, as compared to selling to classical brokers.

An outstanding aspect of the intervention is the fact that due to travel restrictions under the COVID19 pandemic no foreign expert was able to actually travel to the area of intervention. This case study shows the successful combination of classical teaching material, innovative communication tools, and social media to promote a small-scale rural development project in a remote area. Due to the political crises in Myanmar, project activities were ended. A return to military dictatorship will reverse Myanmar's economic development of the past years.

Keywords: IPM, remote teaching, rural development , vegetable nursery

Contact Address: Wolfram Spreer, Maejo University, International College, Chiang Mai, Thailand, e-mail: wolfram.spreer@gmx.net

The Changing Nature of Icts for Agricultural Extension in Developing Countries - A Review

Rashid Khan[1], Thomas Daum[1], Saurabh Gupta[2], Regina Birner[1], Claudia Ringler[3]

[1]*University of Hohenheim, Inst. of Agric. Sci. in the Tropics (Hans-Ruthenberg-Institute), Germany*

[2]*Indian Institute of Management Udaipur, Centre for Development Policy and Management,*

[3]*International Food Policy Research Institute (IFPRI), Environment and Production Technology Division, United States*

Information is essential for accessing key inputs to agriculture, including climate information, for improving farming practices, and for providing a better reach to markets. Information and Communication Technologies (ICTs) and their applications in agricultural extension have witnessed dramatic changes over the last seven decades (1950–2020): moving from an era of "transfer of technology" and no use of ICT–by largely involving face-to-face meetings with extension staff and farmer field days–to the current era of "co-innovation" using a series of advanced ICT tools, including smartphones and internet-based services. This evolution was spurred by advances in ICT and the need for more complex responses to agricultural challenges given climate change, and growing pressures on natural resources. Changing access to information has been a key to rural livelihoods and a significant enabling factor for improving agricultural practices, strengthening market access, and overcoming weather uncertainties. It is clear that the move toward modern forms of communication has helped change extension approaches from a top-down model toward a more networked and interactive format. However, many questions remain regarding the extent to which new ICT methods of transmitting agricultural information have positively affected the livelihoods of farmers in developing countries. This is particularly important for smallholder farmers in Sub-Saharan Africa and South and Southeast Asia that are less likely to have direct access to advanced ICT tools. To address this question, this review analysed 196 papers that focused on the use of various extension methods in the target regions. The results suggest that extension information disseminated through mobile phones has the potential to democratize diversified and complex agriculture practices.

Keywords: Agriculture extension, information communication technology, mobile, video

---

Contact Address: Rashid Khan, University of Hohenheim, Inst. of Agric. Sci. in the Tropics (Hans-Ruthenberg-Institute), 70593 Stuttgart, Germany, e-mail: rkhan.p@gmail.com

Energy Audit in Tofu Industry: Evidence from Indonesia

Lydia Mawar Ningsih[1], Jana Mazancová[1], Udin Hasanudin[2], Hynek Roubik[1]

[1]*Czech University of Life Sciences Prague, Fac. of Tropical AgriSciences, Dept. of Sustainable Technologies, Czech Republic*

[2]*University of Lampung, Department of Agro-industrial Technology, Faculty of Agriculture, Indonesia*

Tofu industry production in Indonesia is generally producing on small- and medium scale levels, which requires a lot of energy inputs in the production process. In general, the fuels used are not environmentally friendly, such as firewood and liquid petroleum gas (LPG). In addition, the tofu industry also uses large amounts of water for the production process, generate wastewater in large quantities. It is resulting in significant negative effects on the environment. Therefore, it is necessary to study the process systematically and propose ways how to overcome it. This study aimed to audit the energy consumption to obtain the energy consumption and the habits of small and medium-scale level tofu producers. The data collection method involved semi-structured interviews and questionnaire surveys carried out in 31 tofu industries in Gunung Sulah District, Bandar Lampung City, Lampung Province, Indonesia, in December 2020. The results showed that energy consumption varies according to the fuel used within the tofu cooking (boiling the soybean and frying tofu) on average; 71.05 MJ kg^{-1} for fuelwood, 16.95 MJ kg^{-1} for LPG, and 6.02 MJ kg^{-1} for wood pellets. The fuel, gasoline or diesel, consumed for milling depend on the type of milling machine used. The gasoline consumption is 1.94 MJ kg^{-1}, and the diesel is 3.22 MJ kg^{-1}. The tofu industry has high energy requirements in the production process. Total energy consumption in tofu industries is 69.33 MJ kg^{-1}, consisting of energy for cooking 66.64 MJ kg^{-1}, soybean milling machine 2.39 MJ kg^{-1}, utilities (electricity) 0.20 MJ kg^{-1}, and human energy 0.11 MJ kg^{-1}. The results of the energy audit make tofu maker aware that the use of energy for the production process is inefficient and not environmentally friendly. Therefore, it is necessary to carry out energy audits more deeply to utilise the wastewater as material for bioenergy, such as biogas to replace the primary fuels.

Keywords: Biogas , energy consumption, tofu industry, wastewater

---

Contact Address: Lydia Mawar Ningsih, Czech University of Life Sciences Prague, Fac. of Tropical AgriSciences, Dept. of Sustainable Technologies, Kamýcká129 16500, Prague, Czech Republic, e-mail: ningsih@ftz.czu.cz

Rice Production among Beneficiaries and Non-Beneficiaries of the Gulf of Mottama Project, Myanmar

No No Aung, Alessandra Giuliani, Urs Scheidegger
Bern University of Applied Sciences, School of Agricultural, Forest and Food Sciences (HAFL), Switzerland

In Myanmar, farmers face so many problems in rice cultivation and get low profits due to low yield with high production cost. The Gulf of Mottama project (GOMP) is conducted in the Gulf of Mottama region, funded by the Swiss Agency for Development and Cooperation (SDC) and implemented by HELVETAS Swiss Intercooperation together with other local NGO and INGOs to conserve biodiversity and develop the local communities sustainably. This study aimed to explore current situation of rice production and socio-economic characteristics of project beneficiary and non-beneficiary households and to assess the rice cultivation practices, their constraints faced in rice production. This study investigated on the influencing factors on rice productivity. A household survey (n=106) by personal interviews with 59 beneficiary households and 47 non-beneficiary households who are randomly selected and focus group discussions were conducted in eight project targeted villages, during the period August to October, 2018. Key informant interviews were also carried out with 10 experts. Descriptive statistics and multiple linear regression analysis were used for data analysis. The study results show that rice production plays a very important role for the interviewed house because about 55 % of their income comes from rice. About 80 % of beneficiary households applied mineral fertilisers and about 14 % of the sampled households used organic fertiliser (compost and farmyard manure). Seed quality problem was mostly found in non-beneficiary households. The rice productivity of the selected fields of sampled households was significantly influenced by being project non-beneficiary households, number of land preparation (tillage). Project beneficiary households get a higher yield and farmers get higher yields if they do only one tillage. The average productivity of the sample households was 39 baskets per acre. Although project farmers get higher yield and have more knowledge about sustainable production than non-beneficiary farmers, it is still needed to try hard to be economically and environmentally sustainable rice production. Most of the farmers have limited knowledge about agricultural inputs and especially on how to use them in the best way. This study indicates the need of further studies on current situation of rice production in different regions of Myanmar.

Keywords: Gulf of Mottama, myanmar, profit, rice production, yield

Contact Address: No No Aung, Bern University of Applied Sciences, School of Agricultural, Forest and Food Sciences (HAFL), Bern, Switzerland, e-mail: nonoaung1992@gmail.com

Social Innovation Initiatives in Agro-Food Systems of the Tropics and Moderate Regions

Eva Hilt, Brigitte Kaufmann

German Institute for Tropical and Subtropical Agriculture (DITSL), Germany

In agriculture, innovation used to be mainly understood as new products, methods of production or processes that are more efficient than previous, thus having a strong technical and economic orientation. However, technical innovation alone are increasingly found to be insufficient to address the complex sustainability challenges that also the current agro-food system faces. Rather, society itself, its norms, values, and paradigms, needs to change. Such changes can be induced by social innovation (SI) which are conceptualised as innovation that are generated through the collaboration of societal actors, serve social needs through a change in social relationships or structures. It is hypothesised that needs for SI differ in agro-food systems under tropical and moderate conditions and consequently, different new relationships are developed. This study systematically reviews of 38 peer-reviewed articles and two book chapters analysing socially innovative cases in the agro-food system. Much more studies are conducted in countries with a moderate climate (n=28) than in tropical. Results show, that in both regions, more than a third of cases address the social need of overcoming corporate control and regain food sovereignty. In tropical countries, another quarter of the initiatives aim to overcome institutional voids, whereas the second most addressed need by initiatives in moderate countries is to reconnect consumers to land, farming and food. Different are also the mechanisms by which the initiatives address the above need. In tropical countries, two thirds of the initiatives establish new relationships between producers, e.g., through self-help groups or farmer cooperatives. In non-tropical countries, six different types of relationship changes between value chain actors are identified, each occurring in quite equal shares. As the needs for SI are rather comparable in both world regions, but cases as well as variety of mechanisms (types of relationship changes) are smaller in tropical countries, it is concluded that currently the potential for SI is underexploited in tropical countries, which offers opportunity for exchange between academic and activist actors in both world regions for increasing knowledge and experience with this new type of innovation for the transformation of agro-food systems.

Keywords: Agro-food system transformation, food sovereignity, social innovation, social learning

Contact Address: Eva Hilt, German Institute for Tropical and Subtropical Agriculture (DITSL), Witzenhausen, Germany, e-mail: e.hilt@ditsl.org

Analyzing Farmers' Socio-economic Conditions by Recording Data through a Mobile Application in Myanmar

Soe Khaing[1], Htun Zaw Htay[2], Alessandra Giuliani[1]

[1]*Bern University of Applied Sciences (BFH), School of Agricultural, Forest and Food Sciences (HAFL), Switzerland*

[2]*Helvetas Myanmar, Gulf of Mottama Project, Myanmar,*

Farmers' mobile phone applications (apps) provide agricultural information including weather, cultivation techniques, and market prices, to improve farm management. By using these apps, farmers can also communicate directly with agricultural technicians, traders, Agri-shop owners, and livestock experts. This study explores the socio-economic status and agricultural knowledge of the households using the farmers' mobile app 'Greenway' in the Gulf of Mottama, Myanmar. The research particularly looks to the monsoon paddy income registered with the use of the App. The major constraints of the Greenway app are also analyzed. The study was conducted through a survey using distance interviews with 46 households trained to use the mobile app in the Gulf of Mottama Project area led by Helvetas Myanmar in Mon State and Bago Region from October to December 2020. Descriptive analysis and multiple linear regression analysis were applied to analyse the data. The descriptive analysis shows that the majority of both regions' households have a male head, aged 49 years old on average. About 40 % of the farmers own primary education level. All the household heads own mobile phones and over half of them possess harrow machines. The average household non-farm income/year is 2,997 USD, with a higher income in the Bago region due to remittance from migrant workers. The average farm size is 15.5 acres. The average monsoon paddy income/acre in 2020 was higher than in 2019 (230 USD), due to higher yield. From the Greenway app, farmers access agricultural information regarding weather, market price, pests& diseases, cropping techniques, livestock, seed information, compose technique, foliar fertiliser, and storage. The result shows that 56 % of the farmers find pests&diseases as the most useful information from the app, followed by weather (28 %) and fertiliser application techniques (26 %). The most appreciated feature of the Greenway app is getting timely information that helps take the right farm management decisions for higher yield. Eighty-seven percent of the respondents indicated the difficulties in using the app due to low education as a major app challenge, followed by poor internet connection (65 %).

Keywords: Farmers' mobile app, ICT in agriculture, Myanmar

Contact Address: Soe Khaing, Bern University of Applied Sciences (BFH), School of Agricultural, Forest and Food Sciences (HAFL), Länggasse 85, 3052 Zollikofen, Switzerland, e-mail: soe.khaing@students.bfh.ch

Impact of Agricultural Market Linkages on Small-Scale Farmers' Welfare: Evidence from Tanzania

André Bueno Rezende de Castro

University of Bonn, Center for Development Research, ZEF, Germany

Subsistence agriculture practised by small-scale farmers (SSFs) is still pervasive in low-income countries (LICs); in Tanzania it plays an important role for food security and income generation in rural areas. In addition to subsistence, the backbone of the local economy is composed by industrial processing of agricultural raw materials, forming the wider agro-industrial sector. In the transition and evolution of food systems, an important step up from subsistence production has been the integration of SSFs into output markets through the strengthening of agricultural value chains – with the potential to lift millions out of poverty. This paper focuses on Tanzania to study how SSFs' participation in vertical market linkages (VMLs) increases household welfare. We contribute to the literature of agricultural economics in two ways. First, by considering a wider variety of welfare indicators (income, poverty, food security, and subjective well-being) in addition to crop yield and commercialisation. Second, by analysing the heterogeneous welfare effects of different VMLs: the first composed by contract farming and cooperatives, the second by processors. The study uses a country-wide representative balanced panel from Tanzania (NPS-LSMS) and employs the fixed effects and the first differences estimators to control for unobserved household time-invariant characteristics. Additionally, propensity score matching (PSM) is used in tandem with these two methods in order to reduce selection bias of farmers; PSM is also employed as a robustness check by estimating average treatment effects. Our main result is that integration to VMLs increases household welfare for all outcome variables in the processors' VML group and less so in the outgrowers' group. However, direct production and income gains do not always convert into better final welfare, including food security. Our findings complement the literature of welfare effects of agricultural markets' integration by showing how different VMLs can improve the lives of SSFs in Tanzania, a case study whose characteristics are also present in similar LICs. We believe that there is scope for further research into the benefits that agricultural value chains can effectively bring to SSFs in the developing world.

Keywords: Agricultural value chains, food security, household welfare, small-scale farmers, Tanzania

Contact Address: André Bueno Rezende de Castro, University of Bonn, Center for Development Research, ZEF, Genscherallee 3, 53113 Bonn, Germany, e-mail: brcandre1@gmail.com

A Discrete Choice Experiment to Measure the Malawian Potential Market for Baobab Fruit Shell Briquettes: Evidence from Consumer Preferences in Mzuzu City

Olha Bovenkerk, Eleydiane Maria Gomes Vale, Dietrich Darr
Rhine-waal University of Applied Sciences, FRac. of Life Sciences, Germany

Our research aims to analyse consumer preferences for baobab fruit shell briquettes, as a potential alternative to substitute current fuel sources for cooking in Mzuzu City, Malawi. The main objective of this work is to determine the parameters for a successful implementation of an alternative energy source for firewood and charcoal. Therefore a discrete choice experiment is designed with 8 attributes and 3 levels in a local conducted survey, where the willingness to pay according to the characteristics of the new product is estimated. Attribute choices were made focusing in important daily use of fuel sources, such as waiting time until cooking can be started, burning time of fuel, help from children in collection, health and comfort when using cooking fuel, price, among others. The results show which characteristics baobab fruit shell briquettes should have in order to meet local consumer needs, and increase the chance of market acceptance. The more useful characteristics the alternative product has, the higher the user adoption, and willingness to pay increases. A positive correlation of income and education with willingness to pay for baobab fruit shell briquettes is expected. The resulting findings may lead to better planning and success assessment of the alternative product. These findings are particularly relevant for investors and policy makers, as a reliable analysis of the accomplishment potential leads to investments and subsidies, which makes a successful establishment of an alternative energy resource more likely. As a result, forest resources may get a higher chance to be secured in the long term, which has a positive effect on the environment, economy and society in Malawi.

Keywords: Baobab fruit shell briquettes, discrete choice experiment, Malawi, willingness to pay

Contact Address: Eleydiane Maria Gomes Vale, Rhine-waal University of Applied Sciences, FRac. of Life Sciences, Marie-Curie-Straße 1, 47533 Kleve, Germany, e-mail: eleydiane.gomes-vale@hochschule-rhein-waal.de

Scaling-Up Agricultural Technologies: Who Should Be Targeted?

Shaibu Mellon Bedi, Lukas Kornher
Center for Development Research, Economic and Technological Change, Germany

Policy directions from most agricultural technology adoption studies in sub-Saharan Africa (SSA) often recommend single solution in addressing challenges associated with already complex farming systems. Thus, previously disseminated agricultural technologies in SSA have either been less adopted or disadopted by farmers or farm households. In this study, we adopt the marginal treatment effect approach in examining how farmers' resource endowment and unobserved factors influence the marginal benefits of adopting sustainable intensification of agricultural practices (SI practices), estimate both the marginal and average benefits of adopting SI practices on maize yield and net returns, and predict which farm households at the marginal entrants will benefit most when targeted during the scaling-up. Our findings suggest that both farmers' resource endowment and unobserved factors affect the marginal benefit of adopting SI practices. Estimates also indicate that the adoption of SI practices increases maize yield and net returns of adopters. Our scaling-up policy prediction estimates suggest that at the marginal entrants, scaling-up SI practices to favour farm households who by observed socioeconomic characteristics appear least likely to adopt would generates the highest marginal benefits. Furthermore, our findings indicate that policies that promotes the formation of farmer-based organisation or coupled with the enhancement of farmers' human capital through extension services can be used to stimulate adoption. Finally, our results caution the use of the average estimate of treated farmers or farm households for scaling-up policy decision, since the average estimate of treated farmers is greater than the average marginal estimate of farmers at the marginal entrants.

Keywords: Adoption, Ghana, marginal treatment effect, scaling-up, sustainable intensification practices

Contact Address: Shaibu Mellon Bedi, Center for Development Research, Economic and Technological Change, Genscherallee 3 - D-53113, Bonn - Germany, Germany, e-mail: sbmellon2005@gmail.com

Adoption of Improved Groundnut Varieties and Output Market Participation among Smallholders in Malawi

Admire Katunga[1], E Wale Zegeye[2], Gerald Ortmann[3]
[1]*Ministry of Agriculture , Agricultural Research Services, Malawi*
[2]*University of Free State, Agricultural Economics, South Africa*
[3]*University of Kwazulu Natal, Agricultural Economics, South Africa*

The quest to increase marketable surplus prompts some farmers to demand improved varieties. Improved groundnut varieties are superior to the conventional ones in terms of yielding potential. Thus, the Malawi Government promotes the adoption of high-yielding improved varieties for smallholders to increase the production of marketable surplus. An increased marketable surplus is the prerequisite for farmers to participate in the market. Increased market participation improves the smallholders' general welfare, such as increased agricultural income earnings and reducing poverty. Increased market participation is also one of the pathways through which the smallholders transform from subsistence to commercialised agriculture. The study investigated the impact of adopting improved groundnut varieties on the intensity to participate in the output market among smallholders in Malawi's selected central and northern districts. Farm household data for 416 groundnut producers were collected in Lilongwe, Salima, Kasungu, and Mzimba. A control function was applied to an endogenous treatment effect model to control endogeneity in the unobservable covariates between treatment and outcome models. The study results indicated that adopting improved varieties positively impacted smallholders' intensity to participate in the output market. Other results revealed that access to seed loans, total land for crop cultivation, and access to extension services positively impacted the smallholders' adoption of improved varieties, while poor conditions of rural roads negatively impacted adoption. To increase the intensity to participate in the market among smallholders, policy strategies that enhance the adoption of improved groundnut varieties should be promoted among them. In this case, access to seed loans and extension services, increasing land productivity, and improving rural roads are critical.

Keywords: Access to extension, endogenous treatment model, increased production, seed loans, yielding potential

Contact Address: Admire Katunga, Ministry of Agriculture , Agricultural Research Services, Chitedze Research Station, 265 Lilongwe, Malawi, e-mail: katungaadmire@gmail.com

Individual Ambient Ware Potato Storage Excels in Uganda

Pieter Wauters[1], Monica Parker[2], Diego Naziri[3], Alice Turinawe[4]

[1]*Centre for International Migration and Development (CIM), Germany & International Potato Center (CIP), Uganda*

[2]*International Potato Center (CIP), Kenya*

[3]*International Potato Center (CIP), Vietnam & Natural Resources Institute (NRI), University of Greenwich, United Kingdom*

[4]*Makerere University, College of Agricultural and Environmental Sciences, Dept. of Agribusiness & Natural Resource Economics, Uganda*

Potato is a key food and cash crop in Uganda, produced by approximately 200,000 smallholder farmers, primarily in the southwestern and eastern highlands of the country. The major production periods are March-July and September-January, with some off-season production in swamps, valley bottoms and irrigated areas. National potato production has steadily grown over time to respond to an increasing demand and consumption. Ugandan potato farmers sell majority (±60 %) of their production immediately after harvest, principally to traders at the farmgate or in nearby local markets. The main reasons are their immediate need for cash, the low volumes of potato harvested, their fear of loss during storage due to pests and diseases, and a lack of adequate storage facilities. Roughly 30 % of the harvested potato is stored as food for later consumption by the household and as seed for planting in the next cropping season. Only a few farmers store small quantities of potato for later sale as ware potato, predominantly in traditional light storage facilities. Due to seasonally fluctuating market prices, most Ugandan potato farmers miss the opportunity to sell at higher price later in the season when potato supply in the market is scarce. To promote potato storage for later sale, improved ambient ware potato storage units were introduced and evaluated in Uganda. Both individual and group stores with a storage capacity of 8 and 50 tonnes, respectively, were piloted. Improved ambient stores ensure that potatoes are kept in the dark and are made from locally available materials. They can maintain marketability of stored potato for up to 3 months by taking advantage of cooler temperatures at night. Only a few of the group storage units generated profits. Furthermore, all of them appeared to present several challenges typical of collective action endeavours. The individual units, however, performed very well with an average payback period of 3–4 years that could even be reduced to less than one year if these stores are used at full capacity. Due to their characteristics, improved individual ambient ware potato stores thus seem to be particularly suitable to increase substantially the income of potato farming households.

Keywords: Cost-benefit analysis, improved individual ambient store, Uganda, ware potato storage

---

Contact Address: Pieter Wauters, Centre for International Migration and Development (CIM), Germany & International Potato Center (CIP), Kampala, Uganda, e-mail: p.wauters@cgiar.org

Marketing Channel and Margin Analysis of Wheat the Case of Debre Elias Woreda, Northwestern Ethiopia

Mezgebu Aynalem Mengstie

Debre Markos University, Burie Campus, Ethiopia

Agriculture is the basic economic sector on which the country relies for its social and economic development. Its contribution to the gross domestic product (GDP), employment, and foreign exchange earnings of the country is about 35.8, 72.7 and 90 percent, respectively. This research attempted to analyse the market chain of wheat in Debre Elias woreda. Specifically aims to analyse the market structure-conduct-performance of wheat market.
The result of this study was analysing the market chain of wheat in Debre Elias woreda. Specifically aims to analyse the market structure-conduct-performance of wheat market. To collect primary data, 154 wheat producers were selected using simple random sampling method and 31 traders were selected purposively. To address the objectives of the study, descriptive statistics were used. The result indicated that Debre Elias wheat market was inefficient, characterised by oligopolistic market structure. The major barrier to enter into the market was shortage of capital. Licensing and years of trading experience did not hinder entry into wheat trading activities. Moreover, the markets were overwhelmed by information asymmetry with low degree of market transparency. Although trading of wheat is profitable across all sample farmers and traders, problems like oligopolistic market structure and information asymmetry made the trading business uncompetitive and inefficient.
The enhancement of wheat producers' bargaining power through cooperatives is the best measure that should target at reducing the oligopolistic market structure in the Debre Elias regional market. The measure also favours the sustainable supply of wheat at reasonable price to consumers. The study recommends the agricultural sector should improve wheat production capacity by providing new technologies and create stable market system for farmer's surplus production.

Keywords: Conduct, market, performance and oligopoly, structure, wheat

Contact Address: Mezgebu Aynalem Mengstie, Debre Markos University, Burie Campus, Debre Markos, Ethiopia, e-mail: mezgebu12aynalem@gmail.com

The Role of Sustainable Value Chain Management in Mainstreaming SCP Practices in Kenyan Horticulture Sector

Joshua Aseto, Kartika Anggraeni, Marianne Magnus-Melgar
Collaborating Centre on Sustainable Consumption and Production (CSCP), Sustainable Business and Entrepreneurship (SBE), Germany

Adoption of sustainable consumption and production (SCP) practices by micro small and medium enterprises (MSMEs) in agriculture contributes to higher produce yields, food security, improved livelihoods, and resilience to climate change. This notwithstanding, adoption of SCP by agricultural MSMEs in Africa is still low due to inadequate knowhow on SCPs, limited access to agri-finance, over-reliance on rain-fed production, and weak and non-transparent value chain with informal and fragmented local markets that involve myriad middlemen. This study discusses and analyses the outcomes of the EU-funded Switch Africa Green 'Green hOrticulture At LAke Naivasha' (GOALAN) project (2018–2021) that seeks to promote the adoption of SCP practices among MSMEs in the horticulture sector in Lake Naivasha Basin in Kenya through training and skill provision, and to improve the MSMEs access to finance, market opportunities and infrastructure. The methods of data collection used in this study include desk review, key informant interviews, focus group discussions, and semi-structured interviews. The findings show that MSMEs face challenges in accessing markets due to their remote locations and lack of access to appropriate and reliable means of transportation. This poses a major challenge considering the fragility of fresh fruits and vegetables (FFVs), which can be easily damaged during transit. The challenge is exacerbated by the lack of well-maintained roads connecting production sites and markets, and lack of cooling storage to extend the life of FFVs. This has resulted in post-harvest losses, including the discarding of 'less appealing' FFVs. These post-harvest losses can be reduced by establishing produce aggregation hubs and strategic partnerships with key value chain actors; supporting the youth to establish centralised and well-coordinated digital marketing platforms for MSMEs; and by ensuring that MSMEs have access to simple machinery to add value to their produce. Project interventions, particularly through infrastructural and training support, have resulted in an increased capacity among MSMEs to achieve the Kenyan Standard (KS)1758 for horticulture allowing them access to markets (contract farming). However, MSMEs can only benefit from such interventions fully, if the requisite infrastructure i.e., roads, aggregation hubs, cold storage facilities, reliable means of transportation are available and functional.

Keywords: Consumption, horticulture, Kenyan, markets, MSMEs, production, sustainable, value addition, value chain

---

Contact Address: Joshua Aseto, Collaborating Centre on Sustainable Consumption and Production (CSCP), Sustainable Business and Entrepreneurship (SBE), Hagenauer Str. 30, 42107 Wuppertal, Germany, e-mail: joshua.aseto@scp-centre.org

Innovating Productivity for Small Farmers: A Case Study from Arid Zone

Dheeraj Singh, M K Chaudhary, Chandan Kumar

Central Arid Zone Research Institute, Krishi Vigyan Kendra, India

Almost 50 % of India's total population consists of small farmers and their families, and 85 % of all farms are less than two hectares. Eventually the situation of small farms is of enormous importance to the overall social wellbeing of India but the questions regarding the role of small farm in agricultural growth is still not very clear. Keeping the above facts in view CAZRI KVK made an attempt to access and improve the productivity of small farms in Pali district through their participation in technological development innovations. Under this programme front line demonstrations (FLD) were organised in its selected villages during 2017–20. For accessing the productivity, a total of 600 demonstrations were laid on 200 hectare in 10 villages across five blocks of Pali district. Under this programme, quality seeds of improved varieties of prominent crops of the area were distributed to the identified farmers and a number of trainings on scientific production technology to the identified farmers in the villages were also arranged for technology empowerment of farmers. In order to harness the synergy between technologies and the community participation, special emphasis was given to build farmer's capacity to produce quality produce with enhanced yield. The farmers adopted the concept and undertook the programme in operational area which showed a considerable spread of selected variety in nearby villages with a considerable improvement in yield of all crops. Fine-tuning of the production technology based on the location specific conditions and resources available with the farmers enhanced the adoption rate. The results indicate higher additional returns and effective yield under demonstrations which were due to improved variety, scientific proven technology, non-monetary factors, timely operations of crop cultivation and scientific monitoring. From an initial start of 600 farmers the variety and innovation spread to 140 villages covering 790 hectares of area. Certainly, the greater use of inputs, improved varieties and making more intensive use of land and new technology helps to gain higher productivity in small farms. Above all, future work will require a mult-disciplinary approach that involves not just soil scientists, agronomists, and farmers, but also ecologists, policymakers, and social scientists.

Keywords: Arid region, economics, front line demonstrations (FLD), production technology, yield gap

Contact Address: Dheeraj Singh, Central Arid Zone Research Institute, Krishi Vigyan Kendra, Jodhpur Road, 306401 Pali, India, e-mail: dheerajthakurala@yahoo.com

Farmers Market Integration: The Importance of Trust Perception and Competence Signals

Dalel Ayari[1], Lokman Zaibet[2], Ghazi Boulila[1]
[1]*Université de Tunis, Ecole Superieure des Sciences Economiques et Commerciales, Tunisia*
[2]*Sultan Qaboos University, Natural Resource Economics, Oman*

Agriculture development and exports of agricultural products would benefit by increased competitiveness by means of improved product quality and reduced uncertainty along the value chain. The dairy sector in Tunisia is exclusively dominated by informal distribution channels with failure in delivering quality milk. Formal marketing channels are seen as a means to create value and contribute to the development of the agricultural sector. The main objective underlining this study is to investigate the factors that might affect dairy farmers' market decisions regarding the selection of milk marketing channels and to consider trust perception as key to shaping breeders' decision and behaviour in the local economy.

The data used in this study were collected from a random sample of 45 smallholder breeders in northern Tunisia. Multinomial logit model was used to estimate the impacts of trust variables on the dependent variable; selection of milk marketing channel. The empirical results indicated that four factors: trust perception, herd size, price setting and regular communication, are major predictors influencing milk producers to choose formal channels. The results also show that farmers who are able to appreciate the signals of competence are in a position to better judge and trust their partners and therefore chose formal channels rather than the informal one. This leads to the conclusion that there exist an association between trust perception, competence and credibility signals and the choice of formal channels. The implications of these results are the following: collection centres shall develop trust, competence and credibility signals. This will create an enabling environment to develop the sector and integrate farmers in the value chain.

Keywords: Appreciation, competence and credibility signals, farmers market integration, formal marketing channel choice, trust perception

---

Contact Address: Dalel Ayari, Université de Tunis, Ecole Superieure des Sciences Economiques et Commerciales, Tunis, Tunisia, e-mail: ayaridalel@yahoo.fr

Assessing changing systems - paradigm shifts in data acquisition and modelling

---

Oral Presentations

Posters

Reconstructing Uncertain Crop Yield Data from Multiple Sources with a Bayesian Imputation Framework

Bernhard Schauberger[1,2], Abel Chemura[2], Roopam Shukla[2], Christoph Gornott[3,2]

[1]*University of Applied Sciences Weihenstephan-Triesdorf, Fac. of Sustainable Agriculture and Energy Systems, Germany*

[2]*Potsdam Institute for Climate Impact Research (PIK), Germany*

[3]*University of Kassel, Fac. of Organic Agricultural Sciences, Germany*

Climate change may threaten agriculture in West Africa by changing temperature regimes and precipitation patterns. Adapting crop management to these new realities may attenuate potential climate impacts. For planning such adaptation measures, reliable information on past and current crop performance is paramount.

There are several data sources on crop management and performance, which could potentially be used for adaptation planning. We explore five independent data sets of crop performance in Ghana: household surveys, satellite measurements, two crop models and official reported yield data. There are few concordances between the five sources. This disagreement questions the usability of a single data source for studying climate impacts and plan adaptation measures. To overcome issues with data reliability, we propose a solution to reconstruct crop performance data from multiple sources using a Bayesian network. Conditional probability tables are learned from data and lead to probabilistic assessments of crop yields. We test our approach in the US mid West, where yield data are deemed reliable, by introducing artificial data gaps and reporting errors.

Our Bayesian approach is partly able to reconstruct missing data or artificially introduced reporting errors, based on different independent data sources. Yet open questions remain, particularly with respect to extreme yields and the more diverse agricultural landscape in Western Africa compared to the US. Our next steps thus comprise the improvement of the reconstruction in the US and the transfer of the approach to West Africa. A successful reconstruction of crop yield data may then guide adaptation planning under current and future uncertainty.

Keywords: Bayesian network, crop yield, reconstruction, USA, West Africa

---

Contact Address: Bernhard Schauberger, Potsdam Institute for Climate Impact Research (PIK), Telegrafenberg A31, Potsdam, Germany, e-mail: schauber@pik-potsdam.de

UAV Based Nitrogen Management of Maize in China

Yang Zhang [1], Zichong Wang[1], Klaus Spohrer[1], Xiongkui He[2], Joachim Müller[1]

[1]*University of Hohenheim, Inst. of Agricultural Engineering, Tropics and Subtropics Group, Germany*

[2]*China Agricultural University, College of Science, China*

Nitrogen is the main fertiliser of maize in China. However, misuse of nitrogen fertiliser is widespread. To ensure optimal fertilisation and high crop yields as well as to reduce environmental pollution, precise fertiliser application of nitrogen should be implemented. UAV based spectral remote sensing of plant nitrogen status is fast and non-destructive and thus a good method to monitor possible stress in maize with respect to nitrogen fertilisation. After getting the spectral information of the plant canopy, it is important to decide whether the plants are stressed or not. However, most researchers focus on the accurate nitrogen detection of the plants by analysing different vegetation indices (VIs). Since the recommended amount of nitrogen fertiliser for plants is not unique but within a certain range, there are large individual differences in crop growth. There is still not enough research about optimal fertilisation in practice.

In this research, we adopted a different point of view: We do not focus on absolute nitrogen content in leaves but on spatial spectral differences in the field, which are caused by spatially different nitrogen supply. Based on this, areas of nitrogen demand c be identified. In addition to this, focus is also on the vegetative growth stage, which is sensitive to fertiliser deficiency and important for biomass production and yield. Therefore, different nitrogen fertilisation treatments for maize will be conducted. Spectral images will be captured during the plants' vegetative growth stages and compared with nitrogen content in leaves as well as LAI and plant height. The results are expected to provide basic spectral differences between different plant nitrogen status. Based on this it will be tried to give recommendations for additional spatiotemporal nitrogen fertilisation.

Keywords: Maize, nitrogen fertilisation, spectral remote sensing, UAV

Contact Address: Yang Zhang , University of Hohenheim, Inst. of Agricultural Engineering, Tropics and Subtropics Group, Stuttgart, Germany, e-mail: yang.zhang@uni-hohenheim.de

Dynamic Runoff Generation Processes and Models: A Review on Application of Models to Variable Catchment Properties in Ethiopian Highlands

Tadesse Shimels Gebremedhin[1,2], Lutz Breuer[2], Belete Berhanu[1], Solomon Gebreyohannis Gebrehiwot[3]

[1]*Addis Ababa Institute of Technology, School of Civil and Environmental Engineering, Ethiopia*

[2]*Justus-Liebig University Giessen, Inst. for Landscape Ecology and Resources Management (ILR), Germany*

[3]*Addis Ababa University, Ethiopian Institute of Water Resources, Ethiopia*

Hydrological models are developed to answer the question how rainfall becomes runoff. Direct runoff could be generated when either rain falls over saturated area (called saturation excess runoff) or the rainfall intensity is greater than the soil infiltration capacity (called infiltration excess runoff). Many of the model's runoff generation processes (RGP) are formulated based on the concept of either infiltration excess or saturation excess. However, RGPs are highly dynamic in space and time, often changing between the two, and hydrological models should account for that. Hence, this paper aims on highlighting challenges and the way forward in developing models with dynamic RGPs. We review the effect of catchment characteristics on RGPs, commonly used models and how they consider RGPs. Our study focuses on the Ethiopian highlands as an example, where RGPs are known to be very heterogonous. The country is characterized with rugged topography with elevation ranging from 120m bsl to 4620m asl. Furthermore, the soil, land use and other RGP controlling factors are highly variable across the country. However, we find that models with one fixed RGP are applied on catchments with different characteristics and known differences in RGPs, whilst several models with diverse RGPs are applied for the same catchment. This kind of random model application leads to structural model uncertainties which likely result in model outputs that might be right for the wrong reasons. Thus, understanding the RGP dynamics and considering this dynamics in the modeling process is needed to get better and more realistic outputs, which are required to appropriate decisions for water management. We present a blue print to overcome this research gap in current hydrological models.

Keywords: Controlling factors, models, RGP dynamism

Contact Address: Tadesse Shimels Gebremedhin, Addis Ababa Institute of Technology, School of Civil and Environmental Engineering, Addis Ababa, Ethiopia, e-mail: asnie2000@gmail.com

Same Climate, Different Soil: Simulating Maize Yield and Water Use for Smallholder Systems with Different Soil Datasets in the Lowveld Region of Limpopo Province, South Africa

Moshood Owolabi[1], William Nelson[1], Eugenio Diaz-Pines[2], Kingsley Ayisi[3], Reimund P. Rötter[1]

[1]*Georg-August Universität Göttingen, Dept. of Crop Sciences; Tropical Plant Production and Agricultural Systems Modelling, (TROPAGS), Germany*

[2]*University of Natural Resources and Life Sciences (BOKU), Inst. of Soil Research, Austria*

[3]*University of Limpopo, Risk and Vulnerability Science Centre, South Africa*

Maize is the main staple crop in southern African drylands, but smallholder farmers are far from achieving potential yields. Climate change projections indicate that low and variable yields will be exacerbated, making maize farming riskier without adaptation. The assessment of management interventions through Crop Simulation Models (CSM) such as Agricultural Production System sIMulator (APSIM) can help understand these systems and design improvements, but they usually rely on good quality and site-specific soil data. While site-specific soil sampling can be resource-demanding, global soil databases could be a reliable source of information. However, the efficiency and accuracy of such databases in terms of simulated crop yield has not been extensively researched. Using the Limpopo province South Africa as a case study, manually sampled soil data collected across two contrasting villages, Gabaza (fertile soil, high rainfall) and Selwane (marginal soil, low rainfall), were compared with digital soil data from the iSDAsoil database. The CSM APSIM was used to examine the extent to which digital soil data can replace physical soil sampling through its ability to simulate maize yield performance over a 20-year period (1999–2019). Soil parameters used across the two datasheets (observed and gridded data) include the C: N ratio, soil texture and rooting depth. Findings from our analysis revealed that soil parameters from the iSDAsoil map were significantly different ($p < 0.05$) from sampled (observed)data for the fertile site (Gabaza) but not for the marginal soil in Selwane. Simulated maize yield results mirrored these differences. Using a Status quo management scenario (no fertiliser or irrigation, locally recommended plant density of 2.7 plants m^{-2}) showed that average maize yields for Gabaza were 1.7 vs. 1.1 t ha^{-1} (observed vs. iSDA) and for Selwane 0.8 vs. 0.9 t ha^{-1} (observed vs. iSDA). Overall, our study showed that the iSDAsoil map could be used with high confidence to simulate smallholder maize performance only for marginal soils in the study area. Discrepancies for the fertile soils necessitate further investigations.

Keywords: APSIM, smallholder maize systems, sustainable adaptation

Contact Address: Moshood Owolabi, Georg-August Universität Göttingen, Dept. of Crop Sciences; Tropical Plant Production and Agricultural Systems Modelling, (TROPAGS), Grisebachstraße 6, 37077 Göttingen, Germany, e-mail: moshood.owolabi@stud.uni-goettingen.de

Coping with Erratic Rainfall: A Model Approach for Improving Maize Sowing Date

Nuttapon Khongdee[1], Thomas Hilger[1], Wanwisa Pansak[2], Georg Cadisch[1]

[1]*University of Hohenheim, Inst. of Agric. Sci. in the Tropics (Hans-Ruthenberg-Institute), Germany*

[2]*Naresuan University, Dept. of Agricultural Science, Thailand*

The objectives of this study were to (i) evaluate the ability of the Water, Nutrient and Light Capture in Agroforestry Systems (WaNuLCAS) model for predicting maize performance under rainfed conditions, (ii) assess the performance of maize under various sowing date options to sustain maize yields, and iii) identify the best sowing option under irregular rainfall. A two-year data set with various sowing dates from a field experiment in northern Thailand was used to calibrate and validate the model. Results indicated that WaNuLCAS was able to predict maize yield well (Goodness-of-fit statistics: R^2=0.83; EF=-0.61; ME=0.16; CRM=0.02; CD=0.56). An analysis of past rainfall data (1970–2018) of the Phitsanulok province, northern Thailand, indicated that only 27.1 % of the years corresponded with the long-term mean, while the same percentage was either moderately dry or moderately wet. The remaining years (19 %) were very wet or very dry, making sowing date decisions difficult. Five sowing date options were simulated using WaNuLCAS, i.e. farmers' practice (FP), 15, 30, and 45 days before FP, and staggered planting (a combination of them) as a strategy to cope with rainfall variability. Simulations revealed that under current rainfall conditions water was the most limiting factor for growth and yield of maize while nutrients (N and P) had only minor impact. Maize water uptake was significantly correlated with yield formation (R^2: 0.45). Sowing maize 30 days before FP or staggered planting are suitable options for farms prone to irregular rainfall conditions, the later particularly when no distinct weather forecasts are possible. Both options reduced the risk of crop failure while maintaining yields under these conditions.

Keywords: Climate change, decision support tool, growth and yield limitation, staggered planting, upland area

Contact Address: Nuttapon Khongdee, University of Hohenheim, Inst. of Agric. Sci. in the Tropics (Hans-Ruthenberg-Institute), Garbenstr. 13, 70599 Stuttgart, Germany, e-mail: nuttaponkhongdee@gmail.com

Investigation of the Spatial and Temporal Variations of Weather Conditions in a Mesoscale Vineyard

PHILIPP PFISTERER, STEFFEN SCHOCK, IRIS RAMAJ, JOACHIM MÜLLER
University of Hohenheim, Inst. of Agricultural Engineering, Tropics and Subtropics Group, Germany

Climatological conditions and weather variability have a momentous impact on viticulture and vineyard management and can be detrimental for grapevine growth and its yield. Humid weather conditions contribute to the spread of fungal pathogens and diseases, which afterwards degrade the quality of the grapevine and risk the longevity of orchards and vineyards in the tropical and subtropical regions. Therefore, it is critical to monitor the spatial and temporal variations of weather conditions in the vineyards. Despite numerous sensor systems developed in academia and industry to address this problem, a scalable and dense sensor system that guarantees low maintenance, fast and reliable data acquisition is still lacking. Thus, in this study, a low-cost wireless networked system was developed for real-time monitoring of weather parameters namely, temperature, relative humidity and dew point temperature. A capacitive-type sensor SHT31 integrated into an STM32L0 microcontroller was employed as a measuring unit. Data transmittance was empowered via a functional radio network. The sensor housings were designed and manufactured in-house via a 3D printer. The accuracy of data readings was validated by a climatic test chamber CTS-20/1000 under a wide-ranging set of temperatures and relative humidities. As an evasive experimental site, a mesoscale 30-ha vineyard located in Hessigheim, Germany was used to test the monitoring system. A number of 30 sensors were installed irregularly in this area. For the graphical analysis, data collected during the summer and winter periods were compared. From the results, substantial differences in temperature were observed between vineyard sites at $p \leq 0.05$. The spatial temperature gradients altered up to 8°C, which was mainly attributed to the heterogeneous and steeply sloping terrain of the vineyard. These gradients increased over the summer and decreased during the winter. This behaviour was accredited to the diurnal solar orientation, shaded conditions as well as wind direction imposed by a bend in the river. Likewise, significant differences were observed for dew point and relative humidity. In conclusion, the developed network system demonstrated a high capability to track the variability of weather conditions and should be used as a tool for the prediction of infection hotpots in vineyards.

Keywords: Low-cost, mesoscale, monitoring, real-time, sensor, vineyard, wireless

---

Contact Address: Philipp Pfisterer, University of Hohenheim, Inst. of Agricultural Engineering, Tropics and Subtropics Group, Garbenstraße 9, 70599 Stuttgart, Germany, e-mail: p.pfisterer@uni-hohenheim.de

Statistical Analysis of Data from the Field Phenotyping Platform "breedvision"

Md Anisur Rahaman, Hans-Peter Piepho, Hans Peter Maurer
University of Hohenheim, Inst. of Crop Science, Germany

The inclusion of secondary traits and environmental covariates in the prediction model became an important consideration in recent years. Our study was aimed at quantifying the impact of adding secondary traits and covariates in the improvement of the target trait. An experiment was conducted in 2018–19 at Hohenheim for sensor-based non-invasive prediction of yield (biomass and grain) in triticale (× *Triticosecale Wittmack*) field trials with four trial areas t_1, t_2, t_3 and t_4 and four nitrogen levels N_1= 40 %, N_2= 70 %, N_3= 100 %, and N_4= 130 % in each trial area. A trial area contains 25 triticale genotypes in an α-lattice design with ten incomplete blocks within two replicates and the data were recorded for dry matter yield (DMY) and a secondary trait canopy temperature (CT). CT was measured with a sensor machine using the field phenotyping platform "BreedVision" from the Senselgo project during the vegetation period by using hyperspectral cameras and the sensor machine ran twice in each plot for the data recording. Mixed models are effective in handling repeated measures on the same statistical units and are widely used in biological sciences. Therefore, considering repeated measure issues and measurement of two traits in parallel, a mixed linear bivariate model was developed to predict the target trait 'DMY' without measuring it in the field. Radiation intensity and ambient temperature are two covariates also considered in the model. The model with covariates showed reasonable improvement in prediction performance. There was no gain from the model with secondary trait CT (bivariate model), however, we recommend using this model where it is difficult to measure the target trait DMY due to extreme weather conditions and limited seed supply. Thus, our model allows early selections to be made and saving considerable resources in breeding experiments.

Keywords: Mixed modelling, phenotyping, prediction performance, repeated measures

Contact Address: Md Anisur Rahaman, University of Hohenheim, Institute of Crop Science, Stuttgarterstr. 28, 73760 Ostfildern, Germany, e-mail: anis307@gmail.com

Land Acquisition for Industrial Parks Anh Household Income: A Study from the Mekong Delta, Vietnam

PHI HO DINH[1], HUONG DIEN PHAM[2]

[1]*University of Phan Thiet, Vietnam*

[2]*Banking University of Ho Chi Minh City, Business Administration Dept., Vietnam*

In recent years, the process of forming and developing industrial parks (IPs) has created a new modern infrastructure system, contributing to the rapid expansion of capital and becoming an important contributor to GDP growth, economic restructuring, job and income generating in Vietnam. To serve this process, land acquisition has been taking place strongly throughout the country. Up to February 20, 2021, the whole country has 33.215 projects with a total registered capital of 388,8 billion USD. The country had 326 industrial zones established with a total natural land area of nearly 93,000 hectares, of which the industrial land area was nearly 64,000 hectares, accounting for about 68 % of the total natural land area (Economic Zone Management Department, 2021). If each hectare of acquired land for industrial zone development affects 10 agricultural workers (Dinh Long, 2010), the acquisition of agricultural land affects the lives of 640,000 workers. Industrial parks are crucial to attract investment, however, for sustainable economic development, attention needs to be paid to the quality of life and income of farming households after land acquisition. Although many studies have been done, changes in life of land users after land acquisition have not been fully noticed, especially income change. Therefore, identifying the factors influencing income change of people whose land was acquired is neccesary for researchers and policy makers. This study leaves its focus on (i) the factors affecting the changes of farm household's income after land acquisition; and (ii) policy implications to improve farm household income. This study conducted a direct survey of 280 households whose land was acquired in industrial parks in Can Tho City, Tien Giang, Ben Tre and Vinh Long in the Mekong Delta to apply and test the model. Binary Logistic regression model is applied to analyse the data. The results show that there are six factors or activities influencing income changes; they are (i) whether or not investing the compensation money on production and business, (ii) whether or not borrowing from formal financial institutions, (iii) working in the industrial parks, (iv) educational attainment, (v) dependency ratio and (vi) age of household head.

Keywords: Binary logistic regression, land acquisition, household income

Contact Address: Huong Dien Pham, Banking University of Ho Chi Minh City, Business Administration Dept., 36 Ton That Dam, District 1, Ho Chi Minh City, Vietnam, e-mail: dienhuongpham@gmail.com

Understanding the Determinants of Soil C Stocks and Water-Use Efficiency in the Semi-Arid Tropics of India Using APSIM

Margarethe Sophie Karpe[1], Andrew Smith[2], Thomas Falk[2], Klaus Dittert[1], GV Anupama[2], Dakshina Murthy[3]

[1]*Georg-August University of Göttingen, Dept. of Crop Sciences, Germany*

[2]*International Crops Research Institute for the Semi-Arid Tropics (ICRISAT), Innovation Systems for the Drylands, India*

[3]*ANGR Agricultural University Guntur, Agronomy, India*

Low soil organic carbon (SOC) concentrations and variable rainfall are major limitations to Indian dryland agriculture that threaten yield stability and food production for a growing population. However, increasing SOC sequestration has the potential to positively influence soil productivity, improve farmers' food security and livelihoods, and mitigate climate change. The impact of management on SOC depends on a wide range of factors. Trends in shifting towards more water-intensive cash crops such as cotton and rice and changing climatic patterns necessitate the exploration of site-specific management options. Our study focuses on the state of Maharashtra, India, and uses output data from the crop model APSIM. The model input for 92 grids (0.25 ° × 0.25°) was obtained from observed climate data, global circulation models (GCMs at RCP 8.5), crop and soil maps, and crop production surveys (160 households). The data represent an agro-ecological gradient in terms of temperature and precipitation (610 – 998 mm annual average precipitation on district-level over 30 years). Simulations were constructed to compare different cropping systems under five management scenarios that included the use of inorganic and organic fertilisers as well as irrigation management. Simulations were run for three climate scenarios. We used conditional inference tree analysis (CTree) for data mining. The results concur with findings of other studies that soil type together with land use management are influential determinants for changes in SOC stocks. SOC concentrations can be significantly influenced by management, especially through changes in systems (crop rotation choice). The two systems that sequestered most SOC were sorghum-chickpea (viz. rainy season crop then post-monsoon crop) and sorghum-wheat. Under rising temperatures, mungbean-sorghum gained in importance.

Keywords: Classification and regression tree, site-specific management, soil degradation

---

Contact Address: Margarethe Sophie Karpe, Georg-August University of Göttingen, Dept. of Crop Sciences, Göttingen, Germany, e-mail: margarethekarpe@gmail.com

Interactive Validation of an Agent-Based Model in the Time of COVID-19

Habtamu Demilew Yismaw, Christian Troost, Thomas Berger
University of Hohenheim, Inst. of Agric. Sci. in the Tropics (Hans-Ruthenberg-Institute), Germany

This study developed an agent-based model (ABM) to examine the effect of smallholder farmers' investment in agroforestry to adapt to extreme climate and price variability in Ethiopia. The agent-based simulation package Mathematical Programming-based Multi-Agent Systems (MPMAS) was used to build the model. Validation is an essential step during the modeling process. In a first step, simulated agent production decisions were compared with cross-sectional survey data collected in the study area in 2018. Nevertheless, comparing simulation results with cross-sectional survey data obtained at a single point of time is prone to overfitting and, more importantly, cannot provide insights into the appropriate dynamic behavior of the model. To complement the results obtained using survey-based validation, a participatory model validation method was designed. The participatory validation was intended to be conducted in the field with farmers who had participated in the survey. However, it was not possible to administer farmer-to-model interactive validation due to the COVID-19 pandemic without inflicting exposure to contagion. As a result, an alternative online participatory platform was designed to validate the model with agricultural experts instead of farmers. Experts were selected based on relevance of their expertise to the objective of the model and their experience working in the area. To undertake the participatory validation, an interactive web application was developed using R Shiny. The app was designed as a web page. Participants were sent the link and could open it using any device they wished to use. The interactive session was guided by the researcher through a video call. All necessary simulations were run in advance and results uploaded to the server. The app operates with minimal bandwidth requirements suitable for the poor internet connectivity in Ethiopia. The interactive session has been successful and provided valuable insights to improve the modeling process. In this way, the interactive model validation was undertaken while keeping all participants safe from COVID-19.

Keywords: Farm-level model, MPMAS, participatory modelling, R Shiny, web-based interface

Contact Address: Habtamu Demilew Yismaw, University of Hohenheim, Inst. of Agric. Sci. in the Tropics (Hans-Ruthenberg-Institute), 70593 Stuttgart, Germany, e-mail: h.demilewyismaw@uni-hohenheim.de

Institutions

ILRI session

Environmental Footprint of Livestock Farming in Sub-Saharan Africa – Local Evidence

Sonja Leitner[1], Phyllis Ndung'u[2,1], Mike W. Graham[1], Yuhao Zhu[3], David Pelster[4], Taro Takahashi[5], Linde Du Toit[2], Klaus Butterbach-Bahl[3], J.P. Goopy[2], P. Chelanga[1], N.D. Jensen[1], F. Fava[1], C. Arndt[1], Lutz Merbold[6]

[1]*International Livestock Research Institute (ILRI), Mazingira Centre, Kenya*

[2]*University of Pretoria, Dept. of Animal and Wildlife Sciences, South Africa*

[3]*Karlsruhe Institute of Technology (KIT), Inst. of Meteorology and Climate Research, Atmospheric Environmental Research, Germany*

[4]*Agriculture and Agri-Food Canada, Canada*

[5]*Rothamsted Research, United Kingdom*

[6]*Agroscope, Research Division Agroecology and Environment, Switzerland*

Livestock are essential to livelihoods and nutrition security of millions of people in sub-Saharan Africa (SSA). At the same time, livestock farming creates environmental challenges such as greenhouse gas (GHG) emissions. The magnitude of these challenges in SSA is not well understood due to a lack of reliable in-situ data. Here, we present three examples of studies conducted at ILRI's Mazingira Centre for Environmental Research that address this knowledge gap and contribute to an informed discussion about the environmental footprint of livestock.

The first study quantified farm-level GHG emissions intensities (EIs) for cattle using primary data on animal productivity from smallholder farms in Western Kenya. Data from individual animals demonstrated an extremely heterogenous EI status amongst ostensibly similar farms and provides indicators on how low EIs may be achieved in these environments. Contrary to common belief, our data show that industrial-style intensification is not always required to achieve a low EI.

The second study assessed the indirect effects of the COVID-19 pandemic on GHG emissions from livestock systems in Northern Kenya using proxy data and a framework based on changes in herd size, feed availability, and livestock movement. We found that overall livestock GHG emissions in Northern Kenya have decreased due to the pandemic because of reductions in herd size and decreased livestock movement during the lockdown.

The third study investigated GHG emissions from cattle manure in Kenya. We found that GHG emissions were lower than the IPCC default emission factors because of lower N concentration and higher C:N ratio of the manure due to poor feed quality.

Contact Address: Sonja Leitner, International Livestock Research Institute (ILRI), Mazingira Centre, Kabete Campus, Old Naivasha Rd., 00100 Nairobi, Kenya, e-mail: s.leitner@cgiar.org

Healthy Animals for Healthy Lives in Low and Middle-Income Countries

VISH NENE, HUNG NGUYEN
International Livestock Research Institute (ILRI), Kenya

The animal and human health programme at the International Livestock Research Institute (ILRI) seeks to effectively manage or eliminate livestock, zoonotic and food-borne diseases that matter to the poor through the generation and use of knowledge, technologies, products and approaches, leading to higher farmer incomes and better health and nutrition for consumers and livestock. We present 4 examples from the program's research showcasing the importance of livestock health research in low- and middle-income countries in Sub-Sahara Africa and Asia, and how we apply the One Health approach.

Many transboundary animal diseases, such as African swine fever (ASF), do not cause disease in humans, but they have a devastating effect on the livelihoods of poor farmers – ASF was once limited to Sub-Sahara Africa, over the past decade its impacts have been felt around the globe; at ILRI we have been working on diagnostics and vaccine candidates for more than 15 years as well as on controlling ASF in Africa and Asia by strengthening biosecurity, movement controls and communication.

Three-quarters of new human diseases emerge from animals; for most of those with pandemic potential, domestic animals are involved. During the COVID-19 ILRI supported the Government of Kenya in routine testing and surveillance by offering the state-of-the art veterinary laboratory because a veterinary laboratory does not discriminate between a human or an animal sample. Rift Valley fever was first reported from East Africa and we have contributed to developing a decision-support framework protecting around 50 million people in East Africa – since the World Health Organisation considers it another viral disease with the potential for a pandemic, our research may contribute to the health of many more people. We have also helped test the efficacy of a new RVF vaccine in cattle, sheep and goats. The same vaccine is suitable for use in humans.

At the same time neglected zoonoses continue to impose a huge health burden on poor people, reduce the value of their livestock assets and may foster the use of antimicrobials which lead to new problems such as antimicrobial resistance (AMR). Our programme is studying these drivers and is developing interventions for surveillance and control of neglected tropical diseases and reducing AMR such as the use bacteriophages, viruses that infect and kill bacteria.

Contact Address: Vish Nene, International Livestock Research Institute (ILRI), Naivasha, Kenya, e-mail: v.nene@cgiar.org

Most of the food in low- and middle-income countries is sold in unregulated markets. Instead of banning these markets, our research generates evidence on how to gradually improve them. The introduction of technologies, policies, training and certification for informal milk traders benefiting 6.5 million consumers in India and Kenya and generating tens of millions of dollars in economic benefits. We support national food safety working groups in countries to translate research into interventions to improve food safety. However, what works in one place, may not work elsewhere; therefore, we also look at interventions that failed and how to avoid this.
The programme outputs have been instrumental in global initiatives, for example, in the World Health Organisation strategy for cysticercosis control and the Global Alliance for Livestock Medicines. The ILRI Animal and Human Health programme thanks all donors & organisations which globally support its work through their contributions to the CGIAR Trust Fund.

BMZ/GIZ Session: Bringing innovations into practice – for a healthy and sustainable future

Oral Presentations

Innovation Promotion as a Focus of German Development Cooperation

Sebastian Lesch

Federal Ministry for Economic Cooperation and Development (BMZ), Sustainable agricultural supply chains, international agricultural policy, agriculture, innovation (122), Germany

Developing innovative approaches is important, but not sufficient to achieve the Sustainable Development Goals. To generate impact, roll-out of innovative approaches needs to be improved for innovations to be put into use.

BMZ supports research, impact-oriented innovation promotion and scaling of innovative approaches in various ways.

BMZ has been a strong supporter of the ongoing reform of the global research partnership CGIAR, which strives for more impact orientation in international agricultural research. The new "One CGIAR" research initiatives, starting from January 2022, are designed to contribute to five impact areas for more sustainable and resilient food, land, and water systems, and meet SDG targets. Germany provides strategic advice and support in priority setting for a coherent research portfolio, robust M&E systems and rigorous impact assessments. BMZ funding modality shifts from bilateral to pooled funding for „One CGIAR" research initiatives. Germany encourages research institutions not being part of "One CGIAR" to actively engage and participate as co-developers in the design and then implementation of the initiatives. Integrated experts associated in a Task Force on Scaling advise and support outreach and impact at broader scale.

A particularly agile tool for the promotion of innovations is the newly established "Fund for the Promotion of Innovation in Agriculture (i4Ag)". The fund allows for flexible promotion of user-centreed and sustainable innovations at different maturity levels. In order to ensure developmental benefits, i4Ag promotes innovations that have broad and positive impacts on income generation, food security, resource conservation and climate resilience, with a particular focus on women and youth. To achieve this, i4Ag is working through innovation partnerships with private and public organisations that promote ownership and provide the basis for further up-scaling.

Innovations depend on ideas and resources. Therefore, funding genebanks is a valuable complementary investment in the context of innovation promotion: they conserve the genetic base of critical crop diversity needed to improve

Contact Address: Sebastian Lesch, Federal Ministry for Economic Cooperation and Development (BMZ), Sustainable agricultural supply chains, international agricultural policy, agriculture, innovation (122), Stresemannstraße 94, 10963 Berlin, Germany, e-mail: sebastian.lesch@bmz.bund.de

yield, nutrition and resistance to pests and diseases. Germany has been funding the Global Crop Diversity Trust and genebanks over a long period of time. Currently, BMZ is supporting the Crop Trust with identifying possibilities to improve operations with a view to more sustainable impact generation.

Keywords: Agricultural research, impact-orientation, innovation promotion, scaling

GIZ's Fund for the Promotion of Innovations in Agriculture (i4Ag)

Evelin Ayadi-Krenzer, Helmut Albert
Deutsche Gesellschaft für Internationale Zusammenarbeit (GIZ), Germany

In many developing countries, agricultural production and processing remain below their capacities. Innovations are therefore constantly needed to fully unlock the great potential of the agriculture and food sector to contribute to the sustainable development of rural areas. Such innovations change existing routines or introduce new ones. They should boost income and employment, improve food security as well as mitigate and adapt to the impacts of climate change and the consequences of inappropriate land use. However, the successful introduction of innovations is often hindered by a lack of access to knowledge, financing tailored to target groups and a suitable political environment. Especially women and young people often have difficulty accessing innovations for structural and cultural reasons. In many cases, innovations are also not fully developed or are insufficiently adapted to the local context. This is where i4Ag provides a contribution: i4Ag is a fund that promotes innovations in the agriculture and food sector, focusing on the five key areas of mechanisation, digitalisation, renewable energy, research and extension and cooperation with the private sector. The objective of i4Ag is to promote gender-sensitive and sustainable innovations for the agriculture and food sector that are self-sustaining and can be leveraged by the fund's target groups, such as smallholder farmers and food processors. Lessons learned from innovation partnerships that have received support are applied by multipliers, such as international cooperation projects, companies, public organisations and research institutions. One of the first funded innovation partnership is the four-year Academy of International Agricultural Research (ACINAR) which aims to strengthen the scientific cooperation between German Universities and the CGIAR++ Centersby providing PhD training and stipends. They are awarded to research topics with a developmental relevance. Other examples from the current portfolio of already funded innovation partnerships will be presented.

Keywords: Agricultural research and extension and private sector cooperation, digitalisation, innovation promotion, mechanisation, renewable energy

Contact Address: Evelin Ayadi-Krenzer, Deutsche Gesellschaft für Internationale Zusammenarbeit (GIZ), Eschborn, Germany, e-mail: evelin.ayadi-krenzer@giz.de

Harnessing Innovations in Aquatic Food Systems for Nourishing Nations

SHAKUNTALA THILSTED

WorldFish, Malaysia

With nine harvest seasons left before 2030, we are still far from attaining the targets outlined in the Sustainable Development Goals, especially in addressing Zero Hunger and Life Below Water. Dialogues on food systems transformation are acknowledging the role and potential of aquatic foods in addressing food and nutrition security, and are being reflected in the design of innovations, solutions and action plans globally. This presentation draws upon my experience of aquatic food innovations implemented in Asia and Africa, and the potential to adopt, adapt and scale up these innovations to meet food and nutrition security challenges,improve livelihoods and social equity, while leaving no one behind.

Keywords: Aquatic food systems, food and nutrition security, livelihoods, social equity

Contact Address: Shakuntala Thilsted, WorldFish, Kuala Lumpur, Malaysia, e-mail: S.Thilsted@cgiar.org

Global initiative for the Secure Long-Term Preservation of Clonal and Recalcitrant Crop Genetic Resources Collections via Cryopreservation

Badara Gueye

International Institute of Tropical Agriculture (IITA), Genetic Resources Centre, Nigeria

There is global agreement on the necessity of diverse and available genetic resources for crop improvement, plant production, enhanced and sustained nutrition and welfare in the face of our changing climate. *Ex situ* conservation initiatives have been untaken globally, yet clonally propagated and recalcitrant crop germplasm collections are relatively hard and expensive to conserve. Most of these collections encounter challenges such as sub-optimal storage conditions and a lack of secure safety back-up. Cryopreservation has been reported to be the best long-term conservation option for these crops as it offers options for safety backup, virus elimination, long-term cost-effective preservation, increased longevity, and greater genetic stability. Although cryopreservation offers huge benefits, up to 100,000 Annex 1 accessions are in danger of loss in field and *in vitro* genebanks. The COVID-19 pandemic has further threatened these collections while also clearly demonstrating the urgent need for a global initiative for the preservation of these invaluable collections for humankind. Therefore, considering the high initial cost of a cryobank, the necessary high technical skills and difficulties related to the validation of uniqueness and virus cleanliness of germplasm, there is need for a global effort that will allow to safeguard all those collections at risk of diversity loss. The Global Cryopreservation Initiative has been initiated to safeguard those collections at highest risk and to ensure long-term storage capacity for genetic resources of clonal and recalcitrant seed crops. The Initiative aims to secure endangered crop collections, especially in the developing world, through regional cryo-hubs and ambitions to involve all relevant stakeholders (NARs, ARIs, CGIAR, Universities, etc.). In addition, the initiative will provide capacity building (awareness, know-how, support) and safety cryo back-up in an immerging global plant cryo network. In a dynamic and evolving way, expertise will be formally coalesced to implement methodologies and results into operational protocols, cryopreservation services, cryo back-up facilities, and the formation of a global network and plant cryopreservation Community of Practice. These collections which are now at risk of loss and hard to conserved will be secure and accessible under the ITPGRFA to provide untapped genes to face current and future agricultural challenges.

Keywords: Clonal crops, cryopreservation, genetic resources, global network

Contact Address: Badara Gueye, International Institute of Tropical Agriculture (IITA), Genetic Resources Centre, Oyo Road, Ibadan, Nigeria, e-mail: b.gueye@cgiar.org

BMEL Session - Transformative research approaches for sustainable food production and consumption

Transformative Research Approaches for Sustainable Food Production and Consumption. Sharing Experiences from Research Projects funded by the Federal Ministry of Food and Agriculture

Silvia Dietz

Federal Ministry of Food and Agriculture (BMEL), Division 123, Research and Innovation, Coordination of Research Area, Germany

The BMEL is committed to contribute to the design of sustainable and resilient food systems worldwide and the implementation of the human right to adequate food.
Contributing to the improvement of global food security, BMEL established the funding instrument "Research Cooperation for Global Food Security and Nutrition". The focus of this instrument is on promoting practice oriented trans- and interdisciplinary research projects. The projects contribute to generating needs-oriented knowledge and the further development of capacities on the ground. Funded projects within this scheme thus strengthen the German contribution to sustainable food systems in partner countries through agricultural and nutritional research and by building long-term partnerships between agricultural and nutritional research institutions in Germany, Africa, South and Southeast Asia.
The Federal Office for Agriculture and Food (BLE) acts as project executing agency.
The BMEL-Session aims to illustrate the contribution of its research projects towards a transformation of habitual ways of thinking, aiming at promoting more sustainable production, consumption, nutrition and lifestyle patterns.
The following three research projects will provide insights into their diverse contexts and approaches:

- "Application of new packaging solutions to reduce food losses in West Africa by extending the shelf life of perishable, local food (WALF-Pack)", Dr. D Sylvain Dabade, University of Bonn.
- "Decentralised processing of rarely used plant and animal raw materials into innovative products with high added value to improve the nutritional situation in West Africa (UPGRADE Plus)" and "Enhancing women's agency in navigating changing food environments to improve child nutrition in African drylands (NaviNut)",
  Prof. Dr. Brigitte Kaufmann, German Institute for Tropical and Subtropical Agriculture .

Contact Address: Christine Hbirkou, Federal Office for Agriculture and Food (BLE), International Cooperation and Global Food Security, Deichmanns Aue 29, Bonn, Germany, e-mail: christine.hbirkou@ble.de

- "Strengthening the Resilience of Rural Food Environments in the Context of Disaster Risk and Climate Change in Mozambique (FEMOZ)", Prof. Dr. Samuel Quive, Eduardo Mondlane University, Maputo, Mozambique .

In addition to the presentations on the selected projects, the session will provide the opportunity for an enriching exchange.

Keywords: Food and nutrition security, global South, international research cooperation

Combining practice and higher agricultural education – Challenges and chances

Posters

Managing Natural Resources in a Sustainable Manner for Rural Development in Kosovo

Anika Totojani

The Agricultural University of Tirana, Institute of Plant Genetic Resources (IPGR), Albania

Kosovo has shown an attempt to reform the economy, willingness to further progress and to be integrated politically and economically towards the European accession, to benefit of trade and international markets. Kosovo is one of the most rural and agrarian European countries. Kosovo's agriculture is one of the major contributors to the gross domestic product. Therefore, it represents a key sector whereby most of the rural population engages in. The country is endowed with fertile soils. Wheat production for food and dairy farming were considered as the most important agriculture production of many farmers in Kosovo . This study analysis the mediums of the cereal production and the importance it has for the rural economy. The research used qualitative and quantitative methods. 50 surveys and semi structured interviews were carried out with respondents engaged in the cereal production. Informants were selected purposely in order to deepen the understanding of the sector and to better fill the study's objectives. Results show that farmers engage in cereal production. They cultivate big parcels of land. Wheat dominates the cereal production and is followed by maize and other small quantities of grain. However, the domestic production does not fulfil the domestic demand. The imported cereal production is accompanied with bad quality. Although actors are involved in wheat production, they are not competitive. Farmers face various difficulties. However, wheat production and dairy farming are two sectors that get most attention. Development programmes support the two sectors as they are considered strategic. Therefore, it is encouraged that the farmers should better make use of the resources: fertile soil, good weather condition, farmer's know-how knowledge and cultivation tradition in cereals production. Policies should continue to support cereal production and the good will of rural population to be engaged in agriculture production. Resources should be adequately managed in the most sustainable way. Future studies should be conducted for sectors that have comparative advantage.

Keywords: Cereal production, rural development, sustainable recourse management, wheat

---

Contact Address: Anika Totojani, The Agricultural University of Tirana, Institute of Plant Genetic Resources (IPGR), Siri Kodra Street 132/1, 132/1 Tirana, Albania, e-mail: anikatotojani@yahoo.com

Improving Storage, Processing and Marketing of African Indigenous Vegetables (AIVs) in Western Kenya

Joseph Alulu
University of Nairobi, Agricultural Economics, Kenya

In Kenya, horticulture as a subsector contributes about 40 % to the agricultural GDP thus its relevance to the economy. Vegetables contribute on average 36 % of the total horticultural value. This study project focused on AIVs with a particular interest in Black nightshade, Cowpeas, Spider plant and Amaranth being produced on a small-scale enterprise in Western Kenya. It is estimate that about 60 % of rural households rely on AIVs for food and income as well. The AIVs provide unique opportunities for smallholder producers to diversify farming systems for improved livelihoods. There has been an increasing nutritional and health awareness of AIVs in Kenya, hence increasing demand with urban consumption per capita being 140 kg per annum and 70 kg per annum for rural dwellers. AIVs are rich in vitamins and minerals, hence important components for a nutritionally diversified diet. Extant evidence shows that AIVs provide smallholder farmers with higher returns per unit area as compared to other main crops, e.g. maize. Despite the relevance of vegetables to the economy, smallholder producers encounter several challenges, key among them being post-harvest losses where farmers lose up to 50 % of their produce. This can be attributed to poor storage and inadequate processing activities to increase the shelf life of the vegetables. In addition, due to poor quality and low value addition activities, smallholder farmers fetch low prices at the market. There have been inconsistencies in supply of AIVs to the High value markets. Thus, there is a need for an innovative strategy to profitably meet consumers' needs. Hence, relevance of this Study Project to improve storage, processing and marketing It aims at providing solution to these problems through three interventions; first, improving storage of AIVs through application of a solar powered cooling chamber, secondly, application of solar drying technology for drying the vegetables to increase their shelf life and thirdly through innovative marketing strategy develop an application for enhancing convenience among the busy urban consumers. The enterprise aims at becoming a model farm where other farmers within the community can come, learn and apply the knowledge for improved livelihoods in the long run.

Keywords: AIVs, Marketing, Processing, storage

Contact Address: Joseph Alulu, University of Nairobi, Agricultural Economics, 14942, 00800 Nairobi, Kenya, e-mail: alulujay@gmail.com

Improvement of the Production, Transformation of Noni into Juice and Powder in Mali

Safoura Ousmane Cissé
Promoter and Manager "Health Secret", Mali

Malian agriculture includes an entire area which is agroforestry, that is to say harmoniously combining agriculture and trees. In the current context, Malian agriculture is concerned about the phenomenon of climate change, therefore it is developing its agroforestry component for carbon capture. In fact, Malian agriculture seeks to fight against malnutrition with products rich in vitamins. The cultivation of noni alone fulfils these different functions and constitutes an innovative product of Malian agriculture. Nutrition security involves more than just access to adequate food. It requires access, among other things, to micronutrients, at affordable costs given the poverty level of households. Therefore, credible alternative solutions with high socio-economic and environmental impact must be considered to support national policies and strategies to fight against food and nutritional insecurity. *Morinda citrifolia* is one of those alternatives. *Morinda citrifolia* is planted for its leaves and fruits. It has the particularity of maturing very quickly and contains immuno-stimulating substances and a high concentration of vitamins, minerals, enzymes, alkaloids, antioxidants, amino acids and flavonoids. The fruit also contains many trace elements. This particular composition gives the fruit of *Morinda citrifolia* medicinal properties capable of protecting the body against degenerative diseases such as cancer and delaying aging (anti glycation). Consuming the fruit helps to maintain the functions of the immune system, and an acceptable level of cholesterol in the body. It regulates sleep through its ability to fight stress and provide relaxing effects. Knowledge of all these virtues of *Morinda citrifolia* (leaves, fruits and seeds) is currently earning him great interest in the food industry in particular and in the processing industry in general. It is with this in mind that the present Noni production and processing project has been initiated in Mali and is mainly aimed at the production of Noni juice and powder.

Keywords: Mali, *Morinda citrifolia*, noni

Contact Address: Safoura Ousmane Cissé, Promoter and Manager "Health Secret", Segou, Mali, e-mail: safouraousmanecisse@gmail.com

Valuation in the Ginger Value Chain: Extraction of Essential Oil from Ginger Roots (rhizomes)

Ayao Dzigbodi Midodji

IFAD, Togo

Ginger is a flowering plant that originated in Southeast Asia. It's among the healthiest spices on the planet. The rhizome (underground part of the stem) is the part commonly used as a spice. The total production of ginger in the world is 1683.00 thousand tons with the total acreage of 310.43 thousand ha. China, India, Nepal and Thailand are the major producers of ginger in the world, having production of 396.60 thousand tons, 385.33 thousand tons, 210.79 thousand tons and 172.68 thousand tons respectively. Africa as a whole accounts for 5 % of the world ginger market and remains a price taker. Among Africa countries, Nigeria is the largest producer (152,106 tons), followed by Cameroon, Côte d'Ivoire, Ethiopia and Fiji (FAO, 2009). Worldwide, the demand for ginger is constantly increasing, especially in these times of Covid 19 and the demand for ginger oil has been estimated at 517.68 tons once a year. This demand is expected to reach 4,212 tons by 2022. In Togo, annual ginger production is estimated at 60,000 tons at the national level. The western part of the Plateaux region is its main production area. On the local market, the demand for ginger is constantly increasing and national production is still insufficient. The deficit is filled by imports from other countries in the sub-region such as Nigeria. Despite this insufficient production, losses occur throughout the chain since some actors cannot even sell their products to maintain production. The objective of our project is first to regroup the producers, then by extracting the essential oil of ginger which will be commercialised on the local and international market, to quickly valorize the harvests to avoid the losses and thus relaunch the production.

Keywords: Ginger, ginger oil, ginger root, rhizome, Togo

Contact Address: Ayao Dzigbodi Midodji, IFAD, Lomé, Togo, e-mail: midodjiayao24@gmail.com

Enhancing the Value Chain of Poultry Production by Using Solar Energy in Ethiopia

Mezgebu Aynalem Mengstie
Debre Markos University, Burie Campus, Ethiopia

The world poultry population has been estimated to be about 16.2 billion, with 71.6 % in developing countries, producing 67,718,544 tons of chicken meat and 57,861,747 tons of hen eggs. Chicken production system is an appropriate and locally available resource in livestock populations. In Africa, village poultry contributes over 70 % of poultry products and 20 % of animal protein intake. Ethiopian indigenous chicken population is estimated to be 40.6 million and producing about 78 million of eggs per year. In Ethiopia, the average flock size under rural chicken production system ranges from 7 – 10 birds in each house hold consisting of 2 – 4 adult hens, one cock and some growers of different age groups. The egg production is estimated to be 40 to 60 eggs per birds per year with an average egg weight of 40 grams. The main problem of poultry production is electricity, market linkage, high production cost and poor management practices. Today, it is widely recognised that electricity is needed for society mobility. Electricity is not able to address supply in rural areas but solar power is one of the solutions for those problems concerning environmental impact and emissions. So by using Solar power in rural areas it will increase chicken production. The major inputs and auxiliary raw materials required are day old chickens, commercial formula feed, and high quality vaccines which have to be imported. The mean goal of this project is enhancing the value chain of poultry production by using solar energy in Ethiopia. Specifically to get economic profit which will help to expand the capacity of farm, create employment opportunity for town residents, by using solar energy to enhancing poultry production, to use manure/ excrement of poultry to biogas energy, to enhance biogas slurry used to fertiliser and to provide products to consumer at reasonable price.

Keywords: Ethiopia, poultry value chain, solar energy

Contact Address: Mezgebu Aynalem Mengstie, Debre Markos University, Burie Campus, Debre Markos, Ethiopia, e-mail: mezgebu12aynalem@gmail.com

Construction and Validation of Photovoltaic Powered Fan Supported Solar Drier Prototypes in Asella, Ethiopia

Bezayit Seifu Woldegiorgis
Arsi University, Ethiopia

The food industry is responsible for supplying the food demand of the ever-increasing global population. The food chain is one of the major contributors to greenhouse gas (GHG) emissions, and global food waste accounts for one-third of produced food. A solution to this problem is preserving crops, vegetables, and fruits with the help of an ancient method of sun drying. For drying agricultural products, several types of dryers are also being developed. However, they require a large amount of energy supplied conventionally from pollutant energy sources. The environmental concerns and depletion risks of fossil fuels persuade researchers and developers to seek alternative solutions. In order to perform drying applications, sustainable solar power may be effective because it is highly accessible in most regions of the world. Solar dryers are simple facilities that can provide large capacities for drying agricultural products. This study project objective is to construct and validate indirect solar dryers' prototypes (Icaro1.5 model and Tunnel driers) with solar technologies, photovoltaic (PV). To investigate the effectiveness of the prototypes different vegetables will be dried in the driers. Additionally, the effectiveness of the prototypes will be assessed by taking into consideration different parameter which will determine the performance of the dryers. In this regard, this study if it is implemented will contribute in the reduction of postharvest loss of agricultural produce and allow usage of renewable energy sources in the food processing. Additionally, the introduction of these simple drying technologies will have economic benefit for producers, processors as well as end consumers.

Keywords: Agricultural products, economic benefit, postharvest loss reduction, solar drying

---

Contact Address: Bezayit Seifu Woldegiorgis, Arsi University, Asella, Ethiopia, e-mail: bezaseifu28@gmail.com

Improvement of Linkages between Research-Producers in Tef Value Chains in Ethiopia

Yazachew Genet Ejigu

Ethiopian Institute of Agricultural Research, Debre Zeit Agricultural Research Centre, Ethiopia

Tef (*Eragrostis tef*) is the most important cereal crops in Ethiopia in terms of food and feed (straw), medicine, cash and foreign currency earnings. In a country of nearly 114.9 million people, Tef is staple food over 77 % Ethiopian population (114.9 million). Tef covered 3.1 million hectares (30 % of cereals (10.5 million ha)), over 7.1 million households grow tef and its production is 5.7 million tons (20 % of cereals 29.7MT) annually (CSA, 2020). Tef is nutritional & healthy crops. Importantly, tef is also gluten-free so it is well-suited to address growing global gluten-free demand. It is considered as the "latest super food of the 21st century However, it has been historically neglected compared to other staple grain crops, yields are relatively low (1.85 t/ha compared to 6 t/ha at research plot), and long and complex value chain, limited number of traders and market outlets, lack of access to improved seed, poor availability of farm implements, poor extension linkage and export of tef grain is banned due to low production and volatility of the price of tef. In addition, while wholesale prices for tef are relatively high making the crop attractive to some producers as a cash crop, the production costs are also high as reflected by the high fertiliser prices and the labour intensity of cultivation, weeding, harvesting, and threshing. Knowledge sharing, training, follows up of tef value interventions, and actors linkages contribute to improve the skill and knowledge of value chain actors and service providers, including women. Availability of improved technologies (varieties and farm implements) results in expansion of tef production areas. Farmers' group seed enterprise can multiply and be source of improved varieties for smallholder farmers. Farmer to farmer exchange, strengthen existing cooperatives and farmer groups can make up the market for improved different option can be used depending on the stage of development. Thus, the aim of the project is to improve tef productivity, profitability, and sustainability by improving research- producers' linkage and implementing actionable value chain interventions.

Keywords: Research- producer linkage, tef, value chain

Contact Address: Yazachew Genet Ejigu, Ethiopian Institute of Agricultural Research, Debre Zeit Agricultural Research Centre, Addis Ababa, Ethiopia, e-mail: yazachewgenet@gmail.com

Reducing Post-harvest Losses in Citrus Value Chain in Benue State, Nigeria: Focus on Processing

Agwaza Alfred Iordekighir
Consultatnt GIZ-GIAE-AFC, Nigeria

Agriculture has been considered as one of the major contributing sectors to the Gross Domestic Product (GDP) amongst most of African countries, Nigeria inclusive. To enhance agricultural growth and productivity, an in-depth understanding of both pre and post-harvest managements of agricultural produce is of paramount important. This could go a long way in reducing food losses and wastages, which has great impact on all the four facets of food security of any developing nation. Evident from researches have shown that, post-harvest losses are still high in Africa (20–40 % annually) compared with the other regions of the world. In Africa, the most commonly lost agricultural produces, after harvest, are the fruits and vegetables due to their seasonal glut in production and high perishable nature. In 2019, Nigeria produced over 4 million metric tones of Citrus fruits. However, the income of the Citrus farmers tends to be relatively low due to several reason such as; high post-harvest losses, unfavourable weather conditions, inadequate infrastructures, lack of value additions, inadequate knowledge on processing technologies among farmers, and inadequate processing factories in the country. It is against this background that, Alfred, one of the 2020 participants from Nigeria in the post-graduate training course, Food Chains in Agriculture, wishes to present his project idea on the topic: Reducing post harvest losses in Citrus fruit Value chain in Benue State, Nigeria with focused on processing. Specifically, the objectives of the project idea are to: train small-scale Citrus farmers on how to process citrus fruits into juice, animal feeds and organic manure, support trained farmers with equipments needed for citrus fruits products through donor funding, create more jobs opportunities to actors in citrus value chain and market citrus fruit products via out-grower scheme with citrus farmers.

Keywords: Citrus, Farmers, Nigeria, post-harvest losses, Processing

Contact Address: Agwaza Alfred Iordekighir, Consultatnt GIZ-GIAE-AFC, 8 Sky Gifted Academy, Mkar, Gboko, Nigeria, e-mail: fredo4all200@gmail.com

Improvement of the Sesame Value Chain by Introducing a Sesame Processing Unit in Bobo Dioulasso/Burkina Faso

Teegwêndé Maggy Ouédraogo
Entrepreneur, Burkina Faso

Sesame is one of Burkina Faso's largest commodities. It is counter-seasonal cash crop, harsh weather tolerant and requiring less input such as fertilisers and pesticides. In Burkina Faso, sesame grains are produced all over the country from organic to conventional and most of the grains are being exported. The main constraints of local sesame oil processing are the capacity, the equipment, the skill and above all, the energy cost. Difficulties in accessing finance, low level of organisation, the low involvement of scientific research, non-compliance with the regulations, Variation of the sesame price low quality of sesame difficulty in stabilising the supply chain are also the challenges that the product has to face. The country has one of the most expensive electricity costs in the world. On the other hand, solar energy is available abundantly. Therefore, development of decentralised solar-powered oil production system at cooperative level in rural areas has a huge potential to address the challenges. As the sesame oil is a very good one, it is open to a large market. Quality parameters of the oil such as free fatty acid, peroxide value, vitamin E, oxidation stability and water content were analysed. It was found that the quality of sesame oil produced from the solar-powered oil press was comparable with the sesame oil sold on the European market and fulfiled the EU standard requirements. The current local market pricing is about 4.89 €/L sesame crude oil. When the system run only during thenday and during harvest season in a year, the fastest payback period was estimated at 2 months. (Pilot Project "Solar-Powered Sesame Oil 2017). The estimations show that this project is affordable and promptly profitable.

Keywords: Sesame, solar, value chain

Contact Address: Teegwêndé Maggy Ouédraogo, Entrepreneur, Bobo-Dioulosso, Burkina Faso, e-mail: maggyouedra@gmail.com

Increase the Revenue of Tomato Smallholders Farmers of Zatta (Côte d'Ivoire) by Linking them to Additional Customers

Audrey Sakhon Dominique Dieket
LONO-CI SARL, Research and development, Ivory Coast (Cote D'Ivoire)

Côte d'Ivoire is a western country of Africa with an economy particularly depending on agricultural outcomes. This sector is 28 % part of the GDP. As in several countries, agricultural products come from the rural areas and are sold in the big cities. Nevertheless, the Ivorian rural population is the poorest with an increased rate of 3.4 % during the five last years. Tomato, for example, is a daily product used in Côte d'Ivoire for each meal. Tomato can be cultivated all over the country but come, mainly, from Abengourou (eastern part) and Zatta (central part). Some farmers interviewed revealed two challenges for small-scale farmers in this last village. First, they undergo the foreign competition (especially Burkina Faso and Niger) in off-season, so they are obliged to reduce the price of the kilogram from 2,000 fcfa – 3,000 fcfa to 1,000 fcfa – 1,500 fcfa. Secondly, because of glut production in favourable season, they can register some losses due to unsold stock. According to FAO, this loss can be estimated to 10 %. Because of low level of knowledges in marketing, they remain focused on the local traditional market and on resellers. So, we think about Hop' vegetables, as a social company that will work for small-scale tomato farmers giving them a marketing support. The main goal is to connect them to additional customers and sell those 10 % of their harvest through an application and decentralised modern vegetable shop. The target additional customers are composed of educated women, employed or with regular income, familiar to online shop and who desire to be delivered their vegetables (17 % of interviewed) or to purchase them in a close and clean market (67 % of interviewed). The cities of Abidjan and Yamoussoukro are chosen for the first step as they register totally around 2, 000, 000 of the target group, and because of many facilities related to company registration and functioning. With a percentage of sales, the company is expected to be amortized in four to five years. The profit will be reinvested in projects for small-scale farmers well-being.

Keywords: Application, Côte d'Ivoire, farmers, marketing, shop, small-scale, support, tomatoes

---

Contact Address: Audrey Sakhon Dominique Dieket, LONO-CI SARL, Research and development, Abidjan, Cocody 2 Plateux mobile, Abidjan, Ivory Coast (Cote D'Ivoire), e-mail: audreydieket@gmail.com

Introducing Solar Cooling Chamber to Reduce Post Harvest Losses of Tomatoes in Machakos County, Kenya

Emmanuel Muasya
Machakos County Government, Kenya

Agriculture is central to Kenya's economy accounting to 24 % of GDP. Horticulture sub sector is a vital source of income for smallholder farmers especially vegetables, fruits, herbs, root crops. Tomato production constitutes 14 % of total vegetable production and 7 % of horticultural crops. Tomato is a short duration crop and gives high yield, hence is economically attractive. Almost all households consume tomatoes and they are rich in vitamin A, C and lycopene. Tomatoes can be eaten fresh, added to salads, cooked as vegetable or processed into tomato paste, jam, sauce, puree and juice. Tomato production in Machakos is through irrigation, rainfed, greenhouses and some farmers rent land. Other vegetables grown are French beans, capsicums, eggplant, onions, kales, black nightshade. The demand and supply of tomatoes varies over the year. Food supply can be improved either by increase in production or reduction in loss. The combination of seasonal glut production and high perishability makes tomatoes more vulnerable to Post harvest losses (PHL). This leads to farmers accepting low prices for their tomatoes. Introducing solar cooling chambers in the high producing areas will reduce the PHL, improve their bargaining power and eliminate middlemen. A group of farmers will manage the solar cooling chamber. Forming groups has its own benefits like contract farming, collection and aggregation of produce/inputs as well as easy access to extension services. Solar cooling chamber is off grid, hence can be installed anywhere. It can be transported in individual parts and assembled easily and has low operational costs after installation. Higher quality products can be marketed over a longer period.

Keywords: Kenya, post harvest loss, solar cooling chamber, Tomato

Contact Address: Emmanuel Muasya, Machakos County Government, Machakos Town, Kenya, e-mail: emmanuelmuasya@gmail.com

Strengthening Rice Value Chain Actors in Nigeria – A Focus on Improving Access to Specific Inputs

Nguemo Achimba
GIZ AFC Nigeria, Extension Service, Nigeria

Rice (*Oryza sativa*) is one of the most consumed staple in Nigeria. taken in different forms across all geopolitical zones, there is also abundant land for production. Nigeria is blessed with FADAMA land for rice production. Despite these potentials, Nigeria is not producing enough rice for its citizens: The consumption rate (7 million metric tones MT) is higher than production rate (5 million metric tones MT). This is largely attributed to poor access to specific farm inputs and communication gap between input suppliers and the farmers and other factors as well. Therefore, the objectives of the project are to strengthen the rice value chain actors, focusing on improving access to specific inputs. If access to farming inputs is improved, it will lead to improvement in food security, reduced importation of rice to Nigeria and improved youth employment. Up till now, a lot of social media marketing content exists in Nigeria, but only a few exist for promotion of access to inputs, especially for rice production.This project is aimed at developing a B2B social media marketing platform that connect inputs dealers and rice producers through advertisement of available inputs and locations on the platform and facilitation of one stop shop installation in rural areas to afford the rural farmers access to inputs for rice production. These will not only improve access to inputs but also strengthen the entire rice value chain since both actors on the value chain are connected. Youths will also be employed to manage the one stop shops established in different locations. They will advice farmers on the right inputs to purchase and use when need be. Also, it will improve the livelihood through improved productivity.

Keywords: B2B, facilitation, inputs, marketing, production, rice, suppliers

Contact Address: Nguemo Achimba, GIZ AFC Nigeria, Extension Service, Makurdi, Nigeria, e-mail: achimbajennifer@gmail.com

Improving Agricultural Extension Services by Formation of Private Extension Groups for Increased Access to Innovation in Nigeria

Joshua Apochi Apochi
GIZ-AFC-AGFIN, Nigeria

Over the years, Nigeria farmers relied on the public Agricultural Extension services through the state Agricultural development programmes (ADP) to access agricultural information and innovation, however, in the past three decades public funding for Agricultural extension dwindled, leading to the drastic reduction in the number of public Extension Agents (EA). A development that has further rubbed off on whatever gains the sector made over the years. In 2013, the National Agricultural Extension Research Liaison Services (NAERLS), and the National Food Reserve Agency (NFRA) conducted a research, which analysed five years (2008–2013) of ADPs and agricultural field situations, especially extension agents' ratio to farm families across the states. Based on the survey, averagely, the EA per farmers' ratio oscillated between 1:1700, 1:2132, 1:3385, 1:2950 and 1:3011 across the country, within the five-year period. The current ratio is between 1:5000 and 1:10,000 which is far above FAO recommendation of 1:800 farm families of developing nation. These development results to limited adoption of research findings and technologies and other factors all combined to keep agricultural productivity low despite the continuous increase in population. On a yearly basis the Universities and other Institutions of higher learnings turn out graduates in Agriculture but neither job nor seed capital to venture into agriculture is provided to support the youth. In line with this challenges, establishing a private extension services in partnership with the ADPs and NGOs to manage dissemination of innovations will not only increase the adoption of innovations but also create job opportunity. The project idea is aimed at creating a cooperative/network of young extension agents in various communities and local government, trained and mentored by an experienced agent to support in stepping down innovations.

Keywords: Innovation, Nigeria, private extension services

Contact Address: Joshua Apochi Apochi, GIZ-AFC-AGFIN, 12 Jato Aka Street Itf Road Wurukum, 970211 Makurdi, Nigeria, e-mail: apochijosh@gmail.com

Enhancing the Participation of Women Groups in Agriculture While Improving their Livelihood through the Promotion of High Quality Cassava Peels (HQCP) in Nigeria

Taiwo Oluwaseun Ayinde
APIN Public Health Initiative, Nigeria

Cassava is one of the most important perennial agricultural food crops in West Africa with Nigeria being the largest producer in the world harvesting over 59Million tonnes and generating 14million a tonne peels annually. About 90 % of the total production is used for human consumption. Women are the key players in the processing and marketing of cassava. In Nigeria, cassava peels tend to end up on dumping sites thereby leading to considerable environmental pollution and most of these cassava processors and women groups lack the technical know-how on how to add value to cassava peels which would have in turn be an additional stream of incomes for them. There is however an opportunity to close the nutrient gap through conversion of waste peels into usable livestock feed which will in turn help to mitigate environmental damage as well as also providing additional source of income to women groups in agriculture if the waste are put into good use. High Quality Cassava Peels is the process of turning Waste(Cassava Peels) into the business of Livestock and Animal's Feed, as a replacement for Corn / Wheat /Grain in Feed Mill productions as well as minimising the hazard caused by environmental pollution through the heap of cassava peels / waste visible all around our cassava value chain industries. However, as a result of the value that has been added to it, when this waste product is processed, the value addition increases the price per pack thereby resulting to enormous financial gain for the people and of course, a ripple effect on the economy. The HQCP if fully harnessed will however hugely subsidise the dependent on maize for animal feeds and thus reduce the competition between livestock and human. This initiative will go a long way in helping women groups have sustainable additional streams of income as well as reducing environmental pollution caused by accumulated cassava peels at dump sites.

Keywords: Cassava peels, Nigeria, women groups

Contact Address: Taiwo Oluwaseun Ayinde, APIN Public Health Initiative, Abeokuta, Nigeria, e-mail: taiwodisu2@gmail.com

Improving the Value Chain of Milk through Solar Milk Cooler in Ethiopia

TIGIST KEFALE MEKONEN
Debre Markos University, Agricultural Economics, Ethiopia

Ethiopia has a large potential in livestock production being the 1st among African countries and 9th in the world. Dairy farming contributes to the livelihoods of 1.7 million livestock farmers. About 4 billion liter of milk is produced in Ethiopia per year among, which smallholder farmers produce about 98 % of the milk. FAO recommends that the per capita consumption of milk be about 200 liters, which means 22 billion liters of milk is required. However, at the current production rate, there is an annual shortage of about 18 billion liters. Hence, in Ethiopia per capita consumption is less than 20 liters. The demand for milk products is rising rapidly in Ethiopia but quality milk supply is not enough due to high post-harvest loss. High perishability of milk and lack of refrigeration in rural areas are leading to millions of tons of fresh produce of milk going to waste every year; even the evening milk is not collected for the market and the income of smallholder farmers are getting low. Therefore, milk cooling facility is a vital technology for storing and transporting milk products. Though, this cooling process is limited or not applied in rural areas due to no access to conventional electric grid and high costs of standalone diesel generators. This problem creates an opportunity for introducing sustainable renewable energy, mainly solar milk cooler. As photovoltaic systems have the potential to be acquired in rural areas throughout the year, this technology is a viable energy source option. The cooling of milk while stored on the farm followed by the cooling during transport has a great role for the different stakeholders along the value chain. Therefore, this project aims to improve the profitability of smallholder dairy farmers and increase access to quality milk for the surrounding community by reducing post-harvest losses of milk through the use of solar milk cooler with the help of different stakeholders and funders.

Keywords: Ethiopia, post-harvest loss, smallholder farmers, solar milk cooler

Contact Address: Tigist Kefale Mekonen, Debre Markos University, Agricultural Economics, Debre Markos, Ethiopia, e-mail: kefale.tigist@yahoo.com

Milk Processing into Cheese in Togo

Wenkonta Pasgo
grassroots development Ministry, Togo

In Togo, the amount of milk and dairy products imported is estimated at 26.9 million $ in 2019 (resourcetrade.earth, 2019). This huge amount of importation is, due to the failure of local production, to respond to the whole need of the population. Also, there is a lack of process units within the country. Cheese is the most important dairy product consumed in Togo (Self survey, 2019). The cheese value chain is affected due to the lack of collection units. Hence, we need to do a group collection for milk and set up a process unit for cheese. The objectives of the project are to create value and wealth through the dairy sector by transforming the raw material which is fresh milk and by increasing the income of producers by ensuring them the market for fresh milk. We want to create cooperatives and share some experiences among the producers. Also, we need to strengthen the link between actors through some contracts. We want to collect the milk from producers, make the market for producers available and make cheese availabe everywhere even in supermarkets. We expect an improvement of producers livelihood up to 30 %, the creation of value and wealth and employement of young people.To conclude, the milk sector has important potentialities because milk contains a lot of proteins which can supplement food diet often based on cereals and tubers in Africa (Adanléhossi et al,2003). Then, this project will contribute to strengthen the links between actors of the value chain. It will also allow people to eat healthy food and create wealth in order to improve the well-being of producers.

Keywords: Cheese, milk processing, Togo

Contact Address: Wenkonta Pasgo, grassroots development Ministry, Lomé, Togo, e-mail: pasgowenkonta@gmail.com

Analyses of Value Chains in Food Production in Africa to Create Start up Ideas

Bernd Müller, Vadym Petrenko, Johanna Menhorn
Weihenstephan-Triesdorf University of Applied Sciences, Germany

A deep understanding of the value chains in agriculture and food production is the basis for entrepreneurial activities and so for income increase and improvement of rural livelihoods in Africa. Success or failure of development in the agricultural sector depends often on the development of strong and effective value chains for the agricultural products. It is not enough to produce agricultural goods but to develop the whole value chain from the input supply over production and processing up to the retail and consumption in order to increase income of farm families. In the past, the scientific focus was often given to single elements of the value chains such as soil, pesticide use or fertilising. Understanding the holistic concept of value chains allows us to identify weak points of a value chain and to overcome the relevant bottlenecks. This leads to optimised resource utilisation and thus income increase. Furthermore, it opens opportunities to identify business start-ups that can be further developed. The presentation deals with the example of such start up ideas created from participants of the Postgraduate Course Food Chains in Agriculture at the Weihenstephan-Triesdorf University of Applied Sciences. The elements of a value chain and their interaction under the holistic concept are shown. The process of elaborating a study project from idea over deep problem analysis and suggestion of a cost efficient solution up to implementation in practice is depicted. A focus is also given to the economic analysis and interpretation of the meaning of the single steps. Pros and cons of the approach as well as challenges for the realisation of the start-ups are analysed and discussed in more detail. The realised start-up gives the opportunity to increase income, create employment in rural areas of African countries, and so improve the living standards of people.

Keywords: Food, start-up, value chains

Contact Address: Bernd Müller, Weihenstephan-Triesdorf University of Applied Sciences, Markgrafenstraße 16, 91746 Weidenbach, Germany, e-mail: bernd.mueller@hswt.de

Improvement of Crucial Aspects of Cereals Value Chains in Ethiopia

Yazachew Genet Ejigu[1], Tigist Kefale Mekonen[2], Mezgebu Aynalem Mengstie[2], Bezawit Seifu[3], Kefala Taye[3]

[1]*Ethiopian Institute of Agricultural Research, Debre Zeit Agricultural Research Centre, Ethiopia*

[2]*Debre Markos University, Agricultural Economics, Ethiopia*

[3]*Arsi University, Ethiopia*

Cereals are source of food and income for Ethiopian farmers. Despite the high cereal production potential, a number of constraints impede the crop cereal sector's and value chain's development. Unbalance of supply and demand in cereal markets have been the source of major concerns for the Government of Ethiopia and its development partners. Increase in cereal price presented serious challenges to the implementation of country's food security programs. Knowledge sharing, training, follows up of interventions, and partner linkages contribute to improve the skill and knowledge of value chain actors and service providers, including women. Availability of improved technologies (varieties and farm implements) results in expansion of cereal production areas. Farmers' group seed enterprise can multiply and be source of improved varieties for smallholder farmers. Farmer to farmer exchange, strengthen existing cooperatives and farmer groups can make up the market for improved different option can be used depending on the stage of development. Small village shop keepers can be effective suppliers of agro chemicals. Hence, the government and other concerned bodies should focus on productivity increasing technologies in the study area in order to boost productivity and thus increase cereal market supply. In addition, to solve the marketing problem, the promotion of value-added practices and the formation of cereal seed enterprise and strengthening of cooperatives are suggested. This Project seeks to contribute to food and nutrition security, and create shared wealth and jobs through improving cereal value chain in Ethiopia. The specific objective is to strengthen linkage between smallholder farmers and input suppliers and increase productivity, on a sustainable basis, the income of rural producers, entrepreneurs that are engaged in the production, processing, storage and marketing of cereals.

Keywords: Cereals, demand and supply, Ethiopia, value chain

Contact Address: Yazachew Genet Ejigu, Ethiopian Institute of Agricultural Research, Debre Zeit Agricultural Research Centre, Addis Ababa, Ethiopia, e-mail: yazachewgenet@gmail.com

Affordable Solutions to Cereal Productivity Increase in West Africa

Joshua Apochi Apochi[1], Nguemo Achimba[1], Bougoum Sayouba[2], Modiere Genevieve Diallo[3], Teegwêndé Maggy Ouédraogo[2]

[1]*GIZ-AFC-AGFIN, Nigeria*

[2]*Entrepreneur, Burkina Faso*

[3]*Agricultural entrepreneur, Mali*

The Economic Community of West African States (ECOWAS) is home to 254 million people in 15 low-income countries. Agriculture accounts for 65 % of employment and 35 % of gross domestic product (GDP) in West Africa, but poverty is highest in rural areas where most of the population depends on agriculture for subsistence, 70 % of rural households depend on agriculture for their livelihood, however, Africa is remains a net importer of food, especially cereal grains. Despite the importance of agriculture and the vast potentials, cereal grain productivity in West Africa has remain low. Agricultural growth remains undermined by low investment in agriculture, poor infrastructure, high population growth rate, and low adoption of technologies. Other challenges to cereal productivity include inherently high climate variability, the looming threat of higher temperatures and more vicious droughts, high incidences of diseases, insect-pests, and parasitic plants, and sub-optimal soil nitrogen. Other factors contributing to food insecurity in Africa are low rural incomes, high poverty levels, high population rate (3 % annually) that is not commensurate to the rate of food production, and low investment in mechanisation. There are also issues with land degradation, climate change, conflict, low use of fertiliser (African average use is 11 kg/ha compared with the world average of 62 kg/ha), limited use of improved seed, limited access to markets, low level of knowledge, food losses due to pests and diseases in the field, high post-harvest losses, policy orientation, trade imbalances, low use of technology, and lately economic shocks due to COVID-19. These are threatening to undermine the continent's pursuit to achieve the Sustainable Development Goals (SDG) number 2 of ending hunger, achieving food security and improved nutrition, and promoting sustainable agriculture by 2030. In other to combat this menace constraining cereal productivity, possible and affordable solutions has been proposed by Genevieve from Mali, Jennifer and Joshua from Nigeria, Sayouba and Maggy from Bokinafaso in their study projects with a common goal of promoting affordable cereal productivity in west Africa.

Keywords: Affordable, agriculture, cereals, solutions, West Africa

Contact Address: Joshua Apochi Apochi, GIZ-AFC-AGFIN, 12 Jato Aka Street Itf Road Wurukum, 970211 Makurdi, Nigeria, e-mail: apochijosh@gmail.com

Value Additions Opportunities for African Fruit and Vegetables Growers

Audrey Sakhon Dominique Dieket[1], Agwaza Alfred Iordekighir[2], Joseph Alulu[3], Safoura Ousmane Cissé[4], John Sélom Amedjonekou[5], Yousra Soua[6]

[1]*LONO-CI SARL, Research and development, Ivory Coast (Cote D'Ivoire)*
[2]*Consultatnt GIZ-GIAE-AFC, Nigeria*
[3]*University of Nairobi, Agricultural Economics, Kenya*
[4]*Promoter and Manager "Health Secret", Mali*
[5]*Entrepreneur, Togo*
[6]*Own Start-up Development in Biological Control Sector, Tunisia*

Many food organisations like FAO mentioned a food crisis for next years, several innovations are shared in developed countries in order to sustain current natural resources and make food available for next generations. Africa, with a third of its population living in rural area and its dynamic growth, is really concerned and must adopt some innovations for food security. Nevertheless, African growers, especially small-scale farmers, faced some challenges decelerating the transition to a secure food system, like in fruits and vegetables production. In Europe, fruits and vegetable growers plan their operations from the production to the marketing phase. In conventional or biological way, cultivated seeds answer to particular climate situation, quantity or quality request. Fertilisation, water supply, sunbeams are calculated to offer only what is needed by plants, under greenhouses for example. The harvest is stored in cold chambers to preserve quality for market and reduce post-harvest losses. Fruits and vegetables are cleaned and sorted according to target customers standards. They may be dried or processed into paste, juice, jam, oil (some parts). Then, European growers ensure their product traceability as they want to penetrate international markets. African growers could adopt these innovations if there were not following difficulties. First, large-scale farmers face the low level of genetic research for efficient seeds, the high costs for imported machines and also, the habits of local population who consume some products only on the fresh form. Additionally, small-scale farmers face more challenges like land access especially for women, high costs of organic fertilisers compared to chemicals ones, low financial support. But we notice particularly a lack of knowledges and zero efforts to know about new practices, the ways to make them traditional and cheaper, the importance of a marketing strategy, the real importance of organic fertilisers beside their costs and their effectiveness manifested after several years and the advantages to be part of farmers organisations such as cooperatives. So, we will recommend African growers to meet in organisations and benefit from the advantages of training and financial support, especially in this context of population growth, and real need of responsible agriculture.

Keywords: African, challenges, fruits, opportunities, value addition, vegetables

Contact Address: Audrey Sakhon Dominique Dieket, LONO-CI SARL, Research and development, Abidjan, Cocody 2 Plateux mobile, Abidjan, Ivory Coast (Cote D'Ivoire), e-mail: audreydieket@gmail.com

Sustainable Approaches for Sustainable Management of African Smallholder Farmer

Hadhami Salem[1], Taiwo Oluwaseun Ayinde[2], Emmanuel Muasya[3], Aményon Akakpo[4], Selamawit Abera Kebede[5]

[1]*Plantméd, Tunisia*
[2]*APIN Public Health Initiative, Nigeria*
[3]*Machakos County Government, Kenya*
[4]*NGO LA Colombe, Togo*
[5]*Hawassa University, Ethiopia*

Agriculture accounts for 30 to 40 % of total Gross domestic product in Africa, and a majority of the continent's population is employed in the sector. In Africa, there are 33 million of smallholder farms. Although agriculture is viewed as a viable activity for rural development for the developing economies, Smallholder farming (i.e. farms which are smaller than five hectares) is the dominant type of agricultural practice, especially in the Sub-Saharan Africa. Smallholder farming as been highlighted as important for rural development, sustained rural livelihoods and an improved economy. However, smallholder farming is faced with a number of challenges which often lead to unsustainable management of their business, which includes, but is not limited to: Low infrastructure, low level of technical skills, low access to finance, land fragmentation, poor value chain links, high post harvest losses, drudgery and low level of technology, climate variability, lack of reliable markets, lack of access to information and a lot more. The following were the suggested approaches and solutions to a sustainable management of the smallholders African farms: extensive trainings on sustainable management for Youths and women in Agriculture, Innovation and use of affordable modern technologies e.g. the use of mobile app for advertisements, digitalisation, Adequate use of the different form of contract farming to ensure a ready market and Creating of a functional cooperative group of smallholder farmers, adoption of different methods of reducing post harvest losses, removing regulatory barriers that inhibit farmers from adopting digital tools, investing in reskilling and upskilling programmes focused on small farms.

Keywords: Agriculture, digitalisation, sustainability

Contact Address: Hadhami Salem, Plantméd, Kef, Tunisia, e-mail: salem123hadhami@gmail.com

Ways of Sustainable Cocoa Value Chain Improvement in West Africa

Frederic Le Roi Abalo[1], Emmanuela Soro[2], Ayao Dzigbodi Midodji[3], Wenkonta Pasgo[4]

[1] *ShARE Professional Training and Consulting, Togo*

[2] *National Polytechnical Institute Houphouet Boigny, Ivory Coast*

[3] *IFAD, Togo*

[4] *grassroots development Ministry, Togo*

Cocoa is one of the major cash crop in West Africa, providing a source of income for about 20 million people. One and half million farms are producing 2.6 million tons of cocoa per year with an average of 3 to 4 hectares size per farm in forested or formerly forested areas. West Africa cocoa production represent 88 % of all Africa production and 63 % of the total world production. In West Africa, Côte d'Ivoire and Ghana respectively with 884,000 and 2.1 million, represent 89 % of West Africa total production. Côte d'Ivoire and Ghana are the two leading producers in the world earning respectively 5 billion and 2 billion dollars from export. The main challenges for farmers are low productivity, low incomes and limited development in farming communities. Many threats are affecting sustainability of cocoa value chain in West Africa at different level. At economical level, challenges are income inequality and inadequate infrastructure and low access to finance are key challenges. Indeed, producing countries retain only a small proportion of the global cocoa market's value, which stood at 45 billion in 2019 and expected to reach 61 billion dollars by 2027. At social level, child labour is the main challenge. In 2021, about 1.6 million children labouring in the Ghana and Côte d'Ivoire cocoa industry, including hazardous tasks such as carrying heavy loads, climbing cocoa trees for harvesting, and using sharp tools to open cocoa pods. The main environmental threats are deforestation and climate change. In the absence of other options to increase efficiency, and overcome poor farm management, old plantations, soil degradation threats, and pressure from pests and diseases, cocoa producers often clear forestland to increase production. In addition, cocoa production is affected by extreme weather conditions, which are becoming more frequent and make certain regions less suitable for growing the crop. Many initiatives in Côte d'Ivoire, Ghana and Liberia aim to improve sustainability of cocoa value chains in West Africa through productivity improvement and processing to create more value. In Cameroon, the Roadmap to Deforestation-free Cocoa initiate by the sustainable trade initiative is a key project to improve sustainability.

Keywords: Cocoa, sustainability, West Africa

Contact Address: Frederic Le Roi Abalo, ShARE Professional Training and Consulting, Lomé, Togo, e-mail: frediabalo@gmail.com

Index of Authors

B

E

F

S

www.ingramcontent.com/pod-product-compliance
Ingram Content Group UK Ltd.
Pitfield, Milton Keynes, MK11 3LW, UK
UKHW022002190726
13853UKWH00004B/1675